LONDON MATHEMATICAL SOCIETY LECTURE NOTE SERIES

Managing Editor: Professor N.J. Hitchin, Mathematical Institute,
University of Oxford, 24–29 St Giles, Oxford OX1 3LB, United Kingdom

The titles below are available from booksellers, or, in case of difficulty, from Cambridge University Press.

46 *p*-adic Analysis: a short course on recent work, N. KOBLITZ
59 Applicable differential geometry, M. CRAMPIN & F.A.E. PIRANI
66 Several complex variables and complex manifolds II, M.J. FIELD
86 Topological topics, I.M. JAMES (ed)
87 Surveys in set theory, A.R.D. MATHIAS (ed)
88 FPF ring theory, C. FAITH & S. PAGE
89 An F-space sampler, N.J. KALTON, N.T. PECK & J.W. ROBERTS
90 Polytopes and symmetry, S.A. ROBERTSON
92 Representations of rings over skew fields, A.H. SCHOFIELD
93 Aspects of topology, I.M. JAMES & E.H. KRONHEIMER (eds)
96 Diophantine equations over function fields, R.C. MASON
97 Varieties of constructive mathematics, D.S. BRIDGES & F. RICHMAN
98 Localization in Noetherian rings, A.V. JATEGAONKAR
99 Methods of differential geometry in algebraic topology, M. KAROUBI & C. LERUSTE
100 Stopping time techniques for analysts and probabilists, L. EGGHE
104 Elliptic structures on 3-manifolds, C.B. THOMAS
105 A local spectral theory for closed operators, I. ERDELYI & WANG SHENGWANG
107 Compactification of Siegel moduli schemes, C.-L. CHAI
109 Diophantine analysis, J. LOXTON & A. VAN DER POORTEN (eds)
113 Lectures on the asymptotic theory of ideals, D. REES
114 Lectures on Bochner-Riesz means, K.M. DAVIS & Y.-C. CHANG
116 Representations of algebras, P.J. WEBB (ed)
119 Triangulated categories in the representation theory of finite-dimensional algebras, D. HAPPEL
121 Proceedings of *Groups - St Andrews 1985*, E. ROBERTSON & C. CAMPBELL (eds)
128 Descriptive set theory and the structure of sets of uniqueness, A.S. KECHRIS & A. LOUVEAU
130 Model theory and modules, M. PREST
131 Algebraic, extremal & metric combinatorics, M.-M. DEZA, P. FRANKL & I.G. ROSENBERG (eds)
132 Whitehead groups of finite groups, ROBERT OLIVER
133 Linear algebraic monoids, MOHAN S. PUTCHA
134 Number theory and dynamical systems, M. DODSON & J. VICKERS (eds)
137 Analysis at Urbana, I, E. BERKSON, T. PECK, & J. UHL (eds)
138 Analysis at Urbana, II, E. BERKSON, T. PECK, & J. UHL (eds)
139 Advances in homotopy theory, S. SALAMON, B. STEER & W. SUTHERLAND (eds)
140 Geometric aspects of Banach spaces, E.M. PEINADOR & A. RODES (eds)
141 Surveys in combinatorics 1989, J. SIEMONS (ed)
144 Introduction to uniform spaces, I.M. JAMES
146 Cohen-Macaulay modules over Cohen-Macaulay rings, Y. YOSHINO
148 Helices and vector bundles, A.N. RUDAKOV *et al*
149 Solitons, nonlinear evolution equations and inverse scattering, M. ABLOWITZ & P. CLARKSON
150 Geometry of low-dimensional manifolds 1, S. DONALDSON & C.B. THOMAS (eds)
151 Geometry of low-dimensional manifolds 2, S. DONALDSON & C.B. THOMAS (eds)
152 Oligomorphic permutation groups, P. CAMERON
153 L-functions and arithmetic, J. COATES & M.J. TAYLOR (eds)
155 Classification theories of polarized varieties, TAKAO FUJITA
156 Twistors in mathematics and physics, T.N. BAILEY & R.J. BASTON (eds)
158 Geometry of Banach spaces, P.F.X. MÜLLER & W. SCHACHERMAYER (eds)
159 Groups St Andrews 1989 volume 1, C.M. CAMPBELL & E.F. ROBERTSON (eds)
160 Groups St Andrews 1989 volume 2, C.M. CAMPBELL & E.F. ROBERTSON (eds)
161 Lectures on block theory, BURKHARD KÜLSHAMMER
162 Harmonic analysis and representation theory, A. FIGA-TALAMANCA & C. NEBBIA
163 Topics in varieties of group representations, S.M. VOVSI
164 Quasi-symmetric designs, M.S. SHRIKANDE & S.S. SANE
166 Surveys in combinatorics, 1991, A.D. KEEDWELL (ed)
168 Representations of algebras, H. TACHIKAWA & S. BRENNER (eds)
169 Boolean function complexity, M.S. PATERSON (ed)
170 Manifolds with singularities and the Adams-Novikov spectral sequence, B. BOTVINNIK
171 Squares, A.R. RAJWADE
172 Algebraic varieties, GEORGE R. KEMPF
173 Discrete groups and geometry, W.J. HARVEY & C. MACLACHLAN (eds)
174 Lectures on mechanics, J.E. MARSDEN
175 Adams memorial symposium on algebraic topology 1, N. RAY & G. WALKER (eds)
176 Adams memorial symposium on algebraic topology 2, N. RAY & G. WALKER (eds)
177 Applications of categories in computer science, M. FOURMAN, P. JOHNSTONE & A. PITTS (eds)
178 Lower K- and L-theory, A. RANICKI
179 Complex projective geometry, G. ELLINGSRUD *et al*
180 Lectures on ergodic theory and Pesin theory on compact manifolds, M. POLLICOTT
181 Geometric group theory I, G.A. NIBLO & M.A. ROLLER (eds)
182 Geometric group theory II, G.A. NIBLO & M.A. ROLLER (eds)
183 Shintani zeta functions, A. YUKIE

184 Arithmetical functions, W. SCHWARZ & J. SPILKER
185 Representations of solvable groups, O. MANZ & T.R. WOLF
186 Complexity: knots, colourings and counting, D.J.A. WELSH
187 Surveys in combinatorics, 1993, K. WALKER (ed)
188 Local analysis for the odd order theorem, H. BENDER & G. GLAUBERMAN
189 Locally presentable and accessible categories, J. ADAMEK & J. ROSICKY
190 Polynomial invariants of finite groups, D.J. BENSON
191 Finite geometry and combinatorics, F. DE CLERCK *et al*
192 Symplectic geometry, D. SALAMON (ed)
194 Independent random variables and rearrangement invariant spaces, M. BRAVERMAN
195 Arithmetic of blowup algebras, WOLMER VASCONCELOS
196 Microlocal analysis for differential operators, A. GRIGIS & J. SJÖSTRAND
197 Two-dimensional homotopy and combinatorial group theory, C. HOG-ANGELONI *et al*
198 The algebraic characterization of geometric 4-manifolds, J.A. HILLMAN
199 Invariant potential theory in the unit ball of $\mathbf{C}^n$, MANFRED STOLL
200 The Grothendieck theory of dessins d'enfant, L. SCHNEPS (ed)
201 Singularities, JEAN-PAUL BRASSELET (ed)
202 The technique of pseudodifferential operators, H.O. CORDES
203 Hochschild cohomology of von Neumann algebras, A. SINCLAIR & R. SMITH
204 Combinatorial and geometric group theory, A.J. DUNCAN, N.D. GILBERT & J. HOWIE (eds)
205 Ergodic theory and its connections with harmonic analysis, K. PETERSEN & I. SALAMA (eds)
207 Groups of Lie type and their geometries, W.M. KANTOR & L. DI MARTINO (eds)
208 Vector bundles in algebraic geometry, N.J. HITCHIN, P. NEWSTEAD & W.M. OXBURY (eds)
209 Arithmetic of diagonal hypersurfaces over finite fields, F.Q. GOUVÊA & N. YUI
210 Hilbert C*-modules, E.C. LANCE
211 Groups 93 Galway / St Andrews I, C.M. CAMPBELL *et al* (eds)
212 Groups 93 Galway / St Andrews II, C.M. CAMPBELL *et al* (eds)
214 Generalised Euler-Jacobi inversion formula and asymptotics beyond all orders, V. KOWALENKO *et al*
215 Number theory 1992–93, S. DAVID (ed)
216 Stochastic partial differential equations, A. ETHERIDGE (ed)
217 Quadratic forms with applications to algebraic geometry and topology, A. PFISTER
218 Surveys in combinatorics, 1995, PETER ROWLINSON (ed)
220 Algebraic set theory, A. JOYAL & I. MOERDIJK
221 Harmonic approximation, S.J. GARDINER
222 Advances in linear logic, J.-Y. GIRARD, Y. LAFONT & L. REGNIER (eds)
223 Analytic semigroups and semilinear initial boundary value problems, KAZUAKI TAIRA
224 Computability, enumerability, unsolvability, S.B. COOPER, T.A. SLAMAN & S.S. WAINER (eds)
225 A mathematical introduction to string theory, S. ALBEVERIO, J. JOST, S. PAYCHA, S. SCARLATTI
226 Novikov conjectures, index theorems and rigidity I, S. FERRY, A. RANICKI & J. ROSENBERG (eds)
227 Novikov conjectures, index theorems and rigidity II, S. FERRY, A. RANICKI & J. ROSENBERG (eds)
228 Ergodic theory of $\mathbf{Z}^d$ actions, M. POLLICOTT & K. SCHMIDT (eds)
229 Ergodicity for infinite dimensional systems, G. DA PRATO & J. ZABCZYK
230 Prolegomena to a middlebrow arithmetic of curves of genus 2, J.W.S. CASSELS & E.V. FLYNN
231 Semigroup theory and its applications, K.H. HOFMANN & M.W. MISLOVE (eds)
232 The descriptive set theory of Polish group actions, H. BECKER & A.S. KECHRIS
233 Finite fields and applications, S. COHEN & H. NIEDERREITER (eds)
234 Introduction to subfactors, V. JONES & V.S. SUNDER
235 Number theory 1993–94, S. DAVID (ed)
236 The James forest, H. FETTER & B. GAMBOA DE BUEN
237 Sieve methods, exponential sums, and their applications in number theory, G.R.H. GREAVES *et al*
238 Representation theory and algebraic geometry, A. MARTSINKOVSKY & G. TODOROV (eds)
239 Clifford algebras and spinors, P. LOUNESTO
240 Stable groups, FRANK O. WAGNER
241 Surveys in combinatorics, 1997, R.A. BAILEY (ed)
242 Geometric Galois actions I, L. SCHNEPS & P. LOCHAK (eds)
243 Geometric Galois actions II, L. SCHNEPS & P. LOCHAK (eds)
244 Model theory of groups and automorphism groups, D. EVANS (ed)
245 Geometry, combinatorial designs and related structures, J.W.P. HIRSCHFELD *et al*
246 p-Automorphisms of finite p-groups, E.I. KHUKHRO
247 Analytic number theory, Y. MOTOHASHI (ed)
248 Tame topology and o-minimal structures, LOU VAN DEN DRIES
249 The atlas of finite groups: ten years on, ROBERT CURTIS & ROBERT WILSON (eds)
250 Characters and blocks of finite groups, G. NAVARRO
251 Gröbner bases and applications, B. BUCHBERGER & F. WINKLER (eds)
252 Geometry and cohomology in group theory, P. KROPHOLLER, G. NIBLO, R. STÖHR (eds)
253 The q-Schur algebra, S. DONKIN
254 Galois representations in arithmetic algebraic geometry, A.J. SCHOLL & R.L. TAYLOR (eds)
255 Symmetries and integrability of difference equations, P.A. CLARKSON & F.W. NIJHOFF (eds)
256 Aspects of Galois theory, HELMUT VÖLKLEIN *et al*
257 An introduction to noncommutative differential geometry and its physical applications 2ed, J. MADORE
258 Sets and proofs, S.B. COOPER & J. TRUSS (eds)
259 Models and computability, S.B. COOPER & J. TRUSS (eds)
260 Groups St Andrews 1997 in Bath, I, C.M. CAMPBELL *et al*
261 Groups St Andrews 1997 in Bath, II, C.M. CAMPBELL *et al*
263 Singularity theory, BILL BRUCE & DAVID MOND (eds)
264 New trends in algebraic geometry, K. HULEK, F. CATANESE, C. PETERS & M. REID (eds)

London Mathematical Society Lecture Note Series. 273

Spectral Theory and Geometry

ICMS Instructional Conference, Edinburgh 1998

Edited by

Brian Davies & Yuri Safarov
King's College, University of London

CAMBRIDGE UNIVERSITY PRESS
Cambridge, New York, Melbourne, Madrid, Cape Town, Singapore, São Paulo

Cambridge University Press
The Edinburgh Building, Cambridge CB2 2RU, UK

Published in the United States of America by Cambridge University Press, New York

www.cambridge.org
Information on this title: www.cambridge.org/9780521777490

© Cambridge University Press 1999

This publication is in copyright. Subject to statutory exception
and to the provisions of relevant collective licensing agreements,
no reproduction of any part may take place without
the written permission of Cambridge University Press.

First published 1999

A catalogue record for this publication is available from the British Library

ISBN-13 978-0-521-77749-0 paperback
ISBN-10 0-521-77749-6 paperback

Transferred to digital printing 2005

Cambridge University Press has no responsibility for the persistence or accuracy
of email addresses referred to in this publication.

Spectral Theory and Geometry

ICMS Instructional Conference

Edinburgh, 30 March – 9 April 1998

Contents

Preface

The ICMS Instructional Conference between 30 March and 9 April 1998 was one of a series of highly successful such conferences held under the auspices of the International Centre of Mathematical Sciences in Edinburgh, Scotland. Some of these meetings have been held in the house where Maxwell was born, which is now the ICMS administrative headquarters. The lectures for larger meetings, such as this one, were in the James Clerk Maxwell Building of the University of Edinburgh in a different part of the city. Accommodation for the meeting was provided for all speakers and audience in the Pollock Halls of Residence belonging to the University and providing a refreshing twenty minute walk to and from the lectures.

The subject of this meeting was Spectral Theory and Geometry, and the format followed that of earlier meetings. Twelve principal speakers were invited to give lecture courses of three hours each, graded into three levels, Introductory, Medium and Advanced (see the list below). There were several exchanges before the meeting about the contents of the introductory lectures, since these had to provide the core material for all of the later courses. These were all given by world experts, who provided organised surveys of their fields, with proofs in many cases. This volume contains articles by most of the lecturers. Many of these go beyond what they could present within the time limits of the conference, and we are most grateful to them for undertaking the considerable amount of work involved in producing the lecture notes.

The organisers of the meeting, Professors E B Davies and Y Safarov, acknowledge with thanks the advice received from the Scientific Panel, namely Professors I Chavel, P Sarnak and J Sjöstrand. We would also particularly like to thank Mrs Tracey Dart, who took over as Administrator of the ICMS at a crucial stage in the organisation, and Mrs Julie Brown; both of them made an enormous contribution to the success of the meeting. Finally we must thank Professor E G Rees, without whom it would never have got started.

We acknowledge with thanks the several organisations which provided funding for the meeting. These were the European Union, The Engineering and Physical Sciences Research Council (grant number GR/L52536), The London Mathematical Society, and the ICMS itself.

Brian Davies
Yuri Safarov

Department of Mathematics
King's College
Strand
London WC2R 2LS

List of lecture courses

Introductory courses:

F E Burstall, *Basic Riemannian geometry*
I Chavel, *The Laplacian on Riemannian manifolds*
E B Davies, *Computational spectral theory*

Medium level courses:

M Ashbaugh, *Isoperimetric and universal inequalities for eigenvalues*
A Grigor'yan, *Estimates of heat kernels on Riemannian manifolds*
M Shubin, *Spectral theory of the Schrödinger operators on non-compact manifolds: qualitative results*
D Vassiliev, *Spectral asymptotics of fractals*

Advanced courses:

Y Colin de Verdière, *Spectral theory of graphs*
R Melrose, *Pseudodifferential operators, contact manifolds and index theory*
A Voros, *Quantum resurgence*
S Zelditch, *Lectures on wave invariants*
M Zworski, *Resonances in geometry and mathematical physics: an introduction*

3 List of participants

Name	Affiliation	Email
M Agranovich	MGIEM, Moscow	msa@funcan.msk.su
P Almeida	Instituto Superior Technico, Lisboa	palmeida@math.ist.utl.pt
M Ashbaugh	University of Missouri	mark@math.missouri.edu
A Aslanyan	King's College London	aslanyan@mth.kcl.ac.uk
G Barbatis	University of Crete	gbarbati @anaximenis.math.uch.gr
C Batty	Oxford University	charles.batty@sjc.ox.ac.uk
S Blachere	University Paul Sabatier, Toulouse	blachere@cict.fr
L Boulton	King's College London	lboulton@mth.kcl.ac.uk
F Burstall	University of Bath	feb@maths.bath.ac.uk
J Butler	Université Paris–Sud	jonathan.butler @math.u-psud.fr
E Buzano	Universita Di Torino	buzano@dm.unito.it
G Carron	École Normale Supérieure de Lyon	gcarron @umpa.ens-lyon.fr
I Chavel	CUNY	ichavel@email.gc.cuny.edu
N Claire	King's College London	nclaire@mth.kcl.ac.uk
T Coulhon	Cergy–Pontoise University	coulhon@paris.u-cergy.fr
Y Colin de Verdiere	Institut Fourier, Grenoble	yves.colin-de-verdiere @ujf-grenoble.fr
E B Davies	King's College London	e.brian.davies@kcl.ac.uk
E-M Delicha	National Technical University of Athens	delicha@math.ntua.gr
T Delmotte	Cergy–Pontoise University	delmotte@u-cergy.fr
S Doukakis	University of Patras	doukakis@hotmail.com
D Drivabiaris	University of Edinbirgh	
D Elton	University of Sussex	d.m.elton@sussex.ac.uk
B Fairfax	King's College London	benjamin.fairfax @kcl.ac.uk
G Garello	Universita Di Torino	garello@dm.unito.it
P Giannopoulou	University of Athens	nagia@math.ntua.gr
D Gioev	Royal Institute of Technology, Stockholm	gioev@math.kth.se
I Goldsheid	Queen Mary and Westfield College, London	i.goldsheid@qmw.ac.uk
V Gontcharenko	Loughborough University	v.m.gontcharenko @lboro.ac.uk

Name	Affiliation	Email
G Green	University of Newcastle	g.m.green@ncl.ac.uk
A Grigoryan	Imperial College, London	a.grigoryan@ic.ac.uk
G Gudmundsdottir	University of Lund	gudrun@maths.lth.se
T Hausel	Oxford University & Cambridge University	hausel@maths.ox.ac.uk
L Hermi	University of Missouri	hermi@math.missouri.edu
L Hillairet	Institut Fourier, Grenoble	luc.hillairet @ujf-grenoble.fr
M Hitrik	Lund Institute of Technology	mike@math.lth.se
A Holst	University of Lund	ah@maths.lth.se
A Jensen	Aalborg University	matarne@math.auc.dk
M Joshi	Cambridge University	joshi@dpmms.cam.ac.uk
K Karagatsos	Gothenburg University	konst@math.chalmers.se
F Klopp	Université Paris–Nord	klopp @math.univ-paris13.fr
S Krusch	Cambridge University	sk211@damtp.cam.ac.uk
Y Kurylev	Loughborough University	y.v.kurylev@lboro.ac.uk
A Laptev	Royal Institute of Technology, Stockholm	laptev@math.kth.se
D Levin	Weizmann Institute	levdan @wisdom.weizmann.ac.il
M Levitin	Heriot–Watt University	m.levitin@ma.hw.ac.uk
V Liskevich	University of Bristol	v.liskevich@bristol.ac.uk
N Mandache	University of Bristol	niculae.mandache @bristol.ac.uk
C Mason	King's College London	mason@mth.kcl.ac.uk
N Matzakos	National Technical University of Athens	nikmatz@math.ntua.gr
J Maubon	Institut Fourier, Grenoble	maubon@ujf-grenoble.fr
O Mazet	University Paul Sabatier, Toulouse	mazet@cict.fr
I McGillivray	University of Bristol	i.mcgillivray @bristol.ac.uk
R Melrose	Massachusetts Institute of Technology	rbm@math.mit.edu
F Menendez-Conde	University of Sussex	mmpd7@sussex.ac.uk

Name	Affiliation	Email
S Micciche	Loughborough University	s.micciche@lboro.ac.uk
J Nath	King's College London	jnath@mth.kcl.ac.uk
L Nedelec	Université Paris–Nord	nedelec @math.univ-paris13.fr
Y Netrusov	King's College London	netrusov@mth.kcl.ac.uk
J Neuheisel	John Hopkins University	
T Nikulainen	Heriot–Watt University	t.nikulainen@ma.hw.ac.uk
A Noll	TU Clausthal Institute for Mathematics	noll@math.tu-clausthal.de
L Notarantonio	Heriot–Watt University	lino@ma.hw.ac.uk
I Polterovich	Weizmann Institute	iossif @wisdom.weizmann.ac.il
O Post	Technische Universität Braunschweig	o.post@tu-bs.de
M Qafsaoui	LAMFA	mahmoud.qafsaoui @u-picardie.fr
Z Qian	Imperial College, London	z.qian@ic.ac.uk
P Redparth	King's College London	redparth@mth.kcl.ac.uk
S Roussel	University Paul Sabatier, Toulouse	roussel@cict.fr
E Russ	Cergy–Pontoise University	russ@u-cergy.fr
B Rynne	Heriot–Watt University	bryan@ma.hw.ac.uk
Y Safarov	King's College London	ysafarov@mth.kcl.ac.uk
G Salomonsen	University of Bonn	gorm @styx.iam.uni-bonn.de
G Scheffler	University Paul Sabatier, Toulouse	scheffler@cict.fr
K-M Schmidt	Ludwig-Maximilians- -Universität München	kschmidt@rz-mathematik. uni-muenchen.de
E Shargorodsky	University of Sussex	e.shargorodsky @sussex.ac.uk
A Shkalikov	Moscow State University	shkal@mech.math.msu.su
M Shubin	Northeastern University	shubin@neu.edu
O Sick	University of Bonn	sick@math.uni-bonn.de
Z Sobol	University of Bristol	z.sobol@bristol.ac.uk
S Srisatkunarajah	Heriot–Watt University	sri@ma.hw.ac.uk
A Valette	Neuchatel University	alain.valette @maths.unine.ch

Name	Affiliation	Email
D Vassiliev	University of Sussex	d.vassiliev@sussex.ac.uk
A Voros	Centre d'Etudes Nucleaire de Saclay	voros@spht.saclay.cea.fr
S Vungoc	University of Utrecht	vungoc@math.vun.nl
J Walthoe	University of Sussex	j.m.walthoe@sussex.ac.uk
T Weidl	Royal Institute of Technology, Stockholm	weidl@math.kth.se
S Wood	University of Sussex	mmpd3@sussex.ac.uk
M Yourkine	Moscow State University	
S Zelditch	John Hopkins University	zel@chow.mat.jhu.edu
A Ziggioto	Universita Di Torino	ziggioto@dm.unito.it
M Zworski	University of California, Berkeley	zworski@math.berkeley.edu

Basic Riemannian Geometry

F.E. Burstall
Department of Mathematical Sciences
University of Bath

Introduction

My mission was to describe the basics of Riemannian geometry in just three hours of lectures, starting from scratch. The lectures were to provide background for the analytic matters covered elsewhere during the conference and, in particular, to underpin the more detailed (and much more professional) lectures of Isaac Chavel. My strategy was to get to the point where I could state and prove a Real Live Theorem: the Bishop Volume Comparison Theorem and Gromov's improvement thereof and, by appalling abuse of OHP technology, I managed this task in the time alloted. In writing up my notes for this volume, I have tried to retain the breathless quality of the original lectures while correcting the mistakes and excising the out-right lies.

I have given very few references to the literature in these notes so a few remarks on sources is appropriate here. The first part of the notes deals with analysis on differentiable manifolds. The two canonical texts here are Spivak [5] and Warner [6] and I have leaned on Warner's book in particular. For Riemannian geometry, I have stolen shamelessly from the excellent books of Chavel [1] and Gallot–Hulin–Lafontaine [3]. In particular, the proof given here of Bishop's theorem is one of those provided in [3].

1 What is a manifold?

What ingredients do we need to do Differential Calculus? Consider first the notion of a continuous function: during the long process of abstraction and generalisation that leads from Real Analysis through Metric Spaces to Topology, we learn that continuity of a function requires no more structure on the domain and co-domain than the idea of an open set.

By contrast, the notion of differentiability requires much more: to talk about the difference quotients whose limits are partial derivatives, we seem to require that the (co-)domain have a linear (or, at least, affine) structure.

However, a moment's thought reveals that differentiability is a completely *local* matter so that all that is really required is that the domain and co-domain be *locally* linear, that is, each point has a neighbourhood which is homeomorphic to an open subset of some linear space. These ideas lead us to the notion of a *manifold*: a topological space which is locally Euclidean and on which there is a well-defined differential calculus.

We begin by setting out the basic theory of these spaces and how to do Analysis on them.

1.1 Manifolds

Let M be a Hausdorff, second countable[1], connected topological space.

M is a C^r *manifold* of dimension n if there is an open cover $\{U_\alpha\}_{\alpha\in I}$ of M and homeomorphisms $x_\alpha : U_\alpha \to x_\alpha(U_\alpha)$ onto open subsets of $\mathbb{R}^n$ such that, whenever $U_\alpha \cap U_\beta \neq \emptyset$,

$$x_\alpha \circ x_\beta^{-1} : x_\beta(U_\alpha \cap U_\beta) \to x_\alpha(U_\alpha \cap U_\beta)$$

is a C^r diffeomorphism.

Each pair (U_α, x_α) called a *chart.*

Write $x_\alpha = (x^1, \dots, x^n)$. The $x^i : U_\alpha \to \mathbb{R}$ are *coordinates.*

1.1.1 Examples

1. Any open subset $U \subset \mathbb{R}^n$ is a C^∞ manifold with a single chart $(U, 1_U)$.
2. Contemplate the unit sphere $S^n = \{v \in \mathbb{R}^{n+1} : \|v\| = 1\}$ in $\mathbb{R}^{n+1}$. Orthogonal projection provides a homeomorphism of any open hemisphere onto the open unit ball in some hyperplane $\mathbb{R}^n \subset \mathbb{R}^{n+1}$. The sphere is covered by the $(2n+2)$ hemispheres lying on either side of the coordinate hyperplanes and in this way becomes a C^∞ manifold (exercise!).
3. A good supply of manifolds is provided by the following version of the Implicit Function Theorem [6]:

   Theorem. *Let $f : \Omega \subset \mathbb{R}^n \to \mathbb{R}$ be a C^r function $(r \geq 1)$ and $c \in \mathbb{R}$* a regular value, *that is, $\nabla f(x) \neq 0$, for all $x \in f^{-1}\{c\}$.*
   Then $f^{-1}\{c\}$ is a C^r manifold.

   Exercise. Apply this to $f(x) = \|x\|^2$ to get a less tedious proof that S^n is a manifold.

[1] This means that there is a countable base for the topology of M.

4. An open subset of a manifold is a manifold in its own right with charts $(U_\alpha \cap U, x_\alpha|_{U_\alpha \cap U})$.

1.1.2 Functions and maps

A continuous function $f : M \to \mathbb{R}$ is C^r if each $f \circ x_\alpha^{-1} : x_\alpha(U_\alpha) \to \mathbb{R}$ is a C^r function of the open set $x_\alpha(U_\alpha) \subset \mathbb{R}^n$.

We denote the vector space of all such functions by $C^r(M)$.

Example. Any coordinate function $x^i : U_\alpha \to \mathbb{R}$ is C^r on U_α.

Exercise. The restriction of any C^r function on $\mathbb{R}^{n+1}$ to the sphere S^n is C^r on S^n.

In the same way, a continuous map $\phi : M \to N$ of C^r manifolds is C^r if, for all charts (U, x), (V, y) of M and N respectively, $y \circ \phi \circ x^{-1}$ is C^r on its domain of definition.

A slicker formulation[2] is that $h \circ \phi \in C^r(M)$, for all $h \in C^r(M)$.

At this point, having made all the definitions, we shall stop pretending to be anything other than Differential Geometers and henceforth take $r = \infty$.

1.2 Tangent vectors and derivatives

We now know what functions on a manifold are and it is our task to differentiate them. This requires some less than intuitive definitions so let us step back and remind ourselves of what differentiation involves.

Let $f : \Omega \subset \mathbb{R}^n \to \mathbb{R}$ and contemplate the derivative of f at some $x \in \Omega$. This is a linear map $\mathrm{d}f_x : \mathbb{R}^n \to \mathbb{R}$. However, it is better for us to take a dual point of view and think of $\mathbf{v} \in \mathbb{R}^n$ is a linear map $\mathbf{v} : C^\infty(M) \to \mathbb{R}$ by

$$\mathbf{v}f \stackrel{\text{def}}{=} \mathrm{d}f_x(\mathbf{v}).$$

The Leibniz rule gives us

$$\mathbf{v}(fg) = f(x)\mathbf{v}(g) + \mathbf{v}(f)g(x). \tag{1.1}$$

Fact. *Any linear* $\mathbf{v} : C^\infty(\Omega) \to \mathbb{R}$ *satisfying* (1.1) *arises this way.*

Now let M be a manifold. The preceding analysis may give some motivation to the following

[2] It requires a little machinery, in the shape of bump functions, to see that this *is* an equivalent formulation.

Definition. A *tangent vector* at $m \in M$ is a linear map $\xi : C^\infty(M) \to \mathbb{R}$ such that

$$\xi(fg) = f(m)\xi(g) + \xi(f)g(m)$$

for all $f, g \in C^\infty(M)$.

Denote by M_m the vector space of all tangent vectors at m.

Here are some examples

1. For $\gamma : I \to M$ a (smooth) path with $\gamma(t) = m$, define $\gamma'(t) \in M_m$ by
$$\gamma'(t)f = (f \circ \gamma)'(t).$$
Fact. *All $\xi \in M_m$ are of the form $\gamma'(t)$ for some path γ.*

2. Let (U, x) be a chart with coordinates $x^1, \ldots, x^n$ and $x(m) = p \in \mathbb{R}^n$. Define $\partial_{i|m} \in M_m$ by
$$\partial_{i|m} f = \left.\frac{\partial(f \circ x^{-1})}{\partial x^i}\right|_p$$
Fact. $\partial_{1|m}, \ldots, \partial_{n|m}$ *is a basis for* M_m.

3. For $p \in U \subset \mathbb{R}^n$ open, we know that U_p is canonically isomorphic to $\mathbb{R}^n$ via
$$\mathbf{v}f = \mathrm{d}f_p(\mathbf{v})$$
for $\mathbf{v} \in \mathbb{R}^n$.

4. Let $M = f^{-1}\{c\}$ be a regular level set of $f : \Omega \subset \mathbb{R}^n \to \mathbb{R}$. One can show that M_m is a linear subspace of $\Omega_m \cong \mathbb{R}^n$. Indeed, under this identification,
$$M_m = \{\mathbf{v} \in \mathbb{R}^n : \mathbf{v} \perp \nabla f_m\}.$$

Now that we have got our hands on tangent vectors, the definition of the derivative of a function as a linear map on tangent vectors is almost tautological:

Definition. For $f \in C^\infty(M)$, the *derivative* $\mathrm{d}f_m : M_m \to \mathbb{R}$ *of* f *at* $m \in M$ is defined by

$$\mathrm{d}f_m(\xi) = \xi f.$$

We note:

1. Each $\mathrm{d}f_m$ is a linear map and the Leibniz Rule holds:
$$\mathrm{d}(fg)_m = g(m)\mathrm{d}f_m + f(m)\mathrm{d}g_m.$$
2. By construction, this definition coincides with the usual one when M is an open subset of $\mathbb{R}^n$.

Exercise. If f is a constant map on a manifold M, show that each $\mathrm{d}f_m = 0$.

The same circle of ideas enable us to differentiate maps between manifolds:

Definition. For $\phi : M \to N$ a smooth map of manifolds, the *tangent map* $\mathrm{d}\phi_m : M_m \to N_{\phi(m)}$ *at* $m \in M$ is the linear map defined by
$$\mathrm{d}\phi_m(\xi)f = \xi(f \circ \phi),$$
for $\xi \in M_m$ and $f \in C^\infty(N)$.

Exercise. Prove the chain rule: for $\phi : M \to N$ and $\psi : N \to Z$ and $m \in M$,
$$\mathrm{d}(\psi \circ \phi)_m = \mathrm{d}\psi_{\phi(m)} \circ \mathrm{d}\phi_m.$$

Exercise. View $\mathbb{R}$ as a manifold (with a single chart!) and let $f : M \to \mathbb{R}$. We now have two competing definitions of $\mathrm{d}f_m$. Show that they coincide.

The *tangent bundle of* M is the disjoint union of the tangent spaces:
$$TM = \coprod_{m\in M} M_m.$$

1.3 Vector fields

Definition. A *vector field* is a linear map $X : C^\infty(M) \to C^\infty(M)$ such that
$$X(fg) = f(Xg) + g(Xf).$$

Let $\Gamma(TM)$ denote the vector space of all vector fields on M.

We can view a vector field as a map $X : M \to TM$ with $X(m) \in M_m$: indeed, we have
$$X_{|m} \in M_p$$

where

$$X_{|m}f = (Xf)(m).$$

In fact, vector fields can be shown to be exactly those maps $X : M \to TM$ with $X(m) \in M_m$ which satisfy the additional smoothness constraint that for each $f \in C^\infty(M)$, the function $m \mapsto X(m)f$ is also C^∞.

The *Lie bracket* of $X, Y \in \Gamma(TM)$ is $[X, Y] : C^\infty(M) \to C^\infty(M)$ given by

$$[X, Y]f = X(Yf) - Y(Xf).$$

The point of this definition is contained in the following

Exercise. Show that $[X, Y] \in \Gamma(TM)$ also.

The Lie bracket is interesting for several reasons. Firstly it equips $\Gamma(TM)$ with the structure of a Lie algebra; secondly, it, and operators derived from it, are the only differential operators that can be defined on an arbitrary manifold without imposing additional structures such as special coordinates, a Riemannian metric, a complex structure or a symplectic form.

There is an extension of the notion of vector field that we shall need later on:

Definition. Let $\phi : M \to N$ be a map. A *vector field along* ϕ is a map $X : M \to TN$ with

$$X(m) \in N_{\phi(m)},$$

for all $m \in M$, which additionally satisfies a smoothness assumption that we shall gloss over.

Denote by $\Gamma(\phi^{-1}TN)$ the vector space of all vector fields along ϕ.

Here are some examples:

1. If $c : I \to N$ is a smooth path then $c' \in \Gamma(\phi^{-1}TN)$.
2. More generally, for $\phi : M \to N$ and $X \in \Gamma(TM)$, $\mathrm{d}\phi(X) \in \Gamma(\phi^{-1}TN)$. Here, of course,

$$\mathrm{d}\phi(X)(m) = \mathrm{d}\phi_m(X_{|m}).$$

3. For $Y \in \Gamma(TN)$, $Y \circ \phi \in \Gamma(\phi^{-1}TN)$.

1.4 Connections

We would like to differentiate vector fields but as they take values in different vector spaces at different points, it is not so clear how to make difference quotients and so derivatives. What is needed is some extra structure: a *connection* which should be thought of as a "directional derivative" for vector fields.

Definition. A *connection on TM* is a bilinear map

$$\begin{aligned} TM \times \Gamma(TM) &\to TM \\ (\xi, X) &\mapsto \nabla_\xi X \end{aligned}$$

such that, for $\xi \in M_m$, $X, Y \in \Gamma(TM)$ and $f \in C^\infty(M)$,

1. $\nabla_\xi X \in M_m$;
2. $\nabla_\xi(fX) = (\xi f)X_{|m} + f(m)\nabla_\xi X$;
3. $\nabla_X Y \in \Gamma(TM)$.

A connection on TM comes with some additional baggage in the shape of two multilinear maps:

$$\begin{aligned} T_m &: M_m \times M_m \to M_m \\ R_m &: M_m \times M_m \times M_m \to M_m \end{aligned}$$

given by

$$\begin{aligned} T_m(\xi, \eta) &= \nabla_\xi Y - \nabla_\eta X - [X, Y]_{|m} \\ R_m(\xi, \eta)\zeta &= \nabla_\eta \nabla_X Z - \nabla_\xi \nabla_Y Z - \nabla_{[Y,X]_{|m}} \end{aligned}$$

where X, Y, $Z \in \Gamma(TM)$ with $X_{|m} = \xi$, $Y_{|m} = \eta$ and $Z_{|m} = \zeta$.

T_m and R_m are, respectively, the *torsion* and *curvature* at m of ∇.

Fact. *R and T are well-defined—they do not depend of the choice of vector fields X, Y and Z extending ξ, η and ζ.*

We have some trivial identities:

$$\begin{aligned} T(\xi, \eta) &= -T(\eta, \xi) \\ R(\xi, \eta)\zeta &= -R(\eta, \xi)\zeta. \end{aligned}$$

and, if each $T_m = 0$, we have the less trivial *First Bianchi Identity*:

$$R(\xi, \eta)\zeta + R(\zeta, \xi)\eta + R(\eta, \zeta)\xi = 0.$$

A connection ∇ on TN induces a similar operator on vector fields along a map $\phi : M \to N$. To be precise, there is a unique bilinear map

$$TM \times \Gamma(\phi^{-1}TN) \to TN$$
$$(\xi, X) \mapsto \phi^{-1}\nabla_\xi X$$

such that, for $\xi \in M_m$, $X \in \Gamma(TM)$, $Y \in \Gamma(\phi^{-1}TN)$ and $f \in C^\infty(M)$,

1. $\phi^{-1}\nabla_\xi Y \in N_{\phi(m)}$;
2. $\phi^{-1}\nabla_\xi(fY) = (\xi f)Y_{|\phi(m)} + f(m)\phi^{-1}\nabla_\xi Y$;
3. $\phi^{-1}\nabla_X Y \in \Gamma(\phi^{-1}TN)$ (this is a smoothness assertion);
4. If $Z \in \Gamma(TN)$ then $Z \circ \phi \in \Gamma(\phi^{-1}TN)$ and
$$\phi^{-1}\nabla_\xi(Z \circ \phi) = \nabla_{d\phi_m(\xi)} Z.$$

$\phi^{-1}\nabla$ is the *pull-back of* ∇ *by* ϕ. The first three properties just say that $\phi^{-1}\nabla$ behaves like ∇, it is the last that essentially defines it in a unique way.

2 Analysis on Riemannian manifolds

2.1 Riemannian manifolds

A rich and useful geometry arises if we equip each M_m with an inner product:

Definition. A *Riemannian metric* g on M is an inner product g_m on each M_m such that, for all vector fields X and Y, the function

$$m \mapsto g_m(X_{|m}, Y_{|m})$$

is smooth.

A *Riemannian manifold* is a pair (M, g) with M a manifold and g a metric on M.

Here are some (canonical) examples:

1. Let $(\,,\,)$ denote the inner product on $\mathbb{R}^n$.

   An open $U \subset \mathbb{R}^n$ gets a Riemannian metric via $U_m \cong \mathbb{R}^n$:
$$g_m(v, w) = (v, w).$$

2. Let $S^n \subset \mathbb{R}^{n+1}$ be the unit sphere. Then $S^n_m \cong m^\perp \subset \mathbb{R}^{n+1}$ and so gets a metric from the inner product on $\mathbb{R}^{n+1}$.

3. Let $D^n \subset \mathbb{R}^n$ be the open unit disc but define a metric by

$$g_z(v,w) = \frac{4(v,w)}{(1-|z|^2)^2}$$

(D^n, g) is *hyperbolic space.*

Much of the power of Riemannian geometry comes from the fact that there is a *canonical* choice of connection. Consider the following two desirable properties for a connection ∇ on (M,g):

1. ∇ is *metric*: $Xg(Y,Z) = g(\nabla_X Y, Z) + g(Y, \nabla_X Z)$.

2. ∇ is *torsion-free*: $\nabla_X Y - \nabla_Y X = [X,Y]$

Theorem. *There is a unique torsion-free metric connection on any Riemannian manifold.*

Proof. Assume that g is metric and torsion-free. Then

$$\begin{aligned} g(\nabla_X Y, Z) &= Xg(Y,Z) - g(Y, \nabla_X Z) \\ &= Xg(Y,Z) - g(Y,[X,Z]) - g(Y, \nabla_Z X)\ldots \end{aligned}$$

and eventually we get

$$\begin{aligned} 2g(\nabla_X Y, Z) = {} & Xg(Y,Z) + Yg(Z,Y) - Zg(X,Y) \\ & - g(X,[Y,Z]) + g(Y,[Z,X]) + g(Z,[X,Y]). \end{aligned} \tag{2.1}$$

This formula shows uniqueness and, moreover, *defines* the desired connection. □

This connection is the *Levi–Civita connection* of (M,g).

For detailed computations, it is sometimes necessary to express the metric and Levi–Civita connection in terms of local coordinates. So let (U,x) be a chart and $\partial_1, \ldots, \partial_n$ be the corresponding vector fields on U. We now define $g_{ij} \in C^\infty(U)$ by

$$g_{ij} = g(\partial_i, \partial_j)$$

and *Christoffel symbols* $\Gamma^k_{ij} \in C^\infty(U)$ by

$$\nabla_{\partial_i}\partial_j = \sum_k \Gamma^k_{ij}\partial_k.$$

(Recall that $\partial_{1|m}, \dots, \partial_{n|m}$ form a basis for M_m.)

Now let (g^{ij}) be the matrix inverse to (g_{ij}). Then the formula (2.1) for ∇ reads:

$$\Gamma^k_{ij} = \tfrac{1}{2} \sum_l g^{kl}(\partial_i g_{jl} + \partial_j g_{li} - \partial_l g_{ij}) \tag{2.2}$$

since the bracket terms $[\partial_i, \partial_j]$ vanish (exercise!).

2.2 Differential operators

The metric and Levi–Civita connection of a Riemannian manifold are precisely the ingredients one needs to generalise the familiar operators of vector calculus:

The *gradient of* $f \in C^\infty(M)$ is the vector field $\operatorname{grad} f$ such that, for $Y \in \Gamma(TM)$,

$$g(\operatorname{grad} f, Y) = Yf.$$

Similarly, the *divergence* of $X \in \Gamma(TM)$ is the function $\operatorname{div} f \in C^\infty(M)$ defined by:

$$(\operatorname{div} f)(m) = \operatorname{trace}(\xi \to \nabla_\xi X)$$

Finally, we put these together to introduce the hero of this volume: the *Laplacian* of $f \in C^\infty(M)$ is the function

$$\Delta f = \operatorname{div} \operatorname{grad} f.$$

In a chart (U, x), set $\mathrm{g} = \det(g_{ij})$. Then

$$\operatorname{grad} f = \sum_{i,j} g^{ij}(\partial_i f)\partial_j$$

and, for $X = \sum_i X_i \partial_i$,

$$\begin{aligned}\operatorname{div} X &= \sum_i (\partial_i X_i + \sum_j \Gamma^i_{ij} X_j) \\ &= \frac{1}{\sqrt{\mathrm{g}}} \sum_j \partial_j(\sqrt{\mathrm{g}} X_j).\end{aligned}$$

Here we have used $\sum_i \Gamma^i_{ij} = (\partial_j \sqrt{\mathrm{g}})/\sqrt{\mathrm{g}}$ which the Reader is invited to deduce from (2.2) together with the well-known formula for a matrix-valued function A:

$$\mathrm{d} \ln \det A = \operatorname{trace} A^{-1} \mathrm{d}A.$$

In particular, we conclude that

$$\Delta f = \frac{1}{\sqrt{\mathrm{g}}}\sum_{i,j}\partial_i(\sqrt{\mathrm{g}}g^{ij}\partial_j f) = \sum_{i,j} g^{ij}(\partial_i\partial_j f - \Gamma^k_{ij}\partial_k f).$$

2.3 Integration on Riemannian manifolds

2.3.1 Riemannian measure

(M, g) has a canonical measure dV on its Borel sets which we define in steps: First let (U, x) be a chart and $f : U \to \mathbb{R}$ a measureable function. We set

$$\int_U f\,dV = \int_{x(U)} (f \circ x^{-1})\sqrt{\mathrm{g} \circ x^{-1}}\,\mathrm{d}x^1 \ldots \mathrm{d}x^n.$$

Fact. *The change of variables formula ensures that this integral is well-defined on the intersection of any two charts.*

To get a globally defined measure, we patch things together with a *partition of unity*: since M is second countable and locally compact, it follows that every open cover of M has a locally finite refinement. A *partition of unity* for a locally finite open cover $\{U_\alpha\}$ is a family of functions $\phi_\alpha \in C^\infty(M)$ such that

1. $\mathrm{supp}(\phi_\alpha) \subset U_\alpha$;
2. $\sum_\alpha \phi_\alpha = 1$.

Theorem. [6, Theorem 1.11] *Any locally finite cover has a partition of unity.*

Armed with this, we choose a locally finite cover of M by charts $\{(U_\alpha, x_\alpha)\}$, a partition of unity $\{\phi_\alpha\}$ for $\{U_\alpha\}$ and, for measurable $f : M \to \mathbb{R}$, set

$$\int_M f\,dV = \sum_\alpha \int_{U_\alpha} \phi_\alpha f\,dV.$$

Fact. *This definition is independent of all choices.*

2.3.2 The Divergence Theorem

Let $X \in \Gamma(TM)$ have support in a chart (U, x).

$$\begin{aligned}\int_M \operatorname{div} X \, dV &= \int_U \frac{1}{\sqrt{\mathrm{g}}} \partial_i(\sqrt{\mathrm{g}} X_i) \, dV \\ &= \int_{x(U)} (\partial_i \sqrt{\mathrm{g}} X_i) \circ x^{-1} \, \mathrm{d}x^1 \dots \mathrm{d}x^n \\ &= \int_{x(U)} \frac{\partial}{\partial x^i} (\sqrt{\mathrm{g}} X_i) \circ x^{-1} \, \mathrm{d}x^1 \dots \mathrm{d}x^n = 0.\end{aligned}$$

A partition of unity argument immediately gives:

Divergence Theorem I. *Any compactly supported vector field X on M has*

$$\int_M \operatorname{div} X \, dV = 0.$$

Just as in vector calculus, the divergence theorem quickly leads to Green's formulae. Indeed, for $f, h \in C^\infty(M)$, $X \in \Gamma(TM)$ one easily verifies:

$$\operatorname{div}(fX) = f \operatorname{div} X + g(\operatorname{grad} f, X)$$

whence

$$\begin{aligned}\operatorname{div}(f \operatorname{grad} h) &= f\Delta h + g(\operatorname{grad} h, \operatorname{grad} f) \\ \Delta(fh) &= f\Delta h + 2g(\operatorname{grad} h, \operatorname{grad} f) + h\Delta f.\end{aligned}$$

The divergence theorem now gives us Green's Formulae:

Theorem. *For $f, h \in C^\infty(M)$ with at least one of f and h compactly supported:*

$$\begin{aligned}\int_M h\Delta f \, dV &= -\int_M g(\operatorname{grad} f, \operatorname{grad} h) \, dV \\ \int_M h\Delta f \, dV &= \int_M f\Delta h \, dV.\end{aligned}$$

2.3.3 Boundary terms

Supposed that M is oriented and that $\Omega \subset M$ is an open subset with smooth boundary $\partial\Omega$. Thus $\partial\Omega$ is a smooth manifold with

1. a Riemannian metric inherited via $(\partial\Omega)_m \subset M_m$;

2. a Riemannian measure dA;

3. a unique outward-pointing normal unit vector field ν.

With these ingredients, one has:

Divergence Theorem II. *Any compactly supported X on M has*

$$\int_\Omega \operatorname{div} X \, dV = \int_{\partial\Omega} g(X, \nu) \, dA$$

and so Green's Formulae:

Theorem. *For $f, h \in C^\infty(M)$ with at least one of f and h compactly supported:*

$$\int_\Omega h\Delta f + \langle \operatorname{grad} f, \operatorname{grad} h\rangle \, dV = \int_{\partial\Omega} h\langle \nu, \operatorname{grad} f\rangle \, dA$$

$$\int_\Omega h\Delta f - \int_\Omega f\Delta h \, dV = \int_{\partial\Omega} h\langle \nu, \operatorname{grad} f\rangle \, dA - \int_{\partial\Omega} f\langle \nu, \operatorname{grad} h\rangle \, dA$$

where we have written $\langle\, ,\, \rangle$ for $g(\, ,\,)$.

In particular

$$\int_\Omega \Delta f \, dV = \int_{\partial\Omega} \nu f \, dV.$$

3 Geodesics and curvature

In the classical geometry of Euclid, a starring role is played by the straight lines. Viewed as paths of shortest length between two points, these may be generalised to give a distinguished family of paths, the *geodesics*, on any Riemannian manifold. Geodesics provide a powerful tool to probe the geometry of Riemannian manifolds.

Notation. Let (M, g) be a Riemannian manifold. For $\xi, \eta \in M_m$, write

$$g(\xi, \eta) = \langle \xi, \eta \rangle, \qquad \sqrt{g(\xi, \xi)} = |\xi|.$$

3.1 (M, g) is a metric space

A piece-wise C^1 path $\gamma : [a, b] \to M$ has *length* $L(\gamma)$:

$$L(\gamma) = \int_a^b |\gamma'(t)| \, dt.$$

Exercise. The length of a path is invariant under reparametrisation.

Recall that M is connected and so[3] path-connected. For $p,\, q \in M$, set

$$d(p,q) = \inf\{L(\gamma) : \gamma : [a,b] \to M \text{ is a path with } \gamma(a) = p,\ \gamma(b) = q\}.$$

One can prove:

- (M, d) is a metric space.
- The metric space topology coincides with the original topology on M.

The key points here are the definiteness of d and the assertion about the topologies. For this, it is enough to work in a precompact open subset of a chart U where one can prove the existence of $K_1, K_2 \in \mathbb{R}$ such that

$$K_1 \sum_{1 \le i \le n} \xi_i^2 \le \sum_{i,j} g_{ij}\xi_i\xi_j \le K_2 \sum_{1 \le i \le n} \xi_i^2.$$

From this, one readily sees that, on such a subset, d is equivalent to the Euclidean metric on U.

3.2 Parallel vector fields and geodesics

Let $c : I \to M$ be a path. Recall the pull-back connection $c^{-1}\nabla$ on the space $\Gamma(c^{-1}TM)$ of vector fields along c. This connection gives rise to a differential operator

$$\nabla_t : \Gamma(c^{-1}TM) \to \Gamma(c^{-1}TM)$$

by

$$\nabla_t Y = (c^{-1}\nabla)_{\partial_1} Y$$

where ∂_1 is the coordinate vector field on I.

Note that since ∇ is metric, we have

$$\langle X, Y\rangle' = \langle \nabla_t X, Y\rangle + \langle X, \nabla_t Y\rangle,$$

for $X, Y \in \Gamma(c^{-1}TM)$.

Definition. $X \in \Gamma(c^{-1}TM)$ is *parallel* if $\nabla_t X = 0$.

The existence and uniqueness results for linear ODE give:

[3]Manifolds are locally path-connected!

Proposition. *For $c : [a, b] \to M$ and $U_0 \in M_{c(a)}$, there is unique parallel vector field U along c with*

$$U(a) = U_0.$$

If Y_1, Y_2 are parallel vector fields along c, then all $\langle Y_i, Y_j \rangle$ and, in particular, $|Y_i|$ are constant.

Definition. $\gamma : I \to M$ is a *geodesic* if γ' is parallel:

$$\nabla_t \gamma' = 0.$$

It is easy to prove that, for a geodesic γ:

- $|\gamma'|$ is constant.
- If γ is a geodesic, so is $t \mapsto \gamma(st)$ for $s \in \mathbb{R}$.

The existence and uniqueness results for ODE give:

1. For $\xi \in M_m$, there is a maximal open interval $I_\xi \subset \mathbb{R}$ on which there is a unique geodesic $\gamma_\xi : I_\xi \to M$ such that

$$\gamma_\xi(0) = m$$
$$\gamma'_\xi(0) = \xi.$$

2. $(t, \xi) \mapsto \gamma_\xi(t)$ is a smooth map $I_\xi \times M_m \to M$.
3. $\gamma_{s\xi}(t) = \gamma_\xi(st)$.

Let us collect some examples:

1. $M = \mathbb{R}^n$ with its canonical metric. The geodesic equation reduces to:

$$\frac{\mathrm{d}^2\gamma}{\mathrm{d}t^2} = 0$$

and we conclude that geodesics are straight lines.

2. $M = S^n$ and ξ is a unit vector in $M_m = m^\perp$. Contemplate reflection in the 2-plane spanned by m and ξ: this induces a map $\Phi : S^n \to S^n$ which preserves the metric and so ∇ also while it fixes m and ξ. Thus, if γ is a geodesic so is $\Phi \circ \gamma$ and the uniqueness part of the ODE yoga forces $\Phi \circ \gamma_\xi = \gamma_\xi$. Otherwise said, γ_ξ lies in the plane spanned by m and ξ and so lies on a great circle.

To get further, recall that $|\gamma'_\xi| = |\xi| = 1$ which implies:

$$\gamma_\xi(t) = (\cos t)m + (\sin t)\xi.$$

A similar argument shows that the unique parallel vector field U along γ_ξ with $U(0) = \eta \perp \xi$ is given by

$$U \equiv \eta.$$

3. $M = D^n$ with the hyperbolic metric and ξ is a unit vector in $M_0 \cong \mathbb{R}^n$. Again, symmetry considerations force γ_ξ to lie on the straight line through 0 in the direction of ξ and then $|\gamma'_\xi| = 1$ gives:

$$\gamma_\xi(t) = (2\tanh t/2)\xi.$$

Similarly, the parallel vector field along γ_ξ with $U(0) = \eta \perp \xi$ is given by

$$U(t) = \frac{1}{\cosh^2 t/2}\eta.$$

3.3 The exponential map

3.3.1 Normal coordinates

Set $\mathcal{U}_m = \{\xi \in M_m : 1 \in I_\xi\}$ and note that $\mathcal{U}_m$ is a star-shaped open neighbourhood of $0 \in M_m$. We define the *exponential map* $\exp_m : \mathcal{U}_m \to M$ by

$$\exp_m(\xi) = \gamma_\xi(1).$$

Observe that, for all $t \in I_\xi$,

$$\exp_m(t\xi) = \gamma_{t\xi}(1) = \gamma_\xi(t)$$

and differentiating this with respect to t at $t = 0$ gives

$$\xi = \gamma'_\xi(0) = (d\exp_m)_0(\xi)$$

so that $(d\exp_m)_0 = 1_{M_m}$. Thus, by the inverse function theorem, $\exp_m$ is a local diffeomorphism whose inverse is a chart.

Indeed, if $e_1, \dots, e_n$ is an orthonormal basis of M_m, we have *normal coordinates* $x^1, \dots, x^n$ given by

$$x^i = \langle (\exp_m)^{-1}, e_i \rangle$$

for which

$$g_{ij}(m) = \delta_{ij}$$
$$\Gamma^k_{ij}(m) = 0.$$

3.3.2 The Gauss Lemma

Let ξ, $\eta \in M_m$ with $|\xi| = 1$ and $\xi \perp \eta$.

The **Gauss Lemma** says:

$$\langle (d\exp_m)_{t\xi}\eta, \gamma'_\xi(t)\rangle = 0.$$

Thus γ_ξ intersects the image under $\exp_m$ of spheres in M_m orthogonally.

As an application, let us show that geodesics are locally length-minimising. For this, choose $\delta > 0$ sufficiently small that

$$\exp_m : B(0,\delta) \subset M_m \to M$$

is a diffeomorphism onto an open set $U \subset M$. Let $c : I \to U$ be a path from m to $p \in U$ and let $\gamma : I \to U$ be the geodesic from m to p: thus γ is the image under $\exp_m$ of a radial line segment in $B(0,\delta)$.

Write

$$c(t) = \exp_m(r(t)\xi(t))$$

with $r : I \to \mathbb{R}$ and $\xi : I \to S^n \subset M_m$. Now

$$\begin{aligned}\langle c'(t), c'(t)\rangle &= (r')^2 + r^2\langle (\mathrm{d}\exp_m)_{r\xi}\xi', (\mathrm{d}\exp_m)_{r\xi}\xi'\rangle + \quad 2rr'\langle (\mathrm{d}\exp_m)_{r\xi}\xi', \gamma'_\xi\rangle \\ &= (r')^2 + r^2\langle (\mathrm{d}\exp_m)_{r\xi}\xi', (\mathrm{d}\exp_m)_{r\xi}\xi'\rangle\end{aligned}$$

by the Gauss lemma (since $\xi' \perp \xi$). In particular,

$$\langle c'(t), c'(t)\rangle \geq (r')^2.$$

Taking square roots and integrating gives:

$$L(c) \geq \int_a^b |r'|\,\mathrm{d}t \geq \left|\int_a^b r'\,\mathrm{d}t\right| = |r(b) - r(a)| = L(\gamma).$$

From this we conclude:

$$L(\gamma) = d(m,p)$$

and

$$B_d(m,\delta) = \exp_m B(0,\delta).$$

Definition. A geodesic γ is *minimising on* $[a,b] \subset I_\gamma$ if

$$L(\gamma_{|[a,b]}) = d(\gamma(a), \gamma(b)).$$

We have just seen that any geodesic is minimising on sufficiently small intervals.

3.3.3 The Hopf–Rinow Theorem

Definition. (M, g) is *geodesically complete* if $I_\xi = \mathbb{R}$, for any $\xi \in \mathbb{R}$.

This only depends on the metric space structure of (M, d):

Theorem (Hopf–Rinow). *The following are equivalent:*

1. *(M, g) is geodesically complete.*
2. *For some $m \in M$, $\exp_m$ is a globally defined surjection $M_m \to M$.*
3. *Closed, bounded subsets of (M, d) are compact.*
4. *(M, d) is a complete metric space.*

In this situation, one can show that any two points of M can be joined by a minimising geodesic.

3.4 Sectional curvature

Let $\sigma \subset M_m$ be a 2-plane with orthonormal basis ξ, η.
The *sectional curvature* $\mathcal{K}(\sigma)$ of σ is given by

$$\mathcal{K}(\sigma) = \langle R(\xi, \eta)\xi, \eta\rangle.$$

Facts:

- This definition is independent of the choice of basis of σ.
- $\mathcal{K}$ determines the curvature tensor R.

Definition. (M, g) *has constant curvature* κ if $\mathcal{K}(\sigma) = \kappa$ for all 2-planes σ in TM.

In this case, we have

$$R(\xi, \eta)\zeta = \kappa\{\langle \xi, \zeta\rangle\eta - \langle \eta, \zeta\rangle\xi\}.$$

$\mathcal{K}$ is a function on the set (in fact manifold) $G_2(TM)$ of all 2-planes in all tangent spaces M_m of M. A diffeomorphism $\Phi : M \to M$ induces $\mathrm{d}\Phi : TM \to TM$ which is a linear isomorphism on each tangent space and so gives a mapping $\hat{\Phi} : G_2(TM) \to G_2(TM)$. Suppose now that Φ is an *isometry*:

$$\langle \mathrm{d}\Phi_m(\xi), \mathrm{d}\Phi_m(\eta)\rangle = \langle \xi, \eta\rangle,$$

for all $\xi, \eta \in M_m$, $m \in M$. Since an isometry preserves the metric, it will preserve anything built out of the metric such as the Levi–Civita connection and its curvature. In particular, we have

$$\mathcal{K} \circ \hat{\Phi} = \mathcal{K}.$$

It is not two difficult to show that, for our canonical examples, the group of all isometries acts *transitively* on $G_2(TM)$ so that $\mathcal{K}$ is constant. Thus we arrive at the following examples of manifolds of constant curvature:

1. $\mathbb{R}^n$.
2. $S^n(r)$.
3. $D^n(\rho)$ with metric

$$g_{ij} = \frac{4\delta_{ij}}{(1 - |z|^2/\rho^2)^2}.$$

It can be shown that these exhaust all complete, simply-connected possibilities.

3.5 Jacobi fields

Definition. Let $\gamma : I \to M$ be a unit speed geodesic. Say $Y \in \Gamma(\gamma^{-1}TM)$ is a *Jacobi field along* γ if

$$\nabla_t^2 Y + R(\gamma', Y)\gamma' = 0.$$

Once again we wheel out the existence and uniqueness theorems for ODE which tell us:

Proposition. *For $Y_0, Y_1 \in M_{\gamma(0)}$, there is a unique Jacobi field Y with*

$$Y(0) = Y_0$$
$$(\nabla_t Y)(0) = Y_1$$

Jacobi fields are infinitesimal variations of γ through a family of geodesics. Indeed, suppose that $h : I \times (-\epsilon, \epsilon) \to M$ is a variation of geodesics: that is, each $\gamma_s : t \to h(t, s)$ is a geodesic. Set $\gamma = \gamma_0$ and let

$$Y = \left.\frac{\partial h}{\partial s}\right|_{s=0} \in \Gamma(\gamma^{-1}TM).$$

Let ∂_t and ∂_s denote the coordinate vector fields on $I \times (-\epsilon, \epsilon)$ and set $D \doteq h^{-1}\nabla$. Since each γ_s is a geodesic, we have

$$D_{\partial_t}\frac{\partial h}{\partial t} = 0$$

whence

$$D_{\partial_s}D_{\partial_t}\frac{\partial h}{\partial t} = 0.$$

The definition of the curvature tensor, along with the fact that $[\partial_s, \partial_t] = 0$, allows us to write

$$0 = D_{\partial_s}D_{\partial_t}\frac{\partial h}{\partial t} = D_{\partial_t}D_{\partial_s}\frac{\partial h}{\partial t} + R(\frac{\partial h}{\partial t}, \frac{\partial h}{\partial s})\frac{\partial h}{\partial t}.$$

Moreover, it follows from the fact that ∇ is torsion-free that

$$D_{\partial_s}\frac{\partial h}{\partial t} = D_{\partial_t}\frac{\partial h}{\partial s}$$

so that

$$0 = (D_{\partial_t})^2\frac{\partial h}{\partial s} + R(\frac{\partial h}{\partial t}, \frac{\partial h}{\partial s})\frac{\partial h}{\partial t}.$$

Setting $s = 0$, this last becomes

$$(\nabla_t)^2 Y + R(\gamma', Y)\gamma' = 0.$$

Fact. *All Jacobi fields arise this way.*

Let us contemplate an example which will compute for us the (constant) value of $\mathcal{K}$ for hyperbolic space: let (D^n, g) be hyperbolic space and consider a path $\xi : (-\epsilon, \epsilon) \to S^{n-1} \subset D_0$ with $\xi'(0) = \eta \perp \xi(0)$.

We set $h(t, s) = \gamma_{\xi(s)}(t) = (2\tanh t/2)\xi(s)$—a variation of geodesics through 0. We then have a Jacobi field Y along $\gamma = \gamma_{\xi(0)}$:

$$\begin{aligned} Y(t) = \left.\frac{\partial h}{\partial s}\right|_{s=0} &= 2(\tanh t/2)\eta \\ &= \sinh t(\eta/\cosh^2 t/2) \\ &= \sinh t U(t) \end{aligned}$$

where U is a unit length *parallel* vector field along γ.

We therefore have:

$$(\nabla_t)^2 Y = \sinh'' t U(t) = \sinh t U(t)$$

whence

$$U + R(\gamma', U)\gamma' = 0.$$

Take an inner product with U to get

$$\mathcal{K}(\gamma' \wedge U) = -1$$

and so conclude that (D^n, g) has constant curvature -1.

The same argument (that is, differentiate the image under $\exp_m$ of a family of straight lines through the origin) computes Jacobi fields in normal coordinates:

Theorem. *For $\xi \in M_m$, the Jacobi field Y along γ_ξ with*

$$Y(0) = 0$$
$$(\nabla_t Y)(0) = \eta \in M_m$$

is given by

$$Y(t) = (\mathrm{d}\exp)_{t\xi} t\eta.$$

3.6 Conjugate points and the Cartan–Hadamard theorem

Let $\xi \in M_p$ and let $\gamma = \gamma_\xi : I_\xi \to \mathbb{R}$. We say that $q = \gamma(t_1)$ *is conjugate to p along γ* if there is a non-zero Jacobi field Y with

$$Y(0) = Y(t_1) = 0.$$

In view of the theorem just stated, this happens exactly when $(\mathrm{d}\exp_p)_{t_1\xi}$ is singular.

Theorem (Cartan–Hadamard). *If (M, g) is complete and $\mathcal{K} \leq 0$ then no $p \in M$ has conjugate points.*

Proof. Suppose that Y is a Jacobi field along some geodesic γ with $Y(0) = Y(t_1) = 0$. Then

$$\begin{aligned} 0 &= \int_0^{t_1} \langle \nabla_t^2 Y + R(\gamma', Y)\gamma', Y\rangle \,\mathrm{d}t \\ &= -\int_0^{t_1} |\nabla_t T|^2 \,\mathrm{d}t + \int_0^{t_1} \mathcal{K}(\gamma' \wedge Y)|Y|^2 \,\mathrm{d}t \end{aligned}$$

where we have integrated by parts and used $Y(0) = Y(t_1) = 0$ to kill the boundary term. Now both summands in this last equation are non-negative and so must vanish. In particular,

$$\nabla_t Y = 0$$

so that Y is parallel whence $|Y|$ is constant giving eventually $Y \equiv 0$. □

From this we see that, under the hypotheses of the theorem, each $\exp_m$ is a local diffeomorphism and, with a little more work, one can show that $\exp_m : M_m \to M$ is a covering map. Thus:

Corollary. *If (M, g) is complete and $\mathcal{K} \leq 0$ then*

1. *if $\pi_1(M) = 1$ then M is diffeomorphic to $\mathbb{R}^n$.*
2. *In any case, the universal cover of M is diffeomorphic to $\mathbb{R}^n$ whence $\pi_k(M) = 1$ for all $k \geq 2$.*

Analysis of this kind is the starting point of one of the central themes of modern Riemannian geometry: the interplay between curvature and topology.

4 The Bishop volume comparison theorem

Our aim is to prove a Real Live Theorem in Riemannian geometry: the theorem is of considerable interest in its own right and proving it will exercise everything we have studied in these notes.

We begin by collecting some ingredients.

4.1 Ingredients

4.1.1 Ricci curvature

Definition. The *Ricci tensor at $m \in M$* is the bilinear map $\mathrm{Ric} : M_m \times M_m \to \mathbb{R}$ given by

$$\begin{aligned} \mathrm{Ric}(\xi, \eta) &= \mathrm{trace}(\zeta \mapsto R(\xi, \zeta)\eta) \\ &= \sum_i \langle R(\xi, e_i)\eta, e_i \rangle \end{aligned}$$

where $e_1, \ldots, e_n$ is an orthonormal basis of M_m.

Exercise. The Ricci tensor is symmetric: $\mathrm{Ric}(\xi, \eta) = \mathrm{Ric}(\eta, \xi)$.

Example. If (M, g) has dimension n and constant curvature κ then

$$\mathrm{Ric} = (n-1)\kappa g.$$

The Ricci tensor, being only bilinear, is much easier to think about than the curvature tensor. On the other hand, being only an average of sectional curvatures, conditions of the Ricci tensor say much less about the topology of the underlying manifold. For example, here is an amazing theorem of Lohkamp [4]:

Theorem. *Any manifold of dimension at least* 3 *admits a complete metric with* Ric < 0 *(that is* Ric *is neagtive definite).*

4.1.2 Cut locus

Henceforth, we will take M to be complete of dimension n.

For $\xi \in M_m$ with $|\xi| = 1$, define $c(\xi) \in \mathbb{R}^+ \cup \{\infty\}$ by

$$\begin{aligned} c(\xi) &= \sup\{t \colon \gamma_\xi|_{[0,t]} \text{ is minimising}\} \\ &= \sup\{t \colon d(m, \gamma_\xi(t)) = t\}. \end{aligned}$$

The *cut locus* C_m of m is given by

$$C_m = \exp_m\{c(\xi)\xi \colon \xi \in S^{n-1} \subset M_m,\ c(\xi) < \infty\}$$

while $\mathcal{D}_m = \{t\xi \colon \xi \in S^{n-1} \subset M_m,\ t \in [0, c(\xi))\}$ and

$$D_m = \exp_m \mathcal{D}_m.$$

We have:

- $M = D_m \cup C_m$ is a disjoint union.
- $\exp_m : \mathcal{D}_m \to D_m$ is a diffeomorphism.
- $\int_{C_m} dV = 0$.

These facts have practical consequences for integration on M: for $f : M \to \mathbb{R}$ integrable,

$$\begin{aligned} \int_M f\, dV &= \int_{\mathcal{D}_m} f(\exp(x)) \sqrt{\mathrm{g}}\, \mathrm{d}x^1 \dots \mathrm{d}x^n \\ &= \int_{S^{n-1}} \int_0^{c(\xi)} f(\exp(r\xi)) \mathrm{a}(r, \xi)\, \mathrm{d}r \mathrm{d}\xi \end{aligned}$$

where $x^1, \dots, x^n$ are orthonormal coordinates on $\mathcal{D}_m$ and $\mathrm{d}\xi$ is Lebesgue measure on $S^{n-1} \subset M_m$.

Example. For $\kappa \in \mathbb{R}$, let $S_\kappa : \mathbb{R} \to \mathbb{R}$ solve

$$\begin{gathered} S_\kappa'' + \kappa S_\kappa = 0 \\ S_\kappa(0) = 0, \quad S_\kappa'(0) = 1 \end{gathered}$$

Then, if (M, g) has constant curvature κ,

$$\mathrm{a}(r, \xi) = S_\kappa^{n-1}(r).$$

4.2 Bishop's Theorem

4.2.1 Manifesto

Fix $\kappa \in \mathbb{R}$ and $m \in M_m$.

Let $V(m,r)$ denote the volume of $B_d(m,r) \subset M$ and $V_\kappa(r)$ the volume of a radius r ball in a complete simply-connected n-dimensional space of constant curvature κ.

Suppose that $\mathrm{Ric}(\xi,\xi) \geq (n-1)\kappa g(\xi,\xi)$ for all $\xi \in TM$. For each $\xi \in M_m$ of unit length, define $a_\xi : (0, c(\xi)) \to \mathbb{R}$ by

$$a_\xi(t) = \mathbf{a}(t,\xi).$$

We will prove that

$$\frac{a'_\xi}{a_\xi} \leq (n-1)\frac{S'_\kappa}{S_\kappa}.$$

As a consequence, we will see that

$$V(m,r) \leq V_\kappa(r)$$

and even that $V(m,r)/V_\kappa(r)$ is decreasing with respect to r.

4.2.2 Laplacian of the distance function

Our strategy will be to identify the radial logarithmic derivative of $\mathbf{a}$ with the Laplacian of the distance from m. We will then be able to apply a formula of Lichnerowicz to derive a differential inequality for a_ξ.

So view r as a function on M:

$$r(x) = d(m,x).$$

Then

Proposition. $\mathbf{a}^{-1}\partial\mathbf{a}/\partial r = \Delta r \circ \exp_m$.

Here is a fast[4] proof stolen from [3]: for $U \subset S^{n-1} \subset M_m$ and $[t, t+\epsilon]$ such that

$$\Omega_{t,\epsilon} = \{\exp_m(r\xi) : r \in [t, t+\epsilon], \xi \in U\} \subset D_m$$

[4]Isaac Chavel rightly objects that this proof is all a bit too slick. See his contribution to this volume for a more down to earth proof.

we have:

$$\int_{\Omega_{t,\epsilon}} \Delta r \, dV = \int_{[t,t+\epsilon]\times U} (\Delta \circ \exp_m) \mathbf{a} \, dr d\xi.$$

However, the divergence theorem gives

$$\begin{aligned}\int_{\Omega_{t,\epsilon}} \Delta \, dV = \int_{\partial\Omega_{t,\epsilon}} \langle \operatorname{grad} r, \nu \rangle \, dA &= \int_U \mathbf{a}(t+\epsilon) \, d\xi - \int_U \mathbf{a}(t) \, d\xi \\ &= \int_U \int_t^{t+\epsilon} \frac{\partial \mathbf{a}}{\partial r}(r,\xi) \, dr d\xi.\end{aligned}$$

Here we have used that $\langle \operatorname{grad} r, \nu \rangle = \nu r = 1$ along the spherical parts of $\partial\Omega_{t,\epsilon}$ and vanishes along the radial parts.

Thus

$$\int_{[t,t+\epsilon]\times U} (\Delta \circ \exp_m) \mathbf{a} \, dr d\xi = \int_{[t,t+\epsilon]\times U} \frac{\partial \mathbf{a}}{\partial r}(r,\xi) \, dr d\xi$$

and, since t, ϵ and U were arbitrary, we get

$$\mathbf{a}(\Delta \circ \exp_m) = \frac{\partial \mathbf{a}}{\partial r}$$

as required.

4.2.3 Lichnerowicz' formula

For $X \in \Gamma(TM)$, define $|\nabla X|^2$ by

$$|\nabla X|^2(m) = \sum_i |\nabla_{e_i} X|^2$$

where $e_1, \ldots, e_n$ is an orthonormal basis of M_m—this is independent of choices.

We now have

Lichnernowicz' Formula. *Let $f : M \to \mathbb{R}$ then*

$$\tfrac{1}{2}\Delta |\operatorname{grad} f|^2 = |\nabla \operatorname{grad} f|^2 + \langle \operatorname{grad} \Delta f, \operatorname{grad} f \rangle + \operatorname{Ric}(\operatorname{grad} f, \operatorname{grad} f).$$

The proof of this is an exercise (really!) but here are some hints to get you started: the basic identity

$$XYf - YXf = [X,Y]$$

along with the fact that ∇ is metric and torsion-free gives:

$$\langle \nabla_X \operatorname{grad} f, Y\rangle = \langle \nabla_Y \operatorname{grad} f, X\rangle$$

from which you can deduce that

$$\tfrac{1}{2} \operatorname{grad}|\operatorname{grad} f|^2 = \nabla_{\operatorname{grad} f} \operatorname{grad} f$$

whence

$$\begin{aligned} \tfrac{1}{2}\Delta|\operatorname{grad} f|^2 &= \operatorname{div} \nabla_{\operatorname{grad} f} \operatorname{grad} f \\ &= \sum_i \langle \nabla_{e_i} \nabla_{\operatorname{grad} f} \operatorname{grad} f, e_i\rangle. \end{aligned}$$

Now make repeated use of the metric property of ∇ and use the definition of R to change the order of the differentiations ...

As an application, put $f = r$. Thanks to the Gauss lemma, $\operatorname{grad} f = \partial_r$ so that $|\operatorname{grad} f| = 1$ and the Lichnerowicz formula reads:

$$0 = |\nabla \operatorname{grad} r|^2 + \partial_r \Delta r + \operatorname{Ric}(\partial_r, \partial_r). \tag{4.1}$$

On the image of γ_ξ, we have

$$\partial_r \Delta r = (a'_\xi/a_\xi)' = a''_\xi/a_\xi - (\Delta r)^2$$

and plugging this into (4.1) gives

$$0 = a''_\xi/a_\xi - (\Delta r)^2 + |\nabla \operatorname{grad} r|^2 + \operatorname{Ric}(\partial_r, \partial_r)$$

or, defining b by $b^{n-1} = a_\xi$ so that $(n-1)b'/b = a'_\xi/a_\xi$,

$$(n-1)b''/b + \operatorname{Ric}(\partial_r, \partial_r) = -\left(|\nabla \operatorname{grad} r|^2 - \frac{1}{n-1}(\Delta r)^2\right). \tag{4.2}$$

4.2.4 Estimates and comparisons

We now show that the right hand side of (4.2) has a sign: choose an orthonormal basis $e_1, \dots, e_n$ of $M_{\gamma_\xi(t)}$ with $e_1 = \partial_r$. Then

$$\begin{aligned} \Delta r &= \sum \langle \nabla_{e_i} \operatorname{grad} r, e_i\rangle \\ &= \sum_{i\geq 2} \langle \nabla_{e_i} \operatorname{grad} r, e_i\rangle \end{aligned}$$

since $\nabla_{\partial_r} \operatorname{grad} r = \nabla_t \gamma'_\xi = 0$.

Two applications of the Cauchy–Schwarz inequality give

$$(\Delta r)^2 \le \left(\sum_{i\ge 2}|\nabla_{e_i}\operatorname{grad} r|\right)^2 \le (n-1)\sum_{i\ge 2}|\nabla_{e_i}\operatorname{grad} r|^2$$

so that

$$|\nabla \operatorname{grad} r|^2 - \frac{1}{n-1}(\Delta r)^2 \ge 0.$$

Thus (4.2) gives

$$(n-1)b''/b + \operatorname{Ric}(\partial_r, \partial_r) \le 0$$

and, under the hypotheses of Bishop's theorem, we have

$$b''/b \le -\kappa.$$

We now make a simple comparison argument: $b > 0$ on $(0, c(\xi))$ so we have

$$b'' + \kappa b \le 0$$
$$b(0) = 0, \quad b'(0) = 1.$$

On the other hand, set $\bar{b} = S_\kappa$ so that

$$\bar{b}'' + \kappa \bar{b} = 0$$
$$\bar{b}(0) = 0, \quad \bar{b}'(0) = 1$$

We now see that, so long as $\bar{b} \ge 0$, we have

$$\bar{b}b'' - \bar{b}''b \le 0$$

or, equivalently,

$$(b'\bar{b} - \bar{b}'b)' \le 0.$$

In view of the initial conditions, we conclude:

$$b'\bar{b} - \bar{b}'b \le 0. \tag{4.3}$$

Let us pause to observe that at the first zero of $\bar{b}$ (if there is one), $\bar{b}' < 0$ so that, by (4.3), $b \le 0$ also. Since $b > 0$ on $(0, c(\xi))$, we deduce that $\bar{b} > 0$ there also[5].

We therefore conclude from (4.3) that on $(0, c(\xi))$ we have

$$b'/b \le \bar{b}'/\bar{b},$$

or, equivalently,

$$a_\xi'/a_\xi \le (n-1)S_\kappa'/S_\kappa. \tag{4.4}$$

[5] For $\kappa > 0$, this reasoning puts an upper bound on the length of $(0, c(\xi))$ and thus, eventually, on the diameter of M. This leads to a proof of the Bonnet–Myers theorem.

4.2.5 Baking the cake

Equation (4.4) reads

$$\ln(a_\xi/S_\kappa^{n-1})' \leq 0$$

so that, a_ξ/S_κ^{n-1} is decreasing and, in view of the initial conditions,

$$a_\xi \leq S_\kappa^{n-1}.$$

Thus:

$$\begin{aligned} V(m,r) &= \int_{S^{n-1}} \int_0^{\min(c(\xi),r)} a_\xi \, \mathrm{d}r\mathrm{d}\xi \\ &\leq \int_{S^{n-1}} \int_0^{\min(c(\xi),r)} S_\kappa^{n-1} \, \mathrm{d}r\mathrm{d}\xi = V_\kappa(r). \end{aligned}$$

This is Bishop's theorem.

Our final statement is due to Gromov [2] and is a consequence of a simple lemma:

Lemma ([2]). *If $f, g > 0$ with f/g decreasing then*

$$\int_0^r f \Big/ \int_0^r g$$

is decreasing also.

With this in hand, we see that, for $r_1 < r_2$,

$$\int_0^{r_1} a_\xi \, \mathrm{d}r \Big/ \int_0^{r_1} S_\kappa^{n-1} \, \mathrm{d}r \leq \int_0^{r_2} a_\xi \, \mathrm{d}r \Big/ \int_0^{r_2} S_\kappa^{n-1} \, \mathrm{d}r.$$

Integrating this over S^{n-1}, noting that the denominators are independent of ξ, gives finally that $V(m,r)/V_\kappa(r)$ is decreasing.

References

[1] I. Chavel, *Riemannian geometry—a modern introduction*, Cambridge University Press, Cambridge, 1993.

[2] J. Cheeger, M. Gromov, and M. Taylor, *Finite propagation speed, kernel estimates for functions of the Laplace operator, and the geometry of complete Riemannian manifolds*, J. Differential Geom. **17** (1982), no. 1, 15–53.

[3] S. Gallot, D. Hulin, and J. Lafontaine, *Riemannian geometry*, Springer-Verlag, Berlin, 1987.

[4] J. Lohkamp, *Metrics of negative Ricci curvature*, Ann. of Math. (2) **140** (1994), no. 3, 655–683.

[5] M. Spivak, *Calculus on manifolds. A modern approach to classical theorems of advanced calculus*, W. A. Benjamin, Inc., New York-Amsterdam, 1965.

[6] F.W. Warner, *Foundations of differentiable manifolds and Lie groups*, Scott, Foresman and Co., Glenview, Ill.-London, 1971.

The Laplacian on Riemannian Manifolds

Isaac Chavel

Edinburgh Lectures

Contents

§0. Introduction

These are informal notes of talks I gave at the Instructional Conference on Spectral Theory and Geometry, International Centre for Mathematical Sciences, Edinburgh, March 29–April 9, 1998. The first three days featured three introductory mini-courses consisting of three lectures each: (1) E.B. Davies on Friedrichs extensions of densely defined symmetric operators, and max-min methods and their computational aspects, (2) F. Burstall on introductory Riemannian geometry, and (3) myself on the Laplacian on Riemannian manifolds.

Burstall's course started from the definition of a manifold, and surveyed the basic definitions and theorems concerning (with slightly different order) connections, parallel translation, geodesics, exponential map, torsion and curvature, Jacobi fields, Riemannian metrics, Levi-Civita connections, geodesic spherical and Riemann normal coordinates, the conjugate and cut loci of a point, Riemann measure, divergence theorems, and the Laplacian, culminating in an elegant proof of Bishop's volume comparison theorem for Ricci curvature bounded from below (by way of the Lichnerowicz formula).

I will pick up the story from this point, and consider some elementary examples and theorems which are pleasing in their own right, and which are suitable and appropriate for presentation in such a course. We pick and choose in the presentation of detail, if any at all, in the proofs. Also, I have tried, in some strictly Riemannian topics, to complement Burstall's elegant treatment with a more classical approach to some of the same material. In particular, I devote more time than might be warranted to the calculation of the Laplacian in geodesic spherical coordinates. Hopefully, the student will then possess a rich view of the subject. Nearly all the material can be found in either of my books Chavel [**8**] and Chavel [**9**].

No pretense will be made to bibliographic completeness, although I hope to get the reader sufficiently along the road to do both the mathematics and the bibliographic scholarship.

Let M be an n–dimensional Riemannian manifold. A Riemannian metric on M assigns to each pair of tangent vectors ξ, η, with same base point, an inner product $\langle \xi, \eta\rangle$ and associated norm $|\xi| = \sqrt{\langle \xi, \xi\rangle}$. This inner product varies smoothly over M, in the sense that if X, Y are smooth vector fields on M then $\langle X, Y\rangle$ is a smooth function on M.

If $\mathbf{x} : U \to \mathbf{R}^n$ is a chart on M, $\{\partial_1, \ldots, \partial_n\}$ the naturally associated coordinate vector fields on M, then the Riemannian metric is given relative to these vector fields by

$$g_{ij} = \langle \partial_i, \partial_j \rangle, \quad \mathcal{G} = (g_{ij}), \quad \mathcal{G}^{-1} = (g^{ij}), \quad g = \det \mathcal{G} > 0.$$

The Riemannian measure dV on M is given locally by

$$dV = \sqrt{g}\, dx^1 \cdots dx^n,$$

and is turned into a global measure using a partition of unity. For any two functions f, g in L^2 we have the inner product and norm

$$(f,g) = \int_M fg\, dV, \qquad \|f\|^2 = \int_M f^2\, dV.$$

We may also speak of L^2–vector fields, and associated inner product and norm

$$(X,Y) = \int_M \langle X,Y \rangle\, dV, \qquad \|X\|^2 = \int_M |X|^2\, dV.$$

The Riemannian metric induces a natural bundle isomorphism $\theta : TM \to TM^*$ (where TM, TM^* denote the tangent and cotangent bundles of M, respectively) given by

$$\theta(\xi)(\eta) = \langle \xi, \eta \rangle,$$

for all $p \in M$ and $\xi, \eta \in M_p$ (where M_p denotes the tangent space to M at p). For any C^1 function f on M the *gradient vector field of* f, $\operatorname{grad} f$, is defined by

$$\operatorname{grad} f = \theta^{-1}(df).$$

That is, for any $\xi \in TM$, we have $\langle \operatorname{grad} f, \xi \rangle = df(\xi) = \xi f$. For C^1 functions f, h on M we have

$$\begin{aligned} \operatorname{grad}(f+h) &= \operatorname{grad} f + \operatorname{grad} h, \\ \operatorname{grad} fh &= f \operatorname{grad} h + h \operatorname{grad} f. \end{aligned}$$

If $\mathbf{x} : U \to \mathbf{R}^n$ is a chart on M then

$$\operatorname{grad} f = \sum_{j,k} \frac{\partial (f \circ x^{-1})}{\partial x^j} g^{jk} \frac{\partial}{\partial x^k}.$$

For any C^1 vector field X on M we define the *divergence of X with respect to the Riemannian metric*, $\operatorname{div} X$, by

$$\operatorname{div} X = \operatorname{tr}(\xi \mapsto \nabla_\xi X),$$

where ∇ denotes the Levi–Civita connection of the Riemannian metric. For the C^1 function f and vector fields X, Y on M we have

$$\begin{aligned} \operatorname{div}(X+Y) &= \operatorname{div} X + \operatorname{div} Y, \\ \operatorname{div} fX &= \langle \operatorname{grad} f, X\rangle + f\operatorname{div} X. \end{aligned}$$

If $\mathbf{x} : U \to \mathbf{R}^n$ is a chart on M, and

$$X|U = \sum_j \xi^j \frac{\partial}{\partial x^j},$$

then

$$\operatorname{div} X = \frac{1}{\sqrt{g}} \sum_{j=1}^{n} \frac{\partial(\sqrt{g}\xi^j)}{\partial x^j}.$$

One verifies that if X has compact support on M then we have the **Riemannian divergence theorem**

$$\int_M \operatorname{div} X \, dV = 0.$$

In particular, if f is a function, X a C^1 vector field, on M at least one of which has compact support, then

$$\int f\operatorname{div} X \, dV = -\int \langle \operatorname{grad} f, X\rangle \, dV.$$

Let f be a C^2 function on M. Then we define the *Laplacian of* f, Δf, by

$$\Delta f = \operatorname{div} \operatorname{grad} f.$$

Thus, in a chart $\mathbf{x} : U \to \mathbf{R}^n$ we have

$$\Delta f = \frac{1}{\sqrt{g}} \sum_{j,k=1}^{n} \frac{\partial}{\partial x^j} \left\{ \sqrt{g} g^{jk} \frac{\partial (f \circ x^{-1})}{\partial x^k} \right\}.$$

Furthermore, for C^2 functions f and h on M we have

$$\begin{aligned} \Delta(f+h) &= \Delta f + \Delta h, \\ \operatorname{div} f\operatorname{grad} h &= f\Delta h + \langle \operatorname{grad} f, \operatorname{grad} h\rangle \end{aligned}$$

(this last formula only requires that $f \in C^1, h \in C^2$), which implies

$$\Delta\, fh = f\Delta h + 2\langle \operatorname{grad} f, \operatorname{grad} h\rangle + h\Delta f.$$

One immediately has **Green's formulae**: Let $f : M \to \mathbf{R} \in C^2(M)$, $h : M \to \mathbf{R} \in C^1(M)$, with at least one of them compactly supported. Then

$$\int_M \{h\Delta f + \langle \operatorname{grad} h, \operatorname{grad} f\rangle\} \, dV = 0.$$

If both f and h are C^2, then

$$\int_M \{h\Delta f - f\Delta h\}\, dV = 0.$$

Let M be oriented, Ω a domain in M with smooth boundary $\partial\Omega$, ν the outward unit vector field along $\partial\Omega$ which is pointwise orthogonal to $\partial\Omega$ (there is only one such vector field). Then for any compactly supported C^1 vector field X on M we have

$$\iint_\Omega \operatorname{div} X\, dV = \int_{\partial\Omega} \langle X, \nu\rangle\, dA.$$

The corresponding **Green's formulae** are: Given M, Ω, and ν as just described, and given $f \in C^2(M)$, $h \in C^1(M)$, at least one of them compactly supported. Then

$$\iint_\Omega \{h\Delta f + \langle \operatorname{grad} h, \operatorname{grad} f\rangle\}\, dV = \int_{\partial\Omega} h\langle \nu, \operatorname{grad} f\rangle\, dA.$$

If both f and h are C^2, then

$$\iint_\Omega \{h\Delta f - f\Delta h\}\, dV = \int_{\partial\Omega} \{h\langle \nu, \operatorname{grad} f\rangle - f\langle \nu, \operatorname{grad} h\rangle\}\, dA.$$

Notation. We generally write $\partial f/\partial\nu$ for $\langle \nu, \operatorname{grad} f\rangle$.

§1. Spectrum of $-\Delta$

1.1. The elementary eigenvalue problems

Closed eigenvalue problem. Let M be compact, connected. Find all real numbers λ for which there exists a nontrivial solution $\phi \in C^2(M)$ to the equation

$$\Delta\phi + \lambda\phi = 0.$$

Dirichlet eigenvalue problem. For M connected with compact closure and smooth boundary, find all real numbers λ for which there exists a nontrivial solution $\phi \in C^2(M) \cap C^0(\overline{M})$ to

$$\Delta\phi + \lambda\phi = 0, \qquad \phi \mid \partial M = 0.$$

Neumann eigenvalue problem. For M connected with compact closure and smooth boundary, find all real numbers λ for which there exists a nontrivial solution $\phi \in C^2(M) \cap C^1(\overline{M})$ to

$$\Delta\phi + \lambda\phi = 0, \qquad \frac{\partial\phi}{\partial\nu} \mid \partial M = 0.$$

1.2. Basic facts for elementary eigenvalue problems

The desired numbers λ in each of the eigenvalue problems are referred to as *eigenvalues* of Δ, and the vector space of solutions of the eigenvalue problem with given λ — it is a linear problem in all the above instances — its *eigenspace* E_λ. The elements of the eigenspace are called *eigenfunctions.*

Theorem 1. *For each of the above eigenvalue problems, the set of eigenvalues consists of a sequence*

$$0 \le \overline{\lambda}_1 < \overline{\lambda}_2 < \ldots \uparrow +\infty,$$

and each associated eigenspace $E_{\overline{\lambda}_j}$ is finite dimensional. The dimension E_λ is called the multiplicity of λ. We (henceforth) list the eigenvalues as

$$0 \le \lambda_1 \le \lambda_2 \le \cdots \uparrow +\infty,$$

with each eigenvalue repeated according to its multiplicity.

Eigenspaces belonging to distinct eigenvalues are orthogonal in $L^2(M)$, and $L^2(M)$ is the direct sum of all the eigenspaces.

Furthermore, each eigenfunction is in $C^\infty(\overline{M})$.

In all the above eigenvalue problems we have the **Weyl formula**:

$$\sum_{\lambda_j \le \lambda} 1 =: \mathcal{N}(\lambda) \sim \frac{\boldsymbol{\omega}_{\mathbf{n}} V(M)}{(2\pi)^n} \lambda^{n/2}, \qquad as \ \lambda \uparrow +\infty,$$

where $V(M)$ denotes the Riemannian volume of M, and $\boldsymbol{\omega}_{\mathbf{n}} = V(\mathbf{B}^n)$, the n–volume of the unit disk $\mathbf{B}^n$ in $\mathbf{R}^n$. Similarly,

$$\lambda_\ell \sim \frac{(2\pi)^2}{\boldsymbol{\omega}_{\mathbf{n}}{}^{2/n}} \left\{ \frac{\ell}{V(M)} \right\}^{2/n}$$

as $\ell \uparrow +\infty$.

1.3. Rayleigh's Principle and Max–Min methods

For the closed eigenvalue problem, Δ is a symmetric operator on L^2 with domain $\phi \in C^2(M)$ and is essentially self-adjoint. The domain $\mathcal{H}$ of the associated quadratic form

$$\mathcal{Q}[\phi, \phi] = \int |\mathrm{grad}\, \phi|^2 \, dV$$

is the Sobolev subspace $\mathcal{H} = \mathcal{H}(M)$, the domain of the quadratic form of $\Delta_{|C^\infty(M)}$.

For the Dirichlet eigenvalue problem, Δ is a symmetric operator on L^2 with domain $\phi \in C^2(M) \cap C^0(\overline{M}), \phi|\partial\Omega = 0$, and is essentially self-adjoint. The domain $\mathcal{H}$ of the associated quadratic form

$$\mathcal{Q}[\phi, \phi] = \int |\mathrm{grad}\, \phi|^2 \, dV$$

is the Sobolev subspace $\mathcal{H} = \mathcal{H}_c(M)$, the domain of the quadratic form of $\Delta_{|C_c^\infty(M)}$.

For the Neumann eigenvalue problem, Δ is a symmetric operator on L^2 with domain $\phi \in C^2(M) \cap C^1(\overline{M})$, $\partial\phi/\partial\nu|\partial M = 0$, and is essentially self-adjoint. The domain $\mathcal{H}$ of the associated quadratic form

$$\mathcal{Q}[\phi, \phi] = \int |\mathrm{grad}\, \phi|^2 \, dV$$

is the Sobolev subspace $\mathcal{H} = \mathcal{H}(M)$, the domain of the quadratic form of $\Delta_{|C^\infty(M)}$.

Rayleigh's Principle. *Fix an eigenvalue problem, with quadratic form $\mathcal{Q}$ on the domain $\mathcal{H}$. For all $f \in \mathcal{H}$, $f \neq 0$, we have*

$$\lambda_1 \leq \frac{\mathcal{Q}[f, f]}{\|f\|^2},$$

with equality if and only if f is an eigenfunction of λ_1.

If $\{\phi_1, \phi_2, \ldots\}$ is a complete orthonormal basis of $L^2(M)$ such that ϕ_j is an eigenfunction of λ_j for each $j = 1, 2, \ldots$, then for $f \neq 0$ satisfying

$$(f, \phi_1) = \cdots = (f, \phi_{k-1}) = 0,$$

we have the inequality

$$\lambda_k \leq \frac{\mathcal{Q}[f, f]}{\|f\|^2},$$

with equality if and only if f is an eigenfunction of λ_k.

Max–Min Theorem. *Consider one of the eigenvalue problems. Given any $v_1, \ldots, v_{k-1}$ in $L^2(M)$, let*

$$\mu = \inf \mathcal{Q}[f, f]/\|f\|^2,$$

where f varies over the subspace (less the origin) of functions in $\mathcal{H}$ orthogonal to $v_1, \ldots, v_{k-1}$ in $L^2(M)$. Then for the eigenvalue λ_k (the counting is with multiplicity) we have

$$\mu \leq \lambda_k.$$

Of course, if $v_1, \ldots, v_{k-1}$ are orthonormal, with each v_ℓ an eigenfunction of λ_ℓ, $\ell = 1, \ldots, k-1$, then $\mu = \lambda_k$.

Applications.

(1) Given M with both Dirichlet and Neumann eigenvalues. Then all Neumann eigenvalues are less than or equal to their corresponding Dirichlet eigenvalues.

(2) Domain monotonicity of eigenvalues (Dirichlet eigenvalue problem). Let $\Omega_1, \ldots, \Omega_m$ be pairwise disjoint regular domains in M whose boundaries, when intersecting M, do so transversally. Given an eigenvalue problem on M with eigenvalues

$$0 \leq \lambda_1 \leq \lambda_2 \leq \ldots,$$

consider, for each $r = 1, \ldots, m$, the eigenvalues on Ω_r obtained by requiring vanishing Dirichlet data on $\partial\Omega_r \cap M$ and by leaving the original data on $\partial\Omega_r \cap \partial M$ unchanged. Arrange *all* eigenvalues of $\Omega_1, \cdots, \Omega_r$ in an increasing sequence

$$0 \leq \nu_1 \leq \nu_2 \leq \ldots .$$

Then

$$\nu_k \geq \lambda_k \qquad \forall\ k = 1, 2, \ldots .$$

(3) If $\Omega \subset M$, then for the Dirichlet eigenvalue problem on Ω and any eigenvalue problem on M we have

$$\lambda_k(\Omega) \geq \lambda_k \qquad \forall\ k = 1, 2, \ldots .$$

If $M \setminus \Omega$ is open then the inequality is strict.

(4) Domain monotonicity of eigenvalues (Neumann eigenvalue problem). Let $\Omega_1, \ldots \Omega_m$ be pairwise disjoint regular domains in M whose boundaries, when intersecting M, do so transversally; and *assume also* that

$$\overline{M} = \overline{\Omega_1} \cup \cdots \cup \overline{\Omega_m}.$$

Given an eigenvalue problem on M with eigenvalues

$$0 \leq \lambda_1 \leq \lambda_2 \leq \ldots,$$

consider, for each $r = 1, \ldots, m$, the eigenvalue on Ω_r obtained by requiring vanishing Neumann data on $\partial\Omega_r \cap M$ and by leaving the original data on $\partial\Omega_r \cap \partial M$ unchanged. Arrange *all* eigenvalues of $\Omega_1, \ldots, \Omega_r$ in an increasing sequence

$$0 \leq \mu_1 \leq \mu_2 \leq \ldots.$$

Then

$$\mu_k \leq \lambda_k \qquad \forall\ k = 1, 2, \ldots .$$

Definition. Let $f : M \to \mathbf{R}^n \in C^0$. Then the *nodal set of* f is $f^{-1}[0]$ and a *nodal domain of* f is any component of $\overline{M} \setminus f^{-1}[0]$.

(5) Courant's nodal domain theorem. Given a fixed eigenvalue problem with complete orthonormal basis $\{\phi_1, \phi_2, \ldots\}$ of $L^2(M)$ such that ϕ_j is an eigenfunction of the eigenvalue λ_j for each $j = 1, 2, \ldots$, then the number of nodal domains of ϕ_k is less than or equal to k.

(6) Corollary. The eigenfunction of ϕ_1 of λ_1 never vanishes on M.

(7) Corollary. The multiplicity of the eigenvalue λ_1 is precisely equal to 1.

Exercise. Let $j_{n,k}$ denote the k–th Dirichlet eigenvalue of $\mathbf{B}^n$ (with eigenvalues repeated according to multiplicity).

(a) Show that $\lambda_k(\mathbf{B}^n(\epsilon))$, the k–th Dirichlet eigenvalue of $\mathbf{B}^n(\epsilon)$ (the open disk in $\mathbf{R}^n$ of radius ϵ), is given by

$$\lambda_k(\mathbf{B}^n(\epsilon)) = \frac{j_{n,k}}{\epsilon^2}.$$

(b) Let M be an n–dimensional Riemannian manifold, $x \in M$. Use Riemann normal coordinates and the max–min theorem to prove

$$\lambda_k(B(x;\epsilon)) \sim \frac{j_{n,k}}{\epsilon^2} \quad \text{as } \epsilon \downarrow 0.$$

Additional references

Berger–Gauduchon–Mazet [**6**], Bérard [**5**], Courant–Hilbert [**11**], Davies [**12**].

§2. Basic examples

2.1. Tori

Consider $\mathbf{R}^n$, $n \geq 1$, and a *lattice* Γ, that is, a discrete subgroup of $\mathbf{R}^n$. Then Γ acts by translations on $\mathbf{R}^n$ in a properly discontinuous fashion, and determines a covering $\mathbf{R}^n \to \Gamma \setminus \mathbf{R}^n$. We assume the *rank of* Γ is n, that is, there exist n linearly independent vectors $\{v_1, \ldots, v_n\}$ in Γ for which

$$\Gamma = \left\{ \sum_{j=1}^{n} \alpha^j v_j : \ \alpha^j \in \mathbf{Z},\, j = 1, \ldots, n \right\}.$$

Then

$$T_\Gamma =: \Gamma \setminus \mathbf{R}^n$$

is compact and diffeomorphic to the torus $(\mathbf{S}^1)^n$.

We consider the Hilbert space space $L^2(T_\Gamma)$ of complex-valued functions on T_Γ, with standard Hermitian inner product, and the Laplacian acting on real and imaginary parts separately. One checks that the same eigenvalues are obtained, as when only admitting real-valued functions, with the same multiplicity.

A collection of eigenfunctions on T_Γ is obtained as follows: associate with Γ its *dual lattice* Γ^* defined by:

$$\Gamma^* = \{ y \in \mathbf{R}^n : \ \gamma \cdot y \in \mathbf{Z} \ \ \forall\, \gamma \in \Gamma \}.$$

Then Γ^* is itself a lattice of rank n. With the above basis $\{v_1, \ldots, v_n\}$ of Γ is associated the *dual basis* $\{w_1, \ldots, w_n\}$ of Γ^* defined by

$$w_j \cdot v_k = \delta_{jk},$$

where δ_{jk} is the Kronecker delta.

With each $y \in \Gamma^*$ associate the function

$$\phi_y(x) = e^{2\pi i x \cdot y}$$

on $\mathbf{R}^n$. Then

(i) ϕ_y is invariant under Γ,

(ii) ϕ_y is an eigenfunction on T_Γ with eigenvalue $\lambda_y = 4\pi^2|y|^2$,

(iii) The collection of functions $\{\phi_y/\sqrt{V(T_\Gamma)}\ :\ y \in \Gamma^*\}$ is a complete orthonormal basis of $L^2(T_\Gamma)$.

Weyl's Formula for the Torus. *We have*

$$\mathcal{N}(\lambda) \sim \omega_{\mathbf{n}} \lambda^{n/2} \frac{V(T_\Gamma)}{(2\pi)^n}.$$

Proof. One can reduce the question to the *circle problem,* namely, the number of eigenvalues less than or equal to the number λ is equal to the number of elements $y \in \Gamma^*$ in the closed n–disk in $\mathbf{R}^n$, centered at the origin, with radius $\sqrt{\lambda}/2\pi$. Then there is a classical argument of Gauss that derives the formula.

Another method is to use the *Poisson summation formula,*

$$\frac{(2\pi)^n}{V(T_\Gamma)} \sum_{y\in\Gamma^*} \hat{f}(2\pi y) = \sum_{\gamma\in\Gamma} f(\gamma)$$

(where $\hat{f}$ denotes the Fourier transform of f) for Schwartz functions f on $\mathbf{R}^n$, applied to the function

$$f(x) = e^{-|x|^2/4t}.$$

Then

$$\sum_{y\in\Gamma^*} e^{-4\pi^2|y|^2 t} = \frac{V(T_\Gamma)}{(4\pi t)^{n/2}} \sum_{\gamma\in\Gamma} e^{-|\gamma|^2/4t} \quad \text{and} \quad \lim_{t\downarrow 0} \sum_{\gamma\in\Gamma,\gamma\neq 0} e^{-|\gamma|^2/4t} = 0,$$

which implies

$$\sum_{k=0}^{\infty} e^{-\lambda_k t} = \sum_{k=0}^{\infty} e^{-4\pi^2|y|^2 t} \sim \frac{V(T_\Gamma)}{(4\pi t)^{n/2}}$$

as $t \downarrow 0$. Karamata's Tauberian theorem now finishes the job. qed

Comment. The above argument previews two subjects:

(1) The Selberg trace formula

(2) The short-time asymptotics of the heat kernel.

2.2. Local calculations in spherical coordinates

We fix a Riemannian manifold M, $\dim M = n$, with Levi-Civita connection ∇, and with Riemann curvature tensor R. For any tangent vector ξ we denote the geodesic determined by ξ by γ_ξ, namely,

$$\gamma_\xi(t) = \exp t\xi$$

(where exp denotes the exponential map). Recall that for any $p \in M$, $\xi \in M_p$, the map $\eta \mapsto R(\xi, \eta)\xi$ is a self-adjoint linear transformation of M_p.

2.2.1. Jacobi fields

Definition. Given a geodesic γ, a *Jacobi field along* γ is a differentiable vector field Y along γ satisfying *Jacobi's equation*

$$\nabla_t{}^2 Y + R(\gamma', Y)\gamma' = 0. \tag{1}$$

Note that

$$Y \perp \gamma', \quad |\gamma'| = 1 \quad \Rightarrow \quad \langle R(\gamma', Y)\gamma'), Y\rangle = |Y|^2 K(\gamma', Y)$$

(where $K(\gamma', Y)$ denotes the sectional curvature of the 2-plane spanned by γ' and Y).

Theorem 2. *The set $\mathcal{J}$ of Jacobi fields along γ is a vector space over* $\mathbf{R}$ *of dimension equal to $2n$. More particularly, one has: Given any $t_0 \in [\alpha, \beta]$, $\xi, \eta \in M_{\gamma(t_0)}$, there exists a unique $Y \in \mathcal{J}$ satisfying $Y(t_0) = \xi$, $(\nabla_t Y)(t_0) = \eta$.*

For $X, Y \in \mathcal{J}$ the Wronskian $W(X, Y)$ of X and Y is constant, that is,

$$W(X, Y) =: \langle \nabla_t X, Y\rangle - \langle X, \nabla_t Y\rangle = \text{const.}$$

Thus, for any $Y \in \mathcal{J}$ we have constants $a, b \in \mathbf{R}$ for which $\langle Y, \gamma'\rangle = at + b$. In particular,

$$\begin{aligned} \mathcal{J}^\perp =: \{Y \in \mathcal{J} :\ \langle Y, \gamma'\rangle = 0 \text{ on } [\alpha, \beta]\} \quad &\Rightarrow \quad \operatorname{codim} \mathcal{J}^\perp = 2, \\ \dim \{Y \in \mathcal{J}^\perp :\ Y(0) = 0\} \quad &= \quad n - 1. \end{aligned}$$

Jacobi's equation is the linearized geodesic equation, namely, let $\epsilon_0 > 0$, $v : (\alpha, \beta) \times (-\epsilon_0, \epsilon_0) \to M$ be differentiable such that for each ϵ in $(-\epsilon_0, \epsilon_0)$

the path $\omega_\epsilon : (\alpha, \beta) \to M$ given by $\omega_\epsilon(t) = v(t, \epsilon)$ is a geodesic. So $\epsilon \mapsto \omega_\epsilon$ is a 1–parameter family of geodesics. The coordinate vector fields along v will be written as

$$\partial_t v = v_* \partial_t, \quad \partial_\epsilon v = v_* \partial_\epsilon$$

(where v_* is the linear map of tangent spaces associated with v).

Theorem 3. *For each fixed ϵ, the vector field $Y = \partial_\epsilon$ is a Jacobi field along ω_ϵ. Furthermore, all Jacobi fields arise in this manner.*

The philosophy behind studying Jacobi fields is simple. The geodesic equations of the Levi-Civita connection are nonlinear, and extremely difficult to solve explicitly, save a few examples. So one does the next best thing. One assumes that one knows a given geodesic and then attempts, at the infinitesmal level (in the direction of the variation), to gain some insight into the behavior of neighboring geodesics.

The most widely used variation is the following:

Theorem 4. *Let $p \in M$, ξ, η orthonormal vectors in M_p, and*

$$v(t, \theta) = \exp t\{(\cos\theta)\xi + (\sin\theta)\eta\},$$

$$\gamma(t) = v(t, 0) = \exp t\xi$$

the base geodesic of the variation. Then the vector field

$$Y(t) = (\partial_\theta v)(t, 0)$$

is a Jacobi field along γ satisfying

$$Y(0) = 0, \quad (\nabla_t Y)(0) = \eta, \qquad \langle Y, \gamma' \rangle = 0.$$

The orthogonality of Y to the geodesic is the content of Gauss' lemma.

2.2.2. Spaces of constant sectional curvature

For each fixed $\kappa \in \mathbf{R}$ let $\mathbf{M}_\kappa$ denote the simply connected space form of constant sectional curvature κ (we assume the dimension is some fixed $n > 1$).

Thus, for $\kappa = 0$, $\mathbf{M}_\kappa = \mathbf{R}^n$, Euclidean space. The geodesics of Euclidean space are the straight lines, parametrized to a constant multiple of arc length.

that is,

$$\begin{aligned} \Delta f &= \frac{1}{\sqrt{\mathrm{g}}(r;\xi)} \frac{\partial}{\partial r}\left(\sqrt{\mathrm{g}}(r;\xi)\frac{\partial f}{\partial r}\right) + \mathcal{L}_{r;\xi} f \\ &= \frac{\partial^2 f}{\partial r^2} + \frac{1}{\sqrt{\mathrm{g}}(r;\xi)} \frac{\partial \sqrt{\mathrm{g}}(r;\xi)}{\partial r}\frac{\partial f}{\partial r} + \mathcal{L}_{r;\xi} f, \end{aligned}$$

where $\mathcal{L}_{r;\xi}$ denotes the Laplacian of the Riemannian $(n-1)$–submanifold $S(p;r)$ at $q = \exp r\xi$.

Consider the special case: $f(x) = \phi(d(p,x))$ on all M. Then

(9) $$\Delta f = \frac{1}{\sqrt{\mathrm{g}}(r;\xi)} \frac{\partial}{\partial r}\left(\sqrt{\mathrm{g}}(r;\xi)\frac{d\phi}{dr}\right);$$

and:

(10) $$\rho_p(x) = d(p,x) \quad \Rightarrow \quad \Delta \rho_p = \frac{1}{\sqrt{\mathrm{g}}(r;\xi)} \frac{\partial \sqrt{\mathrm{g}}(r;\xi)}{\partial r}.$$

Consider the special case: $M = \mathbf{M}_\kappa$. Then

(11) $$(\Delta f)(\exp r\xi) = \frac{1}{\mathbf{s}_\kappa{}^{n-1}} \frac{\partial}{\partial r}\left(\mathbf{s}_\kappa{}^{n-1}\frac{\partial f}{\partial r}\right) + \frac{1}{\mathbf{s}_\kappa{}^2}\mathcal{L}_\xi f$$

(12) $$= \frac{\partial^2 f}{\partial r^2} + (n-1)\frac{\mathbf{c}_\kappa}{\mathbf{s}_\kappa}\frac{\partial f}{\partial r} + \frac{1}{\mathbf{s}_\kappa{}^2}\mathcal{L}_\xi f,$$

where, when writing $\mathcal{L}_\xi f$ we mean that $f|S(p;r)$ is to be considered as a function on S_p with associated Laplacian $\mathcal{L}$ at ξ.

Exercise. Carry out the above calculations for grad f, with the appropriate interpretations.

2.3. Spheres

Assume $\kappa = 0$, so the manifold is $\mathbf{R}^n$. Assume u is homogeneous of degree k, that is,

$$u(x) = u(r\xi) = r^k G(\xi), \qquad x = r\xi, \quad r > 0, \quad \xi \in \mathbf{S}^{n-1};$$

then (11) implies

$$\Delta_{\mathbf{R}^n} u = r^{k-2}\{\Delta_{\mathbf{S}^{n-1}} G + k(k+n-2)G\}.$$

For $\kappa > 0$, $\mathbf{M}_\kappa = \mathbf{S}^n(\rho)$, the sphere of radius ρ in $\mathbf{R}^{n+1}$, where $\kappa = 1/\rho^2$. The Riemannian metric of $\mathbf{S}^n(\rho)$ is inherited from the ambient $\mathbf{R}^{n+1}$ by restriction of the ambient Riemannian metric. The geodesics are the intersections of 2–dimensional planes in $\mathbf{R}^{n+1}$, which pass through the center of the sphere, with the sphere itself.

For $\kappa < 0$, $\mathbf{M}_\kappa$ is the hyperbolic space of curvature κ. One has two classical models of $\mathbf{M}_\kappa$. The first is the disk model, that is, for $\kappa = -1/\rho^2$ one considers $\mathbf{B}^n(\rho)$ with Riemannian metric

$$ds^2 = \frac{4|dz|^2}{\{1 - |z|^2/\rho^2\}}, \qquad z \in \mathbf{B}^n(\rho).$$

The geodesics are either lines emanating from the origin, or circles intersecting the boundary $\mathbf{S}^{n-1}(\rho)$ orthogonally. The second model is the upper half-plane

$$ds^2 = \frac{1}{\sqrt{-\kappa}}\frac{|dx|^2 + dy^2}{y^2}, \qquad \{(x,y) \in \mathbf{R}^n : \, x \in \mathbf{R}^{n-1}, \, y > 0\}.$$

The geodesics consist of lines and circles which intersect $\mathbf{R}^{n-1} = \{y = 0\}$ orthogonally.

One can give a unified treatment to many calculations in the constant curvature spaces, as follows:

Definition. Given a real constant κ, we let $\mathbf{s}_\kappa$ denote the solution to the ordinary differential equation

$$\psi'' + \kappa\psi = 0,$$

satisfying the initial conditions

$$\mathbf{s}_\kappa(0) = 0, \qquad \mathbf{s}_\kappa{}'(0) = 1.$$

We also let $\mathbf{c}_\kappa$ denote the solution to the above ordinary differential equation satisfying the initial conditions

$$\mathbf{c}_\kappa(0) = 1, \qquad \mathbf{c}_\kappa{}'(0) = 0.$$

Of course, we have

$$\mathbf{s}_\kappa(t) = \begin{cases} (1/\sqrt{\kappa}) \sin \sqrt{\kappa} t & \kappa > 0 \\ t & \kappa = 0 \\ (1/\sqrt{-\kappa}) \sinh \sqrt{-\kappa} t & \kappa < 0 \end{cases},$$

and,

$$\mathbf{c}_\kappa(t) = \begin{cases} \cos\sqrt{\kappa}t & \kappa > 0 \\ 1 & \kappa = 0 \\ \cosh\sqrt{-\kappa}t & \kappa < 0 \end{cases}$$

Furthermore, we have

$$\mathbf{s}_\kappa{}' = \mathbf{c}_\kappa, \qquad \mathbf{c}_\kappa{}' = -\kappa\mathbf{s}_\kappa, \qquad \mathbf{c}_\kappa{}^2 + \kappa\mathbf{s}_\kappa{}^2 = 1,$$

$$(\mathbf{c}_\kappa/\mathbf{s}_\kappa)' = (\mathbf{s}_\kappa{}'/\mathbf{s}_\kappa)' = -\mathbf{s}_\kappa{}^{-2}.$$

The constant curvature of $\mathbf{M}_\kappa$ implies that for the geodesic γ parametrized with respect to arc length, so $|\gamma'| = 1$, one has

$$R(\gamma', Y)\gamma' = \kappa Y \qquad \forall\ Y \perp \gamma',$$

which implies Jacobi's equation on $\mathcal{J}^\perp$ reads as

$$\nabla_t{}^2 Y + \kappa Y = 0.$$

The general solution is

$$Y(t) = A(t)\mathbf{c}_\kappa(t) + B(t)\mathbf{s}_\kappa(t),$$

where

$$\nabla_t A = \nabla_t B = 0 \ \text{ along } \gamma, \qquad A(0) = Y(0), \ \ B(0) = (\nabla_t Y)(0).$$

Then

$$Y(0) = 0 \quad \Rightarrow \quad Y(t) = B(t)\mathbf{s}_\kappa(t).$$

One then has, using spherical coordinates described below, that if $o \in \mathbf{M}_\kappa$, $\xi \in \mathsf{S}_o$ (the unit sphere tangent to M at o), $t > 0$, $q = \exp_o t\xi$, then we may write the Riemannian metric at q as

$$ds^2 = dt^2 + \mathbf{s}_\kappa{}^2(t)|d\xi|^2,$$

where $|d\xi|^2$ denotes the Riemannian metric on the Euclidean $(n-1)$–sphere S_o in M_o.

2.2.3. The Riemannian metric in spherical coordinates

Given a manifold M, a *coordinate system on* M will be a C^∞ map $\varphi : \mathcal{O} \to M$ from an open set $\mathcal{O}$ in $\mathbf{R}^n$.

Given a Riemannian manifold $M, p \in M$, and a coordinate system $\xi : \mathcal{O} \to \mathsf{S}_p$ on the unit tangent sphere S_p at p, a coordinate system v on M is determined by

$$v(r, \theta) = \exp r\xi(\theta).$$

The domain of the map v will consist of the collection $\{(r, \theta)\}$ in $(0, +\infty) \times \mathcal{O}$ for which $r\xi(\theta)$ is in the domain of the exponential map. In general, we restrict r to $[0, c(\xi)]$ (where $c(\xi)$ denotes the distance from p to its cut point along the geodesic γ_ξ).

In what follows, given $\xi \in \mathsf{S}_p$, let $\Im_\xi : M_p \to (M_p)_\xi$ denote the usual identification, and $\tau_{r;\xi} : M_p \to M_{\gamma_\xi(r)}$ denote parallel translation along γ_ξ. Write

$$\partial_\alpha \xi =: \Im_\xi{}^{-1} \circ \xi_*(\partial/\partial\theta^\alpha), \qquad \partial_r v =: v_*(\partial/\partial r), \qquad \partial_\alpha v =: v_*(\partial/\partial\theta^\alpha),$$

$\alpha = 1, \ldots, n-1$, where $\partial/\partial r$ and $\partial/\partial\theta^1, \ldots, \partial/\partial\theta^{n-1}$ are natural coordinate vector fields on $(0, +\infty)$ and $\mathcal{O}$, respectively. Then

$$(\partial_r v)(r; \xi) = \partial_r v_{|\exp r\xi} = \gamma_\xi{}'(r),$$

and

$$(\partial_\alpha v)(r; \xi) = \partial_\alpha v_{|\exp r\xi} = Y_\alpha(r; \xi),$$

where $Y_\alpha(r; \xi)$ is the Jacobi field along γ_ξ determined by the initial conditions

$$Y_\alpha(0; \xi) = 0, \qquad (\nabla_r Y_\alpha)(0; \xi) = \partial_\alpha \xi.$$

Of course,

$$|\partial_r v| = 1; \qquad \langle \partial_r v, \partial_\alpha v \rangle = 0$$

(the second equality follows from Gauss' lemma). So the full knowledge of the Riemannian metric along γ_ξ requires the study of

$$\langle \partial_\alpha v, \partial_\beta v \rangle(\exp r\xi) = \langle Y_\alpha(r; \xi), Y_\beta(r; \xi) \rangle.$$

More generally, the matrix

$$\mathcal{H}(r; \theta) = (\langle \partial_\alpha v, \partial_\beta v \rangle)\,(r; \theta)$$

is the $(n-1)\times(n-1)$–matrix of the induced Riemannian metric on the submanifold $S(p;r)$ at $q = \exp r\xi(\theta)$. The full $n\times n$–matrix $\mathcal{G}$ of the Riemannian metric on M in a neighborhood of q is given by

$$\mathcal{G} = \begin{pmatrix} 1 & 0 \\ 0 & \mathcal{H} \end{pmatrix}. \tag{2}$$

It might be best to think of Jacobi's equation as a matrix equation: Fix $p \in M$, $\xi \in \mathsf{S}_p$, let $\xi^\perp$ denote the orthogonal complement of $\mathbf{R}\xi$ in M_p; and for each $r > 0$, let

$$\mathsf{R}(r) = R(\gamma_\xi{}'(r), \cdot)\gamma_\xi{}'(r).$$

Let $A(r;\xi) : \xi^\perp \to M_{\gamma'(r)}{}^\perp$ be the solution of the matrix (more precisely: linear transformation) Jacobi equation on $\xi^\perp$:

$$\nabla_r{}^2 A + \mathsf{R}(r)A = 0, \tag{3}$$

satisfying the initial conditions

$$A(0;\xi) = 0, \qquad (\nabla_r A)(0;\xi) = I. \tag{4}$$

Then for each $\eta \in \xi^\perp$, the vector field $Y(t)$ along γ_ξ, given by

$$Y(r) = A(r;\xi)\eta, \tag{5}$$

is the Jacobi field, in $\mathcal{J}^\perp$, along γ_ξ determined by the initial conditions

$$Y(0) = 0, \qquad (\nabla_t Y)(0) = \eta. \tag{6}$$

Thus $A(t;\xi)$ is nonsingular for all $t \in (0, \operatorname{conj}\xi)$ (where $\operatorname{conj}\xi$ denotes the distance from p to its first conjugate point along γ_ξ).

We therefore have

$$(\partial_\alpha v)(r;\xi) = Y_\alpha(r;\xi) = A(r;\xi)\partial_\alpha\xi,$$

which implies

$$\langle \partial_\alpha v, \partial_\beta v\rangle(\exp r\xi) = \langle A(r;\xi)\partial_\alpha\xi, A(r;\xi)\partial_\beta\xi\rangle.$$

Therefore, if we denote the matrix of the Riemannian metric on S_p at $\xi(\theta)$ by $\mathcal{H}_p$, that is,

$$\mathcal{H}_p = (\langle \partial_\alpha\xi, \partial_\beta\xi\rangle),$$

then

$$\mathcal{H} = A\mathcal{H}_p A^T \tag{7}$$

(where A^T denotes the transpose of A).

Note that (2) and (7) imply

$$\sqrt{\det \mathcal{G}}(r;\theta) = \sqrt{\det \mathcal{H}}(r;\theta) = \det A(r;\xi)\sqrt{\mathcal{H}_p}(\theta),$$

which implies

$$\text{(8)} \qquad dV = \sqrt{\mathbf{g}}(r;\xi)\, dr d\mu_p(\xi), \qquad \sqrt{\mathbf{g}}(r;\xi) =: \det A(r;\xi),$$

where $d\mu_p$ denotes the $(n-1)$-dimensional measure on S_p.

For M with constant sectional curvature equal to κ, we have

$$\mathcal{A}(r;\xi) = \mathbf{s}_\kappa(r)\tau_{r;\xi}, \qquad \mathcal{H}(r;\xi) = \mathbf{s}_\kappa{}^2(r)\mathcal{H}_p(\xi),$$

which is what we encapsulated above as

$$ds^2 = dr^2 + \mathbf{s}_\kappa{}^2(r)|d\xi|^2.$$

Then

$$dV = \mathbf{s}_\kappa{}^{n-1}(r) dr d\mu_p(\xi).$$

2.2.4. The Laplacian in spherical coordinates

Given our point p, and geodesic spherical coordinates

$$v(r,\theta) = \exp r\xi(\theta),$$

where $\xi : \mathcal{O} \to \mathsf{S}_p$, $\mathcal{O} \subset \mathbf{R}^{n-1}$, as described above. Then for the Laplacian in local coordinates we have

$$\begin{aligned}
\Delta f \;&=\; \frac{1}{\sqrt{\det \mathcal{G}}}\left\{\frac{\partial}{\partial r}\left(g^{rr}\sqrt{\det \mathcal{G}}\frac{\partial f}{\partial r}\right) + \sum_{\alpha=1}^{n-1}\frac{\partial}{\partial r}\left(g^{r\alpha}\sqrt{\det \mathcal{G}}\frac{\partial f}{\partial \theta^\alpha}\right)\right.\\
&\qquad + \sum_{\alpha=1}^{n-1}\frac{\partial}{\partial \theta^\alpha}\left(g^{\alpha r}\sqrt{\det \mathcal{G}}\frac{\partial f}{\partial r}\right)\\
&\qquad \left. + \sum_{\alpha,\beta=1}^{n-1}\frac{\partial}{\partial \theta^\alpha}\left(g^{\alpha\beta}\sqrt{\det \mathcal{G}}\frac{\partial f}{\partial \theta^\beta}\right)\right\}\\
&= \frac{1}{\sqrt{\det A(r;\xi)}}\frac{\partial}{\partial r}\left(\sqrt{\det A(r;\xi)}\frac{\partial f}{\partial r}\right)\\
&\qquad + \frac{1}{\sqrt{\det \mathcal{H}}}\sum_{\alpha,\beta=1}^{n-1}\frac{\partial}{\partial \theta^\alpha}\left(h^{\alpha\beta}\sqrt{\det \mathcal{H}}\frac{\partial f}{\partial \theta^\beta}\right)\\
&= \frac{1}{\sqrt{\mathbf{g}}(r;\xi)}\frac{\partial}{\partial r}\left(\sqrt{\mathbf{g}}(r;\xi)\frac{\partial f}{\partial r}\right) + \mathcal{L}_{r;\xi} f,
\end{aligned}$$

Therefore, a homogeneous harmonic polynomial of degree k, namely,

$$u(x) = \sum_{\alpha_1+\cdots+\alpha_n=k} A_{\alpha_1,\ldots,\alpha_n} {x_1}^{\alpha_1} \cdots {x_n}^{\alpha_n},$$

satisfying

$$\Delta_{\mathbf{R}^n} u = 0,$$

must satisfy, for $G = u|\mathbf{S}^{n-1}$,

$$\Delta_{\mathbf{S}^{n-1}} G + k(k+n-2)G = 0,$$

that is, G is an eigenfunction of the Laplacian on $\mathbf{S}^{n-1}$ with eigenvalue

$$\lambda = k(k+n-2).$$

It is known that for each $\ell = 0, 1, 2, \ldots$, the space of homogeneous harmonic polynomials on $\mathbf{R}^n$, of degree ℓ, restricted to $\mathbf{S}^{n-1}$ constitute the full eigenspace of the distinct ℓ-th eigenvalue

$$\overline{\lambda}_\ell = \ell(\ell+n-2),$$

which has dimension

$$\binom{n-1+\ell}{\ell} - \binom{n-2+\ell}{\ell-1}.$$

Note that all linear polynomials on $\mathbf{R}^n$, when restricted to $\mathbf{S}^{n-1}$, become eigenfunctions of the lowest nonzero eigenvalue $n-1$.

2.4. Metric disks in constant curvature spaces

We now consider the Dirichlet and Neumann eigenvalue problems on

$$M =: \mathbf{B}_\kappa(o; R) \subset \mathbf{M}_\kappa.$$

A standard argument using separation of variables shows that, for each of the eigenvalue problems, a complete set of eigenfunctions is given by functions u of the form

$$u(\exp_o r\xi) = T(r)G(\xi), \qquad r > 0, \;\; \xi \in S_o,$$

where

$$\mathcal{L}G + \nu G = 0,$$

and

$$({\mathbf{s}_\kappa}^{n-1}T')' + \left\{\lambda - \frac{\nu}{{\mathbf{s}_\kappa}^2}\right\} {\mathbf{s}_\kappa}^{n-1}T = 0,$$

where λ is the eigenvalue of u and ν is an eigenvalue of the Laplacian on $\mathbf{S}^{n-1}$.

(1) The boundary conditions are given by

$$\begin{aligned} T(R) &= 0 \qquad &&\text{(Dirichlet)},\\ T'(R) &= 0 \qquad &&\text{(Neumann)}, \end{aligned}$$

respectively.

(2) So for each $\ell = 0, 1, 2, \ldots$, the function G varies over the eigenspace of the eigenvalue $\ell(\ell + n - 2)$ on $\mathbf{S}^{n-1}$, and for each such eigenfunction G we have a sequence of eigenfunctions $\phi_{\ell;j}$ on $\mathbf{B}_\kappa(o; R)$ given by

$$\phi_{\ell,j}(\exp r\xi) = T_{\ell,j}(r)G(\xi), \qquad j = 1, 2, \ldots,$$

where $T_{\ell,j}$ satisfies the *radial equation*,

$$({\mathbf{s}_\kappa}^{n-1}{T_{\ell,j}}')' + \left\{\lambda_{\ell,j} - \frac{\ell(\ell+n-2)}{{\mathbf{s}_\kappa}^2}\right\} {\mathbf{s}_\kappa}^{n-1}T_{\ell,j} = 0,$$

with

$$0 \le \lambda_{\ell,1} < \lambda_{\ell,2} < \cdots \uparrow \infty.$$

For each ℓ, j, $\lambda_{\ell,j}$ determines only a 1-dimensional space of solutions.

(3) Near $t = 0$ the behavior of the radial function $T(t)$ is given by

$$\begin{aligned} T'(0) &= 0 && \text{if } \ell = 0,\\ T(t) &\sim \text{const.}t^\ell \quad \text{as } t \downarrow 0 && \text{if } \ell = 1, 2, \ldots. \end{aligned}$$

(4) For the *Dirichlet eigenvalue problem* the lowest eigenvalue $\lambda(R)$ is given by the lowest eigenvalue corresponding to $\ell = 0$, that is,

$$u(t, \xi) = T(t)$$

is radial, where $T(t)$ satisfies

$$({\mathbf{s}_\kappa}^{n-1}T')' + \lambda {\mathbf{s}_\kappa}^{n-1}T = 0,$$

$$T'(0) = T(R) = 0, \quad T|[0, R) > 0.$$

(5) For the *Neumann eigenvalue problem* the lowest nonzero eigenvalue $\mu(R)$ is given by the lowest eigenvalue for $\ell = 1$. For $\ell = 1$, $\nu = n - 1$, with multiplicity n, as mentioned above. So G_1 varies over the restriction to $\mathbf{S}^{n-1}$ of all linear polynomials on $\mathbf{R}^n$. The radial part $\mathcal{T}(t)$ satisfies

$$(\mathrm{s}_\kappa{}^{n-1}\mathcal{T}')' + \left\{\mu(R) - \frac{n-1}{\mathrm{s}_\kappa{}^2}\right\}\mathrm{s}_\kappa{}^{n-1}\mathcal{T} = 0,$$

$$\mathcal{T}(0) = \mathcal{T}'(R) = 0, \quad \mathcal{T}'|[0,R) \neq 0.$$

2.5. Euclidean and hyperbolic spaces

Henceforth, we speak of hyperbolic space as having constant sectional curvature -1.

Theorem 5. *For* $\mathbf{R}^n$ *we have*

$$\lim_{R\to\infty} \lambda(R) = 0,$$

and for $\mathbf{H}^n$ *we have*

$$\lambda(R) \geq \frac{(n-1)^2}{4} \qquad \forall\ R > 0.$$

Proof for $\mathbf{H}^n$. On $\mathbf{H}^n$ we have

$$ds^2 = dr^2 + \sinh^2 r|d\xi|^2, \qquad r > 0,\ \ \xi \in \mathbf{S}^{n-1}.$$

The volume element is then given by

$$dV = \sinh^{n-1} r\, dr d\mu_{n-1}(\xi).$$

For any $\phi \in C_c^\infty$ we have

$$|\mathrm{grad}\,\phi|^2 = (\partial_r\phi)^2 + \sinh^{-2} r|\partial_\xi\phi|^2,$$

which implies, for each fixed $\xi \in \mathbf{S}^{n-1}$,

$$\begin{aligned}
&\int_0^\infty \phi^2 \sinh^{n-1} r\, dr \\
&\leq \int_0^\infty \phi^2 \sinh^{n-2} r \cosh r\, dr \\
&= \left.\frac{\phi^2 \sinh^{n-1} r}{n-1}\right|_0^\infty - \frac{2}{n-1}\int_0^\infty \phi\phi' \sinh^{n-1} r\, dr \\
&\leq \frac{2}{n-1}\left\{\int_0^\infty \phi^2 \sinh^{n-1} r\, dr\right\}^{1/2} \cdot \left\{\int_0^\infty (\phi')^2 \sinh^{n-1} r\, dr\right\}^{1/2},
\end{aligned}$$

which implies

$$\int_0^\infty (\partial_r\phi)^2 \sinh^{n-1} r\, dr \geq \frac{(n-1)^2}{4} \int_0^\infty \phi^2 \sinh^{n-1} r\, dr.$$

Now integrate over $\mathbf{S}^{n-1}$. Then

$$\begin{aligned}
\int_{\mathbf{H}^n} |\mathrm{grad}\, \phi|^2\, dV &= \int_{\mathbf{S}^{n-1}} d\mu_{n-1} \int_0^\infty |\mathrm{grad}\, \phi|^2 \sinh^{n-1} r\, dr \\
&\geq \int_{\mathbf{S}^{n-1}} d\mu_{n-1} \int_0^\infty (\partial_r\phi)^2 \sinh^{n-1} r\, dr \\
&\geq \frac{(n-1)^2}{4} \int_{\mathbf{S}^{n-1}} d\mu_{n-1} \int_0^\infty \phi^2 \sinh^{n-1} r\, dr \\
&= \int_{\mathbf{H}^n} \phi^2\, dV,
\end{aligned}$$

which implies the theorem, by Rayleigh's principle. qed

Additional references

Berger–Gauduchon–Mazet [**6**], Bérard [**5**], Courant–Hilbert [**11**], Stein–Weiss [**19**].

§3. Spectra of noncompact Riemannian manifolds

Let M denote an arbitrary Riemannian manifold, and Δ some self-adjoint extension of $\Delta|C_c^\infty(M)$. Certainly, spec $-\Delta \subseteq [0, \infty)$.

Definition. A number λ is in the *point spectrum of* $-\Delta$, if λ is an eigenvalue of $-\Delta$, that is, there exists a nontrivial element f of $L^2(M)$ for which $-\Delta f = \lambda f$.

The *discrete spectrum of* $-\Delta$ consists of those $\lambda \in$ spec $-\Delta$ for which there exists $\epsilon > 0$ such that

$$\dim (E_{\lambda+\epsilon} - E_{\lambda-\epsilon}) < \infty,$$

where $\{E_\lambda\}_{\lambda\geq 0}$ denotes the spectral family of projections associated with $-\Delta$. So λ is an eigenvalue of $-\Delta$ of finite multiplicity, *and* is an isolated element of spec $-\Delta$.

The *essential spectrum of* $-\Delta$ consists of the complement of the discrete spectrum. Therefore, the essential spectrum consists of those $\lambda \in$ spec $-\Delta$ such that

$$\dim (E_{\lambda+\epsilon} - E_{\lambda-\epsilon}) = \infty$$

for all $\epsilon > 0$.

Thus $\lambda \in$ spec $-\Delta$ precisely when there exists a sequence $\{\phi_n\} \subseteq -\Delta$, $\|\phi_n\| = 1\ \forall n$, such that $(\Delta + \lambda)\phi_n \to 0$ as $n \to \infty$. And λ is in the essential spectrum precisely when the sequence can be chosen to be orthonormal. For any $\lambda \in$ spec $-\Delta$, the sequence $\{\phi_n\}$ is referred to as a *sequence of normalized approximate eigenfunctions of* λ.

Decomposition principle. *Let $\lambda \in$ ess spec $-\Delta$. Then given any compact $K \subset M$ we may pick the sequence of normalized approximate eigenfunctions to have supports disjoint from K.*

3.1. Spectra of complete Riemannian manifolds

Definition. A function $f \in L^2$ on M has Δf *acting as an* L^2*–distribution on* M if there exists $g \in L^2$ such that

$$(g, \phi) = (f, \Delta\phi), \qquad \phi \in C_c^\infty.$$

We write Δf for g.

Gaffney's Theorem. *If M is a complete Riemannian manifold, then $\mathcal{H}_c(M) = \mathcal{H}(M)$. Given any function $f \in C^\infty$ for which $f, \Delta f \in L^2$, we also have* grad $f \in L^2$ *and*

$$(-\Delta f, f) = (\operatorname{grad} f, \operatorname{grad} f).$$

Finally, $\Delta|C_c^\infty$ is essentially self-adjoint, so the Friedrichs extension of $\Delta|C_c^\infty$ is the unique self-adjoint extension of $\Delta|C_c^\infty$.

Yau's Theorem. *Let M be a complete Riemannian manifold. Then M has no L^2 nonconstant harmonic function, that is, any function f in $L^2(M)$ satisfying $\Delta f = 0$ on all of M must be constant. Moreover, given any $\lambda < 0$, there is no solution $u \in L^2$ to the equation*

$$\Delta u + \lambda u = 0$$

except for $u = 0$ identically on M.

Also, from the general theory of Friedrichs extensions we have

$$\lambda(M) =: \inf \operatorname{spec} -\Delta = \inf_{\phi \in C_c^\infty, \phi \neq 0} \frac{\mathcal{Q}(\phi, \phi)}{\|\phi\|^2}.$$

3.2. The spectrum of hyperbolic space

In particular, for $M = \mathbf{H}^n$ we have

$$\text{spec } -\Delta \subset \left[\frac{(n-1)^2}{4}, \infty\right).$$

Theorem. *The spectrum of* $-\Delta$ *on* $\mathbf{H}^n$ *is equal to its essential spectrum, which is*

$$\text{ess spec } -\Delta = \left[\frac{(n-1)^2}{4}, \infty\right).$$

Proof. For any function ϕ on $\mathbf{H}^n$, its Laplacian is given relative to geodesic spherical coordinates based at some fixed $o \in \mathbf{H}^n$ by (see (11))

$$\Delta\phi = \partial_r{}^2\phi + (n-1)\coth r\partial_r\phi + \sinh^{-2} r\mathcal{L}_\xi\phi.$$

Pick ϕ to be radial, that is, $\phi(\exp_o r\xi) = u(r)$. Then

$$\Delta\phi = u'' + (n-1)u'\coth r.$$

Since for large r we have $\coth r \sim 1$, we also consider the operator

$$Lu = u'' + (n-1)u'.$$

For solutions of

$$Lu + \lambda u = 0$$

we try $u(r) = e^{\alpha r}$. Then

$$\alpha = -\frac{n-1}{2} \pm i\sigma, \qquad \text{where} \qquad \lambda = \frac{(n-1)^2}{4} + \sigma^2.$$

Therefore, for a sequence of approximate eigenfunctions of λ for the operator $-\Delta$ we pick

$$u(r) = e^{\alpha r}\eta(r), . \qquad \eta \in C_c^\infty.$$

We will be more precise about η later on.

Explicit calculation yields

$$\begin{aligned}\Delta u + \lambda u &= e^{\alpha r}\{\alpha^2\eta + 2\alpha\eta' + \eta'' + (n-1)(\alpha\eta + \eta')\coth r + \lambda\eta\} \\ &= e^{\alpha r}\{\alpha(n-1)(\coth r - 1)\eta + (2\alpha + (n-1))\eta'\coth r + \eta''\},\end{aligned}$$

which implies

$$\begin{aligned}\|\Delta u + \lambda u\| &= \mathbf{c_{n-1}} \left\{ \int_0^\infty \{u'' + (n-1)u' \coth r + \lambda u\}^2 \sinh^{n-1} r \, dr \right\}^{1/2} \\ &\le \text{const.} \left\{ \int_0^\infty \{\eta'' + \eta' + (\coth r - 1)\eta\}^2 \, dr \right\}^{1/2}\end{aligned}$$

(where $\mathbf{c_{n-1}}$ denotes the $(n-1)$–dimensional area of the unit sphere $\mathbf{S}^{n-1}$). We now pick η. First fix

$$\psi \in C_c^\infty([1/2, 3/2]), \qquad \psi|[3/4, 5/4] = 1,$$

and then set

$$\eta(r) = e^{-i\sigma R}\psi(r/R), \qquad u_R(r) = \text{Re } e^{-(n-1)r/2} e^{i\sigma(t-R)}\psi(r/R).$$

Since

$$\int |\eta_R|^2 = R \int |\psi|^2, \qquad \int |\eta_R'|^2 = R^{-1} \int |\psi'|^2, \qquad \int |\eta_R''|^2 = R^{-3} \int |\psi''|^2$$

we conclude that for every $\sigma > 0$ we have

$$\left\| \Delta u_R + \left\{ \frac{(n-1)^2}{4} + \sigma^2 \right\} u_R \right\| \to 0$$

in $L^2(\mathbf{H}^n)$ as $R \to +\infty$. On the other hand,

$$\begin{aligned}\int_0^\infty u_R \sinh^{n-1} r \, dr &= \int_{R/2}^{3R/2} u_R \sinh^{n-1} r \, dr \\ &\ge \int_{3R/4}^{5R/4} e^{-(n-1)r} \sinh^{n-1} r \cos^2 \sigma(r-R) \, dr \\ &\ge \text{const.},\end{aligned}$$

which implies the theorem. qed

Exercise. Carry out the corresponding calculations for the hyperbolic horn, that is, the surface of revolution with metric

$$ds^2 = dr^2 + e^{-2r} d\theta^2.$$

An equivalent definition is to consider the hyperbolic metric on the upper half-plane given by

$$ds^2 = \frac{dx^2 + dy^2}{y^2}, \qquad x \in \mathbf{R}, \ y > 0,$$

subject the upper half-plane to the action of the discrete subgroup $(x, y) \mapsto (x + 2\pi\mathbf{Z}, y)$, and consider the covered hyperbolic 2–manifold.

More generally, consider the hyperbolic metric on the upper half-plane of $\mathbf{R}^n$ given by

$$ds^2 = \frac{|dx|^2 + dy^2}{y^2}, \qquad x \in \mathbf{R}^{n-1}, \quad y > 0,$$

subject the upper half-plane to the action of the discrete subgroup $(x, y) \mapsto (x + \Gamma, y)$, where Γ is a discrete subgroup of $\mathbf{R}^{n-1}$ of rank $n-1$, and consider the covered hyperbolic n–manifold. Then again one has

$$\text{ess spec } -\Delta = \left[\frac{(n-1)^2}{4}, \infty\right).$$

Consider the cusp $y > 1$. Show that any sequence of approximate eigenfunctions supported in $\{(x, y) : x \in \mathbf{R}^{n-1}, \ y > 1\}$ must ultimately be radial. HINT: Show that for any $a > 1$ the torus

$$T_a = \{(x, y) : x \in \Gamma \setminus \mathbf{R}^{n-1}, \ y = a\}$$

(with induced metric) has lowest nonzero eigenvalue

$$\lambda_1(T_a) = e^{2a}.$$

Additional References

Baider [**1**], Baider–Feldman [**2**], Davies [**12**], Karp [**15**], Lax–Phillips [**16**].

§4. Hyperbolic manifolds

Definition. A *hyperbolic manifold* is a complete oriented Riemannian manifold of finite volume and constant sectional curvature -1.

We start with compact hyperbolic manifolds. It is customary to label the eigenvalues of a compact Riemannian manifolds by

$$0 = \lambda_0 < \lambda_1 \le \lambda_2 \le \cdots \quad .$$

4.1. Buser's results

Even though the $\lambda(\mathbf{H}^n) = (n-1)^2/4$ we have examples that for a *compact* hyperbolic manifold M it is possible to construct examples with $\lambda_1(M)$ as small as one wishes.

Theorem. *Given any $\epsilon > 0$ and integer $g \geq 2$, there exists a compact Riemann surface of genus g for which*

$$\lambda_{2g-3} < \epsilon.$$

Theorem. *Given positive integers g and k, and any $\epsilon > 0$, there exists a compact Riemann surface M with genus g such that*

$$\lambda_k(M) \leq 1/4 + \epsilon.$$

Theorem. *For any compact Riemann surface of genus g we have*

$$\lambda_{4g-2} > 1/4.$$

Proof of first result. Consider a quadrilateral $ABCD$ in the hyperbolic plane with

$$\measuredangle A = \measuredangle B = \measuredangle C = \pi/2, \qquad \measuredangle D = \pi/3,$$

and

$$AB = r, \qquad BC = t.$$

Then it is an exercise in hyperbolic geometry that

$$\frac{1}{2} = \sinh r \sinh t$$

(see Figure 4, p. 248 of Chavel [**8**]). Now consider six copies of the quadrilateral $ABCD$, wherein we obtain a hyperbolic hexagon $B_1 \cdots B_6$ all of whose angles are

$$\measuredangle B_j = \pi/2 \qquad j = 1, \ldots, 6,$$

and alternating sides have lengths $2r$, $2t$, respectively, that is,

$$B_1B_2 = B_3B_4 = B_5B_6 = 2r, \qquad B_2B_3 = B_4B_5 = B_6B_1 = 2t$$

(see Figure 5, p. 248 of Chavel [**8**]). Now take another copy of $B_1 \cdots B_6$, $E_1 \cdots E_6$, and identify

$$B_2B_3 \equiv E_2E_3, \qquad B_4B_5 \equiv E_4E_5, \qquad B_6B_1 \equiv E_6E_1.$$

Then one obtains a Riemann surface Ω bounded by 3 simple closed geodesics, all of length $4r$. By picking r sufficiently small we may guarantee that $t > 1$.

Now consider the union of the 3 collars, D,

$$D = \{x : \ d(x, \partial\Omega) < 1\}.$$

Then

$$A(D) = 12r \sinh 1.$$

This implies that if

$$\phi = \begin{cases} d(x, \partial\Omega) & d(x, \partial\Omega) \leq 1 \\ 1 & d(x, \partial\Omega) \geq 1, \end{cases}$$

then

$$\begin{aligned} \int_\Omega |\text{grad}\, \phi|^2 \, dA &= 12r \sinh 1, \\ \int_\Omega \phi^2 \geq A(\Omega) - A(D) &= 2\pi - 12r \sinh 1. \end{aligned}$$

Rayleigh's theorem then implies, for the lowest Dirichlet eigenvalue of Ω, $\lambda(\Omega)$,

$$\lambda(\Omega) \leq \frac{12r \sinh 1}{2\pi - 12r \sinh 1}.$$

Next take $2g - 2$ copies of Ω: $\Omega_1, \ldots, \Omega_{2g-2}$, each bounded by simple closed geodesics

$$\gamma_{k,j}, \qquad k = 1, \ldots, 2g - 2, \quad j = 1, 2, 3,$$

each having length $4r$, and identify geodesics as follows:

$$\gamma_{k,2} \equiv {\gamma_{k+1,3}}^{-1}, \qquad k \bmod 2g - 2$$

and

$$\gamma_{2\ell-1,1} \equiv {\gamma_{2\ell,1}}^{-1}, \qquad \ell = 1, \ldots, g - 1$$

(see Figure 6, p. 250 of Chavel [**8**]). We obtain a compact Riemann surface of genus g. Our previous estimate on $\lambda(\Omega)$ of each three-holed sphere, and domain monotonicity of Dirichlet eigenvalues implies

$$\lambda_{2g-3} \leq \frac{12r \sinh 1}{2\pi - 12r \sinh 1},$$

which implies the claim. qed

4.2. The collar lemma

Let M be an n–dimensional Riemannian manifold, Y a k–dimensional submanifold, $k < n$, $\mathbf{n}Y$ the normal bundle of Y in M, $\mathbf{nS}Y$ the unit normal bundle of Y in M, that is, $\mathbf{nS}Y = \{\xi \in \mathbf{n}Y : |\xi| = 1\}$. Also, let

$$\mathbf{nS}_p = \mathbf{nS}Y \cap M_p, \qquad p \in M.$$

Let $\mathrm{Exp} : \mathbf{n}Y \to M$ denote the restriction of the exponential map to $\mathbf{n}Y$. Given $\xi \in \mathbf{nS}Y$, define $c_{\mathbf{n}}(\xi)$ the *distance to the focal cut point of* Y *along* γ_ξ by

$$c_{\mathbf{n}}(\xi) =: \sup\{t > 0 : d(Y, \gamma_\xi(t)) = t\}.$$

The *focal injectivity radius of* Y *in* M is defined by

$$\mathrm{inj}\, Y = \inf_{\xi \in \mathbf{nS}Y} c_{\mathbf{n}}(\xi).$$

We may now consider integration. There exists a density $\sqrt{\mathbf{g}}(r;\xi)$ such that for $f \in L^1(M)$ we have

$$\int_M f\, dV = \int_Y dV_k(p) \int_{\mathbf{nS}_p} d\mu_{n-k-1,p}(\xi) \int_0^{c_{\mathbf{n}}(\xi)} f(\mathrm{Exp}\, r\xi)\sqrt{\mathbf{g}}(r;\xi)\, dr,$$

where dV_k the (k–dimensional) Riemannian measure of Y, and $d\mu_{n-k-1,p}$ the standard $(n-k-1)$–dimensional measure on $\mathbf{nS}_p$. Again, $\sqrt{\mathbf{g}}(r;\xi)$ is the determinant of a solution of the matrix Jacobi equation along γ_ξ, but the initial conditions are now adjusted for the submanifold. Our particular case is as follows:

If M is hyperbolic, and Y is a geodesic γ, then

$$\sqrt{\mathbf{g}}(r;\xi) = \cosh r \sinh^{n-2} r.$$

Randol's Collar lemma. Let M be a compact Riemann surface with simple closed geodesic γ of length ℓ. Then

$$\mathrm{inj}\, \gamma \geq \operatorname{arcsinh} \operatorname{csch} \ell/2.$$

Corollary. *Given a fixed number $\beta > 0$ consider all noncompact Riemann surfaces Ω with compact closure, whose boundary consists of closed geodesic circles with total length $\ell(\partial\Omega)$, satisfying*

$$\ell(\partial\Omega) \leq 2\beta.$$

Then

$$\lambda(\Omega) \le \frac{\ell(\partial\Omega)\cosh\beta}{\operatorname{arcsinh}^2 \operatorname{csch}\beta\{2\pi - \ell(\partial\Omega)\operatorname{csch}\beta\}}$$

Schoen–Wolpert–Yau Theorem. *Given the genus $g \ge 2$, then there exists a constant $\beta(g)$ such that*

$$\lambda_k(M) \le \beta(g)\ell_k(M), \qquad k = 1, \ldots, 2g-3,$$

for every compact Riemann surface M of genus g, where $\ell_k(M)$ denotes the minimum length of the union of simple closed geoedesics which divide M into $k+1$ components.

4.3. The collar lemma in higher dimensions

Kazhdan–Margulis Thick-Thin Decomposition Theorem. *Assume M is compact. Then there exists a positive constant $\mu = \mu_n$ such that for*

$$\mathcal{A} = \{p \in M : \operatorname{inj} p \ge \mu\} = \text{thick}, \qquad \mathcal{B} = \{p \in M : \operatorname{inj} p < \mu\} = \text{thin},$$

we have:

(a) *$\mathcal{A}$ is non-empty and connected.*

(b) *$\mathcal{B}$ is open with finitely many components.*

(c) *Each component $\mathcal{B}_\iota$ is a tubular neighborhood of a simple closed geodesic (referred to as* the core geodesic*) γ_ι of length $< \mu$. More particularly, $\mathcal{B}_\iota$ is a starlike neighborhood (with respect to the exponential map of the normal bundle of γ_ι) within the focal cut locus of γ_ι.*

(d) *Let*

$$R(\gamma) = d(\gamma, \mathcal{A}), \qquad T_\gamma = \{p \in M:\ d(p,\gamma) < R(\gamma)\}.$$

Then T_γ contains a hyperbolic disk of radius $\mu/2$.

(e) *Given a core geodesic γ in $\mathcal{B}$, then*

$$V_{-1}(\mu/2) \le \frac{\mathbf{c}_{n-2}}{n-1}\ell(\gamma)\sinh^{n-1} R(\gamma) \le V(M)$$

where $V_{-1}(r)$ denotes the volume of the geodesic n-disk in $\mathbf{H}^n$ of radius r (for any r), $\mathbf{c}_{n-2}$ denotes the $(n-2)$-dimensional area of the unit sphere in $\mathbf{R}^{n-1}$, and $\ell(\gamma)$ denotes the length of γ.

Remark. Since every component of $\mathcal{B}$ contains a hyperbolic disk of radius $\mu/2$, we have an upper bound on the number of components in terms of the the dimension of M and its volume.

4.4. Lower bounds for $\lambda_1(M)$

Let M be a compact n–dimensional hyperbolic Riemannian manifold.

Schoen–Wolpert–Yau, Dodziuk–Randol Theorem. *If* $n = \dim M = 2$ *with genus* g*, then there exists* $\alpha(g) > 0$ *such that*

$$\lambda_k(M) \geq \alpha(g)\ell_k(M), \qquad k = 1, \ldots, 2g-3.$$

Dodziuk–Randol Theorem. *If* $n = \dim M \geq 3$*, then there exists* $c = c(n) > 0$ *such that*

$$\lambda_1(M) \geq c(n)V^{-2}(M).$$

4.5. 3*–dimensional case*

Let M be 3–dimensional noncompact hyperbolic.

We know

$$\operatorname{ess\,spec} -\Delta = [1, \infty),$$

with "multiplicity" of each $\lambda \in \operatorname{ess\,spec} -\Delta$ equal to the number of cusps.

Kazhdan–Margulis Thick-Thin Decomposition Theorem. *Consider the thick-thin decomposition of* M *into* $\mathcal{A}$ *and* $\mathcal{B}$ *as defined above. Compact components of* $\mathcal{B}$ *have the same structure as before. If a component of* $\mathcal{B}$ *is not compact then it is isometric to the product* $\mathbf{R}^+ \times F$, F *a flat* 2*–dimensional torus, equipped with the metric*

$$ds^2 = d\rho^2 + e^{-2\rho} ds_0^2,$$

where ds_0^2 *is a flat metric on* F.

Thurston Theorem. *There exists a sequence of compact hyperbolic manifolds* $M_i \to M$. *Moreover,* $V(M_i) \leq V(M)$ *for all* i.

Colbois–Courtois Theorem. *The eigenvalues in* $\operatorname{spec} -\Delta$ *on* M *(the limit manifold of the approximating sequence) below* 1 *are limits of eigenvalues of the approximating* M_i.

Chavel–Dodziuk Theorem. *Suppose M has only one cusp. Then, for large i, M_i will contain a metric tubular neighborhood of a short, simple, closed geodesic γ_i of length $\ell_i \to 0$ and of radius $R_i = R(\gamma_i) \to \infty$. Let Δ_i be the Laplacian on M_i,* spec $(-\Delta_i)$ *its spectrum, and*

$$N_i(x) = \#\{\, \lambda \in \operatorname{spec}(\Delta_i) \mid 1 \leq \lambda \leq 1 + x^2 \,\}.$$

Then

$$N_i(x) = \frac{x}{\pi} R_i + O_x(1) = \frac{x}{2\pi} \log \frac{1}{\ell_i} + O_x(1).$$

If M has $q \geq 1$ cusps, then M_i will have q shrinking geodesics $\gamma_i^1, \gamma_i^2, \ldots, \gamma_i^q$ surrounded by disjoint tubes of radii $R_i^1, R_i^2, \ldots, R_i^q$ respectively and the equalities above hold with $R_i = \sum_{j=1}^{j=q} R_i^j$ and with $\log(1/\ell_i)$ replaced by $\sum_{j=1}^{j=q} \log(1/\ell_i^j)$.

Sketch of proof: *Assume that M has exactly one cusp. The modifications required in case of several cusps will be obvious.*

Fix a compact M_i of the approximating sequence, and a core geodesic γ. Let $\tilde{M}_i \cong \mathbf{H}^3$ be the universal covering of M, and A_γ the deck transformation of $\mathbf{H}^3$ determined by γ. Then the cyclic group $\langle A_\gamma \rangle$ generated by γ consists of hyperbolic transformations, with common axis $\tilde{\gamma}$ along which A_γ acts as a translation. Then $\tilde{\gamma}$ is a lift of γ.

Define Fermi coordinates (r, t, θ) in $\mathbf{H}^3$ based on $\tilde{\gamma}$:
$r =$ distance from $\tilde{\gamma}$,
$t =$ arclength along $\tilde{\gamma}$,
$\theta =$ the angular coordinate in the circle of unit vectors perpendicular to $\tilde{\gamma}$ at a point. The angle θ will be determined by a choice of parallel unit vector field along $\tilde{\gamma}$. Any other choice of parallel unit vector field along $\tilde{\gamma}$ changes θ angle by an additive constant.

Then the metric of $\mathbf{H}^3$ is expressed as

$$ds^2 = dr^2 + \cosh^2 r\, dt^2 + \sinh^2 r\, d\theta^2$$

and the deck transformation A_γ is given by

$$A_\gamma(r, t, \theta) = (r, t + \ell(\gamma), \theta + \alpha)$$

for some angle α (there is no a priori reason to assume that A_γ will leave the parallel translation invariant). Then

$$\tilde{T}_\gamma = \{\tilde{p} \in \tilde{M} :\ d(\tilde{p}, \tilde{\gamma}) \leq R\} \qquad \text{and} \qquad T_\gamma = \langle A_\gamma \rangle \setminus \tilde{T}_\gamma$$

are determined up to isometry by R, α, and $\ell = \ell(\gamma)$.

Now introduce global coordinates on the tube T_γ. Consider the mapping

$$f : (r, t, \phi) \mapsto (r, t, \theta), \qquad \theta = \phi + \frac{\alpha}{\ell} t.$$

Note that

$$A_\gamma(r, t, \phi) = (r, t + \ell, \phi).$$

It follows that $0 \leq r \leq R$, $t \in \mathbf{R}/\ell\mathbf{Z}$, and $\phi \in \mathbf{R}/2\pi\mathbf{Z}$ are well-defined functions on $T_\gamma = \langle A_\gamma \rangle \setminus \tilde{T_\gamma}$. The pull-back of the metric on T_γ to $\tilde{T_\gamma}$ is given by

$$ds^2 = dr^2 + \cosh^2 r\, dt^2 + \sinh^2 r\, d\left(\phi + \frac{\alpha}{\ell} t\right)^2.$$

A straightforward calculation (which generalizes the one given earlier for geodesic spherical coordinates) yields the expression for the Laplacian $\Delta = \Delta_i$ on T_γ:

$$\Delta u = Lu + \Delta_r,$$

where

$$Lu = \frac{1}{\sinh r \cosh r} \frac{\partial}{\partial r}\left(\sinh r \cosh r \frac{\partial u}{\partial r}\right) = u'' + 2u' \coth 2r$$

(the primes denote differentiation with respect to r), and Δ_r is the Laplacian on the distance torus

$$F_r = \{p \in T : \ d(p, \gamma) = r\}$$

with respect to the induced metric.

Lemma. *For every $x > 0$ we have*

$$N_i(x) \geq \frac{x}{\pi} R_i + O_x(1)$$

during the degeneration $M_i \to M$.

Proof. Given $\gamma = \gamma_i$ in M_i; let

$$\mathcal{H} = \{\phi \in C^\infty(M_i) : \phi(p) = \psi(d(p, \gamma)), \ \ \operatorname{supp} \psi \subset [1, R)\}.$$

Consider the Sturm-Liouville problem

$$-Lu = \nu\, u, \qquad u(1) = u(R) = 0$$

on the interval $[1, R]$. Its eigenfunctions can be identified with elements of $\mathcal{H}$. Let

$$\operatorname{spec} - L = \{\nu_k : k = 1 \ldots\}, \qquad D_i(x) = \#\{\, k \mid 1 \leq \nu_k \leq 1 + x^2 \,\}.$$

and

$$\operatorname{spec} - \Delta_i = \{\lambda_k : k = 1 \ldots\}, \qquad N_i(x) = \#\{\, k \mid 1 \leq \lambda_k \leq 1 + x^2 \,\}.$$

Then

$$\nu_k \geq \lambda_k \qquad \forall\ k.$$

Upper eigenvalue bounds translate into lower bounds for the counting function. Since the number of eigenvalues of M_i in the interval $[0, 1]$ is bounded by const.$V(M)$ for some positive universal constant, and (one easily shows) $\nu_1 > 1$, we have

$$N_i(x) \geq D_i(x) - O(1).$$

To estimate $D_i(x)$ we transform the Sturm–Liouville problem for $-L$ by

$$u = \sinh^{-1/2} 2r\, f.$$

Then

$$-f'' - \sinh^{-2} 2r\, f = (\nu - 1)\, f, \qquad f(1) = f(R) = 0,$$

which implies, by straightforward max-min methods,

$$\nu_k - 1 \leq \frac{\pi^2 k^2}{(R-1)^2}.\text{qed}$$

The upper bound $N_i(x) \leq xR/\pi + O_x(1)$ is more subtle. The employment of max-min methods will require that we split M_i into the tube T_γ and its complement, and estimate the Neumann eigenvalues of each from below.

STEP 1: Show that the removal of the complement of T_γ has no effect on clustering of eigenvalues. More precisely, one can show that the counting function for Neumann eigenvalues of T_γ differs from $N_i(x)$ by a bounded amount when $i \to \infty$. Beyond some delicate max-min arguments one uses the following:

(i) The geometry of the tori $F_i = \partial T_{\gamma_i}$ are bounded independent of i.

(ii) Near the boundary F_i of γ_i (on the "thin" side) the metric of the tube is a very good approximation of the metric of a cusp. Indeed,

$$\rho = R - r \quad \Rightarrow \quad 4\sinh^2 r \sim 4\cosh^2 r \sim e^{2R} e^{-2\rho} \qquad \text{as } R \to +\infty,$$

which implies

$$ds^2 \sim d\rho^2 + e^{-2\rho} ds_0^2,$$

where ds_0^2 is the metric of F.

STEP 2: One reduces the estimate of the counting function of Neumann eigenvalues on the tube T_γ to an upper bound of the counting function

$$Q_R(x) = \#\{\mu : \ \mu \leq 1 + x^2\}$$

for the Neumann eigenvalue problem

$$-u'' - \sinh^2 ru = \mu - 1, \qquad u'(1) = u'(R - a) = 0,$$

where a is fixed and $R \to \infty$. The ability to restrict to radial functions follows from the exercise at the end of §3.2.

STEP 3: One shows that

$$Q_R(x) = \frac{x}{\pi}(R - a - 1) + O_x(1) = \frac{x}{\pi} R + O_x(1).$$

This is not the usual obvious estimate (the one everyone has in mind is letting x become large). But it is true nonetheless.

Remark . (Dodziuk–Hejhal–Jorgensen) When one considers degeneration of *noncompact* manifolds M_i to M, then one has the weaker estimate

$$N_i(x) = \frac{x}{\pi} \log \frac{1}{l_l} + O_x\left(\left[\log \frac{1}{l_i}\right]^{3/4}\right).$$

Remark. The corresponding question on forms was dealt with by Dodziuk–McGowan.

Remark. The analogous question for surfaces has been studied by Wolpert, Hejhal, Ji. A sharp estimate of the accumulation rate was obtained by Ji and Zworski.

Remark. Problems of this kind do not arise in dimensions greater than or equal to four, since the number of complete hyperbolic manifolds of volume less than or equal to a given constant is finite in this case.

Additional References

Beardon [**3**], Buser [**7**], Chavel [**8**], Benedetti–Petronio [**4**], Gromov [**14**], Ratcliffe [**18**], Thurston [**20**]; Chavel–Dodziuk [**10**], Dodziuk–Randol [**13**], Randol [**17**].

§5. Curvature comparison theorems

Given a complete Riemannian manifold M. Fix $x \in M$. For the moment, fix a unit tangent vector $\xi \in M_x$, and its associated geodesic $\gamma_\xi(r)$. The study of solutions $Y(r)$ to Jacobi's equation (1) along γ_ξ is the study, at the infinitesmal level, of how geodesics emanating from x spread apart one from the other. The larger $|Y(r)|$ is, for all r, for all Y, and for all geodesics γ_ξ emanating from x, the faster M is "opening up" as one leaves x. The basic insight here is that the higher the curvature the slower M "opens up". The first indication of this insight is found in constant curvature, as one checks how $\mathbf{s}_\kappa$ varies as κ varies. The more general answer proceeds in stages.

The simplest case is when $\dim M = 2$, so Jacobi's equation is a *scalar* equation

$$y'' + K(r)y = 0,$$

with initial conditions $y(0) = 0$, $y'(0) = 1$ (recall the variations induced by polar coordinates).

Sturm's Comparison Theorem. *Given the two continuous functions $K(r)$ and $H(r)$, with associated Jacobi equations*

$$\begin{aligned} y'' + K(r)y &= 0, \qquad y(0) = 0, \quad y'(0) = 1, \\ z'' + H(r)z &= 0, \qquad z(0) = 0, \quad z'(0) = 1, \end{aligned}$$

where

$$K(r) \le H(r) \quad \forall\, r \in [0, \beta), \qquad z|(0, \beta) > 0.$$

Then

$$\frac{y'(r)}{y(r)} \ge \frac{z'(r)}{z(r)} \quad \forall\, r \in [0, \beta).$$

Therefore

$$y(r) \ge z(r) \quad \forall\, r \in [0, \beta].$$

One has equality (in either of the above) at any $r_o \in (0, \beta]$ if and only if $K = H$ on all of $[0, r_o]$, in which case $y = z$ on all of $[0, r_o]$.

Proof. Consider the Wronskian

$$W(r) = y'z - yz';$$

then

$$W'(r) = \{H(r) - K(r)\}yz.$$

Since y and z are positive on some interval $(0,\alpha)$, we have $W' \geq 0$ on $(0,\alpha)$, which implies, since $W(0)=0$, $W(r)\geq 0$ on $(0,\alpha)$. Therefore

$$\frac{y'}{y} \geq \frac{z'}{z} \quad \text{on} \quad [0,\alpha],$$

which implies

$$y \geq z \quad \text{on} \quad [0,\alpha].$$

One then has $y>0$ whenever $z>0$, and the inequalities are valid on all of $(0,\beta)$. Given $r_o>0$ at which $y'/y = z'/z$, we have $W(r_o)=0$ which implies, since $W(0)=0$ and $W(r)$ is increasing, that $W(r)=W'(r)=0$ for all $r\in(0,r_o]$, which implies $K=H$ on $[0,r_o]$.

The argument for the case $y(r_o)=z(r_o)$ is similar. qed

In higher dimensions one has (we only consider Jacobi fields orthogonal to the geodesic at all points of the geodesic).

Rauch's Comparison Theorem. *Given $x\in M$, $\xi\in \mathsf{S}_x$, $Y(r)$ a vector solution to Jacobi's equation* (1) *along γ_ξ satisfying* $Y(0)=0$, $|(\nabla_r Y)|(0)=1$.

(a) *If the sectional curvatures along γ_ξ are less than or equal to δ, then*

$$|Y(r)| \geq \mathbf{s}_\delta(r) \quad \forall\, r\in[0,\pi/\sqrt{\delta}]$$

(where, when $\delta\leq 0$, we set $\pi/\sqrt{\delta}=+\infty$).

(b) *If the sectional curvatures along γ_ξ are greater than or equal to κ, then* $\operatorname{conj}\xi \leq \pi/\sqrt{\kappa}$ *and*

$$|Y(r)| \leq \mathbf{s}_\kappa(r) \quad \forall\, r\in[0,\operatorname{conj}\xi].$$

The proof of **(a)** is an easy reduction to the scalar case, but the proof of **(b)** is more difficult. We only present **(a)**.

One has statements concerning equality in either of the above equalities, analogous to those of the scalar equation.

Proof of (a). We start with

$$|Y|' = \langle Y,\nabla_t Y\rangle |Y|^{-1},$$

which implies

$$\begin{aligned} |Y|'' &= |Y|^{-1}\{|\nabla_t Y|^2 - \langle Y, R(\gamma',Y)\gamma'\rangle\} - |Y|^{-3}\langle Y,\nabla_t Y\rangle^2 \\ &\geq -\delta|Y| + |Y|^{-3}\{|\nabla_t Y|^2|Y|^2 - \langle Y,\nabla_t Y\rangle^2\} \\ &\geq -\delta|Y| \end{aligned}$$

by the Cauchy–Schwarz inequality, which implies

$$|Y|'' + \delta|Y| \geq 0.$$

Now one adapts the Wronskian argument, from the Sturm Comparison Theorem, where $y = |Y|$ and $z = \mathbf{s}_\kappa$. qed

Next we consider the volume of M along γ_ξ, that is, we wish to estimate the growth of

$$\sqrt{\mathbf{g}}(r;\xi) = \det A(r;\xi)$$

in r (see (8)). Since $A(r;\xi)$ encapsulates all the information contained in the vector solutions of Jacobi's equation (3)–(6), we are asking for less, in seeking to estimate $\det A(r;\xi)$.

Bishop's Comparison Theorem. *Given $x \in M$, let $\sqrt{\mathbf{g}}(r;\xi)$ denote the volume density in geodesic spherical coordinates based at x.*

(a) *If the sectional curvatures along γ_ξ are less than or equal to δ, then*

$$\frac{\partial_r \sqrt{\mathbf{g}}(r;\xi)}{\sqrt{\mathbf{g}}(r;\xi)} \geq (n-1)\frac{\mathbf{c}_\delta}{\mathbf{s}_\delta}(r) \tag{13}$$

for all $r < \pi/\sqrt{\delta}$.

(b) *Assume $r < \operatorname{conj}(\xi)$. If the Ricci curvatures along γ_ξ are greater than or equal to $(n-1)\kappa$, then*

$$\frac{\partial_r \sqrt{\mathbf{g}}(r;\xi)}{\sqrt{\mathbf{g}}(r;\xi)} \leq (n-1)\frac{\mathbf{c}_\kappa}{\mathbf{s}_\kappa}(r). \tag{14}$$

The proof of **(a)** follows by using the Rauch Comparison Theorem **(a)** to study the logarithmic derivative of

$$r \mapsto \det A^T(r;\xi)A(r;\xi).$$

For **(b)** we have a different proof, one that does not have to use the Rauch Comparison Theorem **(b)**.

Proof of (b). The proof is via the classical Riccati equation, more precisely, Riccati inequality. The idea is as follows: Given the *scalar* Jacobi equation

$$y'' + K(t)y = 0,$$

then $u =: y'/y$ satisfies the Riccati equation

$$u' + u^2 + K = 0. \tag{15}$$

In higher dimensions, even though $t \mapsto (\nabla_t A)(t;\xi)A^{-1}(t;\xi)$ satisfies a *matrix* Riccati equation, the logarithmic derivative of $\det A(t;\xi)$ will only satisfy a scalar Riccati inequality. Nonetheless it will suffice to do the job.

In the notation that follows we drop the ξ (since it is fixed) and replace $(\nabla_t A)$ by A' (for convenience).

Note that the formula for differentiating determinants states

$$\frac{(\det A)'}{\det A} = \operatorname{tr} A'A^{-1}.$$

Next, let

$$\mathbf{Ct}_\kappa(t) =: \mathbf{s}_\kappa{}'(t)/\mathbf{s}_\kappa(t),$$

let $\mathbf{arcCt}_\kappa$ denote the inverse function of $\mathbf{Ct}_\kappa$; and consider

$$\psi =: (n-1)\mathbf{Ct}_\kappa.$$

Then ψ satisfies the scalar Riccati equation

$$\psi' + \frac{\psi^2}{n-1} + (n-1)\kappa = 0. \tag{16}$$

Also, $\psi(t)$ is strictly decreasing with respect to t, for all t; and, when $\kappa \leq 0$, has limiting value, as $t \uparrow +\infty$, equal to $(n-1)\sqrt{-\kappa}$.

Given linear transformations $\mathsf{A}(t), \mathsf{B}(t) : \xi^\perp \to M_{\gamma_\xi(t)}$, depending differentiably on t, we associate their Wronskian $\mathsf{W}(t)$ defined by

$$\mathsf{W}(\mathsf{A},\mathsf{B}) =: \mathsf{A}'^T\mathsf{B} - \mathsf{A}^T\mathsf{B}'.$$

One verifies that for our $A(t)$ we have $\mathsf{W}(A,A) = 0$. Set

$$U =: A'A^{-1}.$$

Then

$$U^T - U = (A^{-1})^T\mathsf{W}(A,A)A^{-1} = 0$$

— so U is self-adjoint. Also, U satisfies the matrix Riccati equation

$$U' + U^2 + \mathsf{R} = 0,$$

which implies

$$(\operatorname{tr} U)' + \operatorname{tr} U^2 + \operatorname{tr} \mathsf{R} = 0$$

($\operatorname{tr} \mathsf{R}$ is the Ricci curvature along γ).

Now the Cauchy–Schwarz inequality implies

$$\operatorname{tr} U^2 \geq \frac{(\operatorname{tr} U)^2}{n-1},$$

which implies, for

$$\phi =: \operatorname{tr} U = \operatorname{tr} A'A^{-1} = \frac{(\det A)'}{\det A},$$

the differential inequality

$$\phi' + \frac{\phi^2}{n-1} + (n-1)\kappa \le 0. \tag{17}$$

We, therefore, wish to compare ϕ with ψ.

As mentioned (16), we have

$$\Psi =: \frac{\psi^2}{n-1} + (n-1)\kappa > 0$$

on all of $(0, \pi/\sqrt{\kappa})$. Next, note that $\phi \sim (n-1)/t$ as $t \downarrow 0$. So there exists $\epsilon_0 > 0$ such that

$$\Phi =: \frac{\phi^2}{n-1} + (n-1)\kappa > 0$$

on $(0, \epsilon_0)$. Then the inequality (17) implies

$$\frac{-\phi'}{\frac{\phi^2}{n-1} + (n-1)\kappa} \ge 1,$$

which implies

$$\int_0^t \frac{-\phi'}{\frac{\phi^2}{n-1} + (n-1)\kappa}(\tau)\, d\tau \ge t.$$

That is,

$$\textbf{arcCt}_\kappa \frac{\phi(t)}{(n-1)} \ge t,$$

which implies

$$\phi \le \psi,$$

which is (14).

So we have an interval $(0, \epsilon_o)$ on which $\phi \le \psi$. Let $\beta > 0$ be the supremum of $\alpha \in (0, \operatorname{conj} \xi)$ for which $\phi \le \psi$ on $(0, \alpha)$. If $\beta = \operatorname{conj} \xi$ then we are done. If not, then $\phi(\beta) = \psi(\beta)$, which implies $\Phi(\beta) = \Psi(\beta) > 0$. Therefore there exists $\delta > 0$ such that $\Phi|(\beta - \delta, \beta] > 0$, which implies $\phi' < 0$ on $(\beta - \delta, \beta]$, which implies $\Phi(\beta - \delta) > 0$. One concludes that $\Phi > 0$ on $(0, \beta]$, and can therefore continue the argument past β. qed

Volume Comparison Theorem. *Given $x \in M$, $V(x; r)$ denote the volume of the metric disk $B(x; r)$. Let $V_\sigma(r)$ denote the volume of the disk in M_σ of radius r.*

(a) *Assume $r < \operatorname{inj} x$. If the sectional curvatures of M are less than or equal to δ, then*

$$V(x;r) \geq V_\delta(r).$$

(b) *If the Ricci curvatures of M are greater than or equal to $(n-1)\kappa$, then*

$$V(x;r) \leq V_\kappa(r).$$

Note that in **(b)** there is no restriction on r, relative to the cut locus of x.

Cheeger's Inequality. *Let M be an arbitrary Riemannian manifold, $\lambda(M)$ the bottom of the spectrum of M. Define $h(M)$ by*

$$h(M) = \inf_{\Omega} \frac{A(\partial\Omega)}{V(\Omega)},$$

where Ω varies over all relatively compact domains in M with smooth boundary. Then

$$\lambda(M) \geq \frac{1}{4}h^2(M).$$

Example. Let M be a complete simply connected Riemannian manifold all of whose sectional curvatures $\leq \kappa < 0$. Then

$$h(M) \geq \sqrt{-\kappa}.$$

Proof. (Yau) One has geodesic polar coordinates centered at some x valid in all of M. Then, by (9) and the Bishop Comparison Theorem **(a)**, we have

$$\Delta r = \frac{\partial_r \sqrt{\mathbf{g}}(r;\xi)}{\sqrt{\mathbf{g}}(r;\xi)} \geq (n-1)\frac{\mathbf{c}_\delta}{\mathbf{s}_\delta}(r) \geq (n-1)\sqrt{-\kappa},$$

which implies for any admissible Ω, by the divergence theorem,

$$(n-1)\sqrt{-\kappa}V(\Omega) \leq \iint_\Omega \Delta r\, dV = \int_{\partial\Omega} \langle \operatorname{grad} r, \nu\rangle\, dA \leq A(\partial\Omega),$$

which implies the claim. qed

Note that by considering $\Omega = B(x;r)$, $r \to +\infty$, one sees that $h(\mathbf{H}^n) = n-1$.

The combination of Cheeger's inequality and the above estimate immediately yields

McKean's Inequality. *Let M be a complete simply connected Riemannian manifold all of whose sectional curvatures $\leq \kappa < 0$. Then*

$$\lambda(M) \geq \frac{(n-1)^2}{4}(-\kappa).$$

We already know that this estimate is sharp.

We now give an application of the Bishop Comparison Theorem **(b)**.

For any x in M and $\rho > 0$ we let $\lambda(x;\rho)$ denote the lowest Dirichlet eigenvalue of $B(x;\rho)$. For $M = \mathbf{M}_\sigma$ we write $\lambda_\sigma(\rho)$.

Cheng Comparison Theorem.

(a) *Assume that the sectional curvatures of $M \leq \delta$. Then for every $x \in M$ we have*

$$\lambda(x;\rho) \geq \lambda_\delta(\rho)$$

for all $\rho \leq \min\{\operatorname{inj} x, \pi/\sqrt{\delta}\}$.

(b) *Assume that the Ricci curvatures of M are all greater than or equal to $(n-1)\kappa$. Then for every $x \in M$, $\rho > 0$, we have*

$$\lambda(x;\rho) \leq \lambda_\kappa(\rho).$$

Note that in **(b)** there is no restriction on r, relative to the cut locus.

Proof of (b). We have

$$\lambda(x;\rho) = \inf_{F \in \mathcal{H}_c(B(x;\rho)), F \neq 0} \frac{\|\operatorname{grad} F\|^2}{\|F\|^2}.$$

The function F is constructed as follows: Let ϕ denote an eigenfunction of $\lambda_\kappa(\rho)$ with $\phi|\mathbf{B}_\kappa(o;\rho) > 0$, where $\mathbf{B}_\kappa(o;\rho)$ denotes the disk in $\mathbf{M}_\kappa$. Then one has

$$\phi(\exp r\xi) = \Phi(r),$$

and Φ satisfies

$$\partial_r{}^2\Phi + (n-1)\frac{\mathbf{C}_\kappa}{\mathbf{S}_\kappa}\partial_r\Phi + \lambda_\kappa(\rho)\Phi = 0,$$

with boundary conditions

$$(\partial_r\Phi)(0) = \Phi(\rho) = 0.$$

Then

$$(\partial_r\Phi)(r) < 0$$

on all of $(0,\rho)$ — so Φ is strictly decreasing with respect to r. Then we define $F : B(x;\rho) \to \mathbf{R}$ by

$$F(\exp r\xi) = \Phi(r);$$

so F is certainly defined for those $r\xi$ inside the tangent cut locus of x.

The function F is certainly well defined on *all* of $\overline{B(x;\rho)}$, since if two minimizing geodesics, emanating from x, intersect at y then both geodesics have the same length. Furthermore, one uses the Bishop Comparison Theorem **(b)** to show that $F \in \mathcal{H}_c(B(x;\rho))$.

Let

$$b(\xi) = \min\{c(\xi), \rho\}.$$

Then it suffices to establish

$$\int_0^{b(\xi)} (\partial_r\Phi)^2 \sqrt{\mathbf{g}}(r;\xi)\,dr \le \lambda_\kappa(\rho) \int_0^{b(\xi)} \Phi^2 \sqrt{\mathbf{g}}(r;\xi)\,dr$$

for every $\xi \in \mathsf{S}_x$. Well,

$$\begin{aligned}
&\int_0^{b(\xi)} (\partial_r\Phi)^2 \sqrt{\mathbf{g}}(r;\xi)\,dr \\
&= \Phi(\partial_r\Phi)\sqrt{\mathbf{g}}(r;\xi)\Big|_0^{b(\xi)} - \int_0^{b(\xi)} \Phi\partial_r\{(\partial_r\Phi)\sqrt{\mathbf{g}}(r;\xi)\}\,dr \\
&= (\Phi\partial_r\Phi)(b(\xi))\sqrt{\mathbf{g}}(b(\xi);\xi) - \int_0^{b(\xi)} \Phi\partial_r\{(\partial_r\Phi)\sqrt{\mathbf{g}}(r;\xi)\}\,dr \\
&\le -\int_0^{b(\xi)} \Phi\{\partial_r{}^2\Phi + (\partial_r\Phi)\frac{\partial_r\sqrt{\mathbf{g}}(r;\xi)}{\sqrt{\mathbf{g}}(r;\xi)}\}\sqrt{\mathbf{g}}(r;\xi)\,dr \\
&\le -\int_0^{b(\xi)} \Phi\{\partial_r{}^2\Phi + (n-1)\frac{\mathbf{C}_\kappa}{\mathbf{S}_\kappa}\partial_r\Phi\}\sqrt{\mathbf{g}}(r;\xi)\,dr \\
&= \lambda_\kappa(\rho)\int_0^{b(\xi)} \Phi^2\sqrt{\mathbf{g}}(r;\xi)\,dr.
\end{aligned}$$

Note that we have used (i) $\Phi|[0,\rho) > 0$, (ii) $\partial_r\Phi|(0,\rho] < 0$, and the Bishop Comparison Theorem **(b)**. Now integrate this last inequality over S_x to obtain the theorem. qed

Remark. A different formulation of Cheng's theorem is that

$$\Delta r \le (n-1)\frac{\mathbf{C}_\kappa}{\mathbf{S}_\kappa}(r)$$

in the sense of distributions, if all Ricci curvatures of $M \ge (n-1)\kappa$.

Other applications of Bishop's theorem.

(1) Maximum and Harnack principle methods.

(2) Dirichlet problem at infinity

(3) Heat kernel comparison theorems

(4) Gromov's relative comparison theorems

(a) Spherical isoperimetric inequality

(b) Lichnerowicz and Obata theorems.

(c) Toponogov' volume comparison theorem.

Bibliography

[1] A. BAIDER. Noncompact Riemannian manifolds with discrete spectra. *J. Diff. Geom* .

[2] A. BAIDER & E.A. FELDMAN. Noncompact Riemannian manifolds with discrete spectra. *Trans. Amer. Math. Soc.* **258** (1980), 495–504.

[3] A.F. BEARDON. *The Geometry of Discrete Groups.* New York: Springer Verlag, 1983.

[4] R. BENEDETTI & C. PETRONIO. *Lectures on Hyperbolic Geometry.* New York: Springer Verlag, 1992.

[5] P.H. BÉRARD. *Spectral Geometry: Direct and Inverse Problems*, Lecture Notes Math. **1207**. Berlin: Springer Verlag, 1986.

[6] M. BERGER, P. GAUDUCHON, & E. MAZET. *Le Spectre d'une Variété Riemanniene*, Lect. Notes Math. **194**. Berlin: Springer Verlag, 1974.

[7] P. BUSER. *Geometry and Spectra of Compact Riemann Surfaces.* Basel: Birkhäuser Verlag, 1992.

[8] I. CHAVEL. *Eigenvalues in Riemannian geometry.* New York: Academic Press, 1984.

[9] ——. *Riemannian Geometry: a modern introduction.* Cambridge: Cambridge U. Press, 1994.

[10] I. CHAVEL & J. DODZIUK. Spectral degeneration of 3–dimensional hyperbolic manifolds. *J. Diff. Geom.* **39** (1994), 123–137.

[11] R. COURANT & D. HILBERT. *Methods of Mathematical Physics.* New York: Wiley (Interscience), Vol. I, 1953; Vol. II, 1967.

[12] E.B. DAVIES. *Spectral Theory and Differential Operators.* Cambridge: Cambridge University Press, 1995.

[13] J. DODZIUK & B. RANDOL. Lower bounds for λ_1 on a finite-volume hyperbolic manifold. *J. Diff. Geom.* **24** (1986), 133–139.

[14] M. Gromov. Hyperbolic manifolds according to Thurston and Jørgensen. *Sem. Bourbaki* **546** (1979).

[15] L. Karp. Noncompact manifolds with purely continuous spectrum. *Mich. Math. J.* **31** (1984), 339–347.

[16] P.D. Lax & R.S. Phillips. *Scattering Theory and Automorphic Functions.* Princeton: Princeton U. Press, 1976.

[17] B. Randol. Cylinders in Riemann surfaces. *Comment. Mat. Helv.* **54** (1979), 1–5.

[18] J.G. Ratcliffe. *Foundations of Hyperbolic Manifolds.* New York: Springer Verlag, 1994.

[19] E. Stein & G. Weiss. *Introduction to Fourier Analysis on Euclidean Spaces.* Princeton: Princeton U. Press, 1971.

[20] W.P. Thurston. *Three-Dimensional Geometry and Topology*, **I**. Princeton, N.J.: Princeton U., 1997. editor, S. Levy.

ICMS LECTURE NOTES ON COMPUTATIONAL SPECTRAL THEORY

E.B. Davies

April 1998

1 Abstract Spectral Theory

The material in this section is standard theory which may be found in many textbooks, for example [4] which I follow closely. Much more comprehensive accounts are given in [11, 8], which are rightly regarded as classic accounts of the subject. My goal in these lectures is not to describe new research, but to provide students with the basic knowledge needed to follow the later and more advanced courses. However, in the third lecture I indulge myself somewhat by describing spectral theory from a computational point of view which will be familiar to numerical analysts, but not to most mathematicians and mathematical physicists. This lecture contains recent research material.

Let $\mathcal{H}$ be a separable Hilbert space, such as $L^2(U)$ where U is a region in $\mathbf{R}^N$, and let A be a differential operator acting in $\mathcal{H}$; it is common to abuse language and say that A acts in U. Since not all functions in $\mathcal{H}$ are differentiable the domain of A cannot be the whole of $\mathcal{H}$, and we assume that it is a dense linear subspace $\mathrm{Dom}(H)$ of $\mathcal{H}$. The precise choice of this subspace is both important and difficult in many cases. One often starts with a domain smaller than the final domain, which consists of suitably regular functions obeying the boundary conditions relevant to the operator in question, and then passes to a slightly larger domain by the following closure procedure.

We say that $A : \mathrm{Dom}(A) \to \mathcal{H}$ is closed (the analogue for differential operators of boundedness) if whenever $f_n \in \mathrm{Dom}(A)$, $f_n \to f \in \mathcal{H}$ and $Af_n \to g \in \mathcal{H}$, one can conclude that $f \in \mathrm{Dom}(A)$ and $Af = g$; equivalently the graph of A is a closed subspace of $\mathcal{H} \times \mathcal{H}$. We say that A is closable if it has a closed extension, and then denote by $\overline{A}$ the smallest closed extension, called the closure of A. Most differential operators are closable.

We say that $z \in \mathbf{C}$ does not lie in the spectrum of the closed operator A if the operator $A - zI$ is one-one with domain $\mathrm{Dom}(A)$ and range $\mathcal{H}$; by the closed graph theorem the inverse operator $(A - zI)^{-1}$ is bounded. The spectrum $\mathrm{Spec}(A)$ is a closed subset of $\mathbf{C}$ which may be empty or equal to the whole of $\mathbf{C}$. Any eigenvalue of A is clearly in the spectrum, but there may also be points of the spectrum, parts of its 'essential spectrum', which are not eigenvalues.

If A is a closed operator it has an adjoint defined by

$$\langle Af, g\rangle = \langle f, A^* g\rangle$$

for all $f \in \mathrm{Dom}(A)$ and $g \in \mathrm{Dom}(A^*)$. The precise domain of A^* is the set of all $g \in \mathcal{H}$ for which there exists $h \in \mathcal{H}$ such that $\langle Af, g\rangle = \langle f, h\rangle$ for all $f \in \mathrm{Dom}(A)$.

We say that H is a symmetric operator if

$$\langle Hf, g\rangle = \langle f, Hg\rangle$$

for all $f, g \in \mathrm{Dom}(H)$. It is an exercise to show that all symmetric operators are closable and that their closures are still symmetric; see [4, Lemma 1.1.4]. We say that H is essentially self-adjoint if its closure is self-adjoint in the sense that $\overline{H} = \overline{H}^*$. Closed symmetric operators need not be self-adjoint, and the condition of self-adjointness is crucial for most important results in spectral theory. In particular the spectrum of a self-adjoint operator is always a closed subset of the real line; see [4, Theorem 1.2.10] and [5] for a correction of the last part of the proof of this theorem.

We start with a simple abstract example, which is also characteristic of many elliptic differential operators in regions with compact closures.

Example 1 Suppose that $\{f_n\}_{n=1}^{\infty}$ is a complete orthonormal set in $\mathcal{H}$ and that H is an operator whose domain contains every f_n and satisfies

$$Hf_n = \lambda_n f_n$$

for all n, where λ_n are real eigenvalues of H. Assume that the sequence λ_n is non-decreasing and that $\lim_{n\to\infty} \lambda_n = +\infty$. It is an exercise to show that if

$$\mathrm{Dom}(H) := \left\{ \sum_{n=1}^{\infty} \alpha_n f_n : \sum_{n=1}^{\infty} \left\{ |\lambda_n \alpha_n|^2 + |\alpha_n|^2 \right\} < \infty \right\}$$

then H is self-adjoint and its spectrum is precisely $\{\lambda_n : n = 1, 2, \ldots\}$; see[4, Lemma 1.2.2]. One can write down an explicit expression for the resolvent operators $(H - z)^{-1}$ whenever $z \notin \mathrm{Spec}(H)$ and deduce that

$$\begin{aligned} \|(H - z)^{-1}\| &= \sup\{|(\lambda_n - z)^{-1}| : n = 1, 2, \ldots\} \\ &= \mathrm{dist}\,(z, \mathrm{Spec}(H))^{-1}. \end{aligned}$$

It can be proved from the spectral theorem below that this identity holds for all self-adjoint operators, but the corresponding statement for other operators may be very far from the truth. We set it as an exercise to prove that for this choice of H the resolvent operators are all compact.

The following theorem is taken from [4, Section 4.5], and is called the variational theorem. Together with its developments, it is the most important method of obtaining general information about the location of the eigenvalues of a self-adjoint operator. It also underlies one of the most important methods of determining eigenvalues of partial differential operators numerically. The following theorem refers to the operator H of Example 1, but we will see a similar result in a more general context later.

Theorem 2 *For all $n \geq 1$ we have*

$$\lambda_n = \min\{\lambda(L) : \dim(L) = n \text{ and } L \subseteq \mathrm{Dom}(H)\}. \tag{1}$$

where

$$\lambda(L) := \sup\{\langle Hf, f\rangle : f \in L \text{ and } \|f\| = 1\}$$

for any finite-dimensional subspace L of $\mathrm{Dom}(H)$.

Proof If $L_n = \mathrm{lin}\{f_r : 1 \leq r \leq n\}$ and $f \in L_n$ then

$$\begin{aligned}\langle Hf, f\rangle &= \sum_{r=1}^{n} \lambda_r |\langle f, f_r\rangle|^2 \\ &\leq \lambda_n \sum_{r=1}^{n} |\langle f, f_r\rangle|^2 \\ &= \lambda_n \|f\|^2.\end{aligned}$$

Therefore $\lambda(L_n) \leq \lambda_n$.

Conversely let L be any n-dimensional linear subspace of $\mathrm{Dom}(H)$ and let P be the orthogonal projection on $\mathcal{H}$ with range L_{n-1}, so that P is given explicitly by

$$Pf := \sum_{r=1}^{n-1} \langle f, f_r\rangle f_r$$

for all $f \in \mathcal{H}$. The range of $P|_L$ has lower dimension than L so there must exist $f \in L$ with norm 1 such that $Pf = 0$. We then have $\langle f, f_r\rangle = 0$ for all $r \leq n-1$. It follows that

$$\begin{aligned}\langle Hf, f\rangle &= \sum_{r=n}^{\infty} \lambda_r |\langle f, f_r\rangle|^2 \\ &\geq \lambda_n \sum_{r=n}^{\infty} |\langle f, f_r\rangle|^2 \\ &= \lambda_n.\end{aligned}$$

We conclude that $\lambda(L) \geq \lambda_n$ for all such subspaces, as required to prove the reverse inequality.

Corollary 3 *The bottom eigenvalue λ_1 is given by*

$$\lambda_1 = \min\{\langle Hf, f\rangle : f \in \mathrm{Dom}(H) \text{ and } \|f\| = 1\}.$$

The key to many of the applications of the variational theorem is the following comparison theorem, whose technical scope will be extended later.

Theorem 4 *Let H and $\tilde{H}$ be two self-adjoint operators with the same domain and suppose that $H \geq \tilde{H}$ in the sense that*

$$\langle Hf, f\rangle \geq \langle \tilde{H}f, f\rangle$$

for all $f \in \mathrm{Dom}(H)$. Then $\lambda_n \geq \tilde{\lambda}_n$ for all n in an obvious notation.

The proof is an elementary consequence of the fact that $\lambda(L) \geq \tilde{\lambda}(L)$ for all subspaces L of $\mathrm{Dom}(H)$. The theorem is not stated in this abstract form in [4], although it is constantly used in the quadratic form context.

The second key idea in these lectures is that of working with quadratic forms rather than operators as far as possible. In the context of Example 1, if we define

$$Q(f) := \langle Hf, f\rangle = \sum_{n=1}^{\infty} \lambda_n |\alpha_n|^2$$

for all $f = \sum_{n=1}^{\infty} \alpha_n f_n$ then we see that the natural domain of Q is

$$\mathrm{Dom}(Q) := \left\{ \sum_{n=1}^{\infty} \alpha_n f_n : \sum_{n=1}^{\infty} \left\{ \lambda_n |\alpha_n|^2 + |\alpha_n|^2 \right\} < \infty \right\}$$

which contains $\mathrm{Dom}(H)$ properly. We define $Q(f) := +\infty$ if $f \notin \mathrm{Dom}(Q)$, so that it is a lower semi-continuous function on $\mathcal{H}$.

We postpone the definition of the quadratic forms associated with general non-negative self-adjoint operators. It has been found that the quadratic form domains of many differential operators are much easier to specify than their operator domains, and that many important theorems can be proved as easily in the quadratic form context, and actually have greater technical scope if one does this. In fact the process of doing everything possible in the quadratic form context has gone so far that one of the key questions of the 1960s, whether a particular operator was self-adjoint on some domain, has come to seem a secondary issue rather than a prerequisite for any serious progress. We may rewrite the variational formula as follows; see [4, Section 4.5] for details. Note that the subspaces L need not be contained within $\mathrm{Dom}(Q)$ because of our extended definition of Q.

Lemma 5 *For all $n \geq 1$ we have*

$$\lambda_n = \min\{\lambda(L) : \dim(L) = n\}.$$

where

$$\lambda(L) := \sup\{Q(f) : f \in L \text{ and } \|f\| = 1\}$$

for any finite-dimensional subspace of $\mathcal{H}$.

The quadratic form comparison theorem does not require the two forms to have the same domain. This is important when comparing two differential operators which only differ in that H satisfies Dirichlet boundary conditions while $\tilde{H}$ satisfies Neumann boundary conditions. The two associated forms Q and $\tilde{Q}$ are given by the same expression, but the form Q has a smaller domain, so the condition of the following theorem is satisfied. We will explain this in more detail in the second lecture.

Theorem 6 *Let H and $\tilde{H}$ be two self-adjoint operators and suppose that $H \geq \tilde{H}$ in the quadratic form sense:*

$$+\infty \geq Q(f) \geq \tilde{Q}(f) > -\infty$$

for all $f \in \mathcal{H}$. Then $\lambda_n \geq \tilde{\lambda}_n$ for all n in an obvious notation.

So far I have given some general definitions, and illustrated them by reference to one example. It is time to write down the spectral theorem for general self-adjoint operators. The quest to obtain an explicit implementation of this theorem for particular operators has been the goal of an enormous amount of research over the last fifty years, much of it led by problems arising from quantum theory or more recently geometry. The spectral theorem has many forms, and also many fundamentally different proofs, and we follow [4, Theorem 2.5.1]. It states that every self-adjoint operator is unitarily equivalent to a multiplication operator.

Theorem 7 *If H is a self-adjoint operator on $\mathcal{H}$ there exists a unitary operator $U : \mathcal{H} \to L^2(X, \mathrm{d}x)$ for some set X equipped with a σ-field of subsets and a measure $\mathrm{d}x$, and a real-valued measurable function h on X such that*

$$(UHf)(x) = h(x)Uf(x)$$

for all $f \in \mathrm{Dom}(H)$. Moreover $f \in \mathrm{Dom}(H)$ if and only if

$$\int_X \left\{|Uf|^2 + |h(Uf)|^2\right\} \mathrm{d}x < \infty.$$

The example we considered earlier corresponds to the case in which X consists of the positive integers, $\mathrm{d}x$ is the counting measure and $h(n) = \lambda_n$ for all n.

Using the spectral theorem essentially any abstract question about a single self-adjoint operator can be reduced to a concrete problem about multiplication operators which can be resolved using measure theoretic methods. For example, [4, Theorem 4.3.1]:

Lemma 8 *The following three conditions on a self-adjoint operator are equivalent:*
(i) $H \geq 0$ in the sense that $\langle Hf, f\rangle \geq 0$ for all $f \in \mathrm{Dom}(H)$,
(ii) $\mathrm{Spec}(H) \subseteq [0, \infty)$,
(iii) $h(x) \geq 0$ almost everywhere in X.

If H is a non-negative self-adjoint operator, its square root $H^{1/2}$ is defined to be the operator corresponding under U to multiplication by $h^{1/2}$. Its precise domain is

$$\mathrm{Dom}(H^{1/2}) := \{f \in \mathcal{H} : \int_X \left\{|Uf|^2 + |h^{1/2}(Uf)|^2\right\} \mathrm{d}x < \infty.$$

We leave it as a non-trivial exercise for the reader to prove that $H^{1/2}$ is the unique operator A acting in $\mathcal{H}$ such that $A = A^* \geq 0$ and $A^2 = H$ in a suitable sense.

If H is a non-negative self-adjoint operator, its quadratic form Q is defined for all $f \in \mathcal{H}$ by

$$Q(f) := \begin{cases} \|H^{1/2}f\|^2 & \text{if } f \in \mathrm{Dom}(H^{1/2}) \\ +\infty & \text{otherwise.} \end{cases}$$

See [4, Theorem 4.4.2] for the proof that Q is a lower semi-continuous function. In many situations it is important that the above definition can be extended to the case in which H is semi-bounded rather than non-negative, but we spare the reader the details.

A quadratic form Q on $\mathcal{H}$ (in the extended sense in which it is assigned the value $+\infty$ outside its natural domain) is associated with a self-adjoint operator if and only if it is lower-semicontinuous; its restriction to its domain is then said to be closed. A quadratic form Q with domain $\mathcal{D} \subseteq \mathcal{H}$ is said to be closable with closure $\overline{Q}$ if $\overline{Q}$ is the least closed extension of Q. See [4, Sect. 4.4] for details.

The spectrum of any operator can be classified into different parts. In the self-adjoint case this is particularly simple. We say that λ is in the discrete spectrum of H if it is an eigenvalue of finite multiplicity and also an isolated point of the spectrum. The essential spectrum of H is by definition the part of the spectrum which is not discrete spectrum, and includes any intervals contained in the spectrum.

Example 9 Let $Hf := -\Delta f$ acting in $L^2(\mathbf{R}^N)$ with domain Schwartz space $\mathcal{S}$, and let $\mathcal{F}$ denote the unitary Fourier transform operator. Then

$$\mathcal{F}Hf(\xi) = |\xi|^2\mathcal{F}f(\xi)$$

for all $f \in \mathcal{S}$. The operator H is not self-adjoint on $\mathcal{S}$ but its closure is self-adjoint with domain

$$\{f \in L^2 : \int_{\mathbf{R}^N} \left\{|\mathcal{F}f(\xi)|^2 + |\xi|^4|\mathcal{F}f(\xi)|^2\right\} \mathrm{d}\xi < \infty\}.$$

The Fourier transform provides an explicit representation of the Spectral Theorem for H, which can be extended to any differential operator acting in $L^2(\mathbf{R}^N)$ with constant coefficients. The closure of H has no eigenvalues, so its spectrum and essential spectrum coincide and are equal to $[0, \infty)$; see [4, Theorem 3.5.3].

Example 10 Let H be defined on the domain $C_c^\infty(\mathbf{R}^N)$ by

$$Hf(x) := -\Delta f(x) + V(x)f(x)$$

where V is a bounded or sufficiently locally regular function which vanishes as $|x| \to \infty$. It is known that H has self-adjoint closure. There are three possiblilities for the spectrum:

(i) $\mathrm{Spec}(H) = [0, \infty)$,
(ii) $\mathrm{Spec}(H) = [0, \infty) \cup \{\mu_n\}_{n=1}^{N}$ where the eigenvalues are negative and are written in increasing order and repeated according to multiplicity,
(iii) $\mathrm{Spec}(H) = [0, \infty) \cup \{\mu_n\}_{n=1}^{\infty}$ where the eigenvalues are negative and are written in increasing order and repeated according to multiplicity, and satisfy $\lim_{n\to\infty} \mu_n = 0$. In all three cases the essential spectrum of H equals $[0, \infty)$.

The variational theorem can be adapted to each of these cases, where we now define λ_n by the variational formula (1). In case (i) one finds that $\lambda_n = 0$ for all n. In case (ii) one has $\lambda_n = \mu_n$ for $1 \leq n \leq N$ and $\lambda_n = 0$ for all $n > N$. In case (iii) one has $\lambda_n = \mu_n$ for all n. Thus the determination of all λ_n distinguishes between the three cases and determines all of the negative eigenvalues. If there are any eigenvalues greater than the smallest number in the essential spectrum (which in this particular case there are not) the variational method is powerless to say anything about their number or location. See [4, Theorem 4.5.2, Theorem 8.5.1].

Example 11 Let $H := -\Delta$ on a compact Riemannian manifold M without boundary. Then H is essentially self-adjoint on $C^\infty(M)$, it has a complete orthonormal set of eigenfunctions f_n exactly as in Example 1, and each eigenfunction lies in $C^\infty(M)$. If M has a smooth boundary ∂M and we impose either Dirichlet or Neumann boundary conditions by restricting the domain appropriately, then the situation is similar. An important issue then is to determine the asymptotic distribution of the eigenvalues, and this will be discussed later in the conference.

If the Riemannian manifold is complete but non-compact then the type of spectrum depends upon the geometry of the manifold. For example hyperbolic space H^{n+1} has spectrum $[n^2/4, \infty)$; [3, p178], while we have already seen that Euclidean space $\mathbf{R}^n$ has spectrum $[0, \infty)$.

Example 12 Suppose that $f_n \in C_c^\infty(M)$ have supports in disjoint balls $B_n \subseteq M$ of varying diameters, and that $\|f_n\|_2 = 1$, $\|\nabla f_n\|_2 \leq c_n$ for all n; if the balls all have the same diameter, and the constants c_n are independent of n, this is known as a bounded geometry assumption. Then by applying the variational formula to the subspaces $L := \mathrm{lin}\{f_{r(1)}, \ldots, f_{r(n)}\}$ for all possible sequences $\{r(1), \ldots, r(n)\}$, one may show that the smallest number in the essential spectrum of H is at most

$$\liminf\{c_n^2 : n = 1, \ldots\}.$$

2 Comparison of Spectra

In the first lecture I discussed some of the abstract tools used in spectral theory and listed a few of the applications without explaining how the results are proved.

In this lecture I wish to go a bit more deeply into the technical details of the applications to elliptic partial differential operators, particularly the Laplacian, acting in regions of Euclidean space. You will have to imagine the modifications to Riemannian manifolds, or listen to the lectures of Burstall and Chavel! Once again most of the proofs may be found in [4]. We say that a function f defined on $U \subseteq \mathbf{R}^N$ lies in $W^{n,p}(U)$ if $D^\alpha f \in L^p(U)$ for all $|\alpha| \leq n$, where $\alpha = (\alpha_1, \ldots \alpha_N)$ is a multi-index and the mixed partial derivatives $D^\alpha f$ are interpreted in the weak sense. If $p = 2$ then this can be rewritten using the Fourier transform map $\mathcal{F}$; see[4, Section 3.7].

Lemma 13 *We have $f \in W^{n,2}(\mathbf{R}^N)$ if and only if*

$$\int_{\mathbf{R}^N} (1 + |\xi|^2)^n |\mathcal{F}f(\xi)|^2 \mathrm{d}x < \infty.$$

The local regularity of a function $f \in W^{n,2}(\mathbf{R}^N)$ increases with n, and f must be continuous and bounded if $n > N/2$, [4, Theorem 3.7.6]. For a much more detailed study of the general theory of Sobolev spaces see [1].

If U is a region in $\mathbf{R}^N$ we define $W_0^{n,p}(U)$ to be the closure of $C_c^\infty(U)$ in $W^{n,p}(U)$ with respect to the obvious norm. Using these ideas we can give the general definition of the Dirichlet Laplacian, without any assumptions of boundary regularity. The operator described in the following theorem coincides with $-\Delta$ on $C_c^\infty(U)$, which lies within its domain; see [4, Section 6.2].

Theorem 14 *There exists a non-negative self-adjoint operator H_D on $L^2(U)$ such that* $\mathrm{Dom}(H_D^{1/2}) = W_0^{1,2}(U)$ *and*

$$\|H_D^{1/2} f\|_2^2 = Q_D(f) = \int_U |\nabla f|^2 \mathrm{d}x$$

for all $f \in W_0^{1,2}(U)$. If U is bounded then H_D has discrete spectrum (and hence compact resolvent).

The final part of the proof depends upon the use of the variational theorem as is explained after Theorem 17 below. It is the case that functions in the domain of H_D vanish at the boundary ∂U provided one interprets this in a sufficiently weak sense. However if ∂U is sufficiently irregular at some point a then even an eigenfunction of H_D need not vanish as $x \to a$. It is no suprise that the situation is more difficult when one considers Neumann boundary conditions, since it is not even clear what one could mean by zero normal derivatives, if the boundary is nowhere differentiable.

Theorem 15 *There exists a non-negative self-adjoint operator H_N on $L^2(U)$ such that* $\mathrm{Dom}(H_N^{1/2}) = W^{1,2}(U)$ *and*

$$\|H_N^{1/2} f\|_2^2 = Q_N(f) = \int_U |\nabla f|^2 \mathrm{d}x$$

for all $f \in W^{1,2}(U)$. If U is bounded with a Lipschitz boundary then H_N has compact resolvent, and hence discrete spectrum.

For the proof see [4, Section 7.2]. The condition that the boundary of U be Lipschitz is stronger than necessary. However, there are by now a number of clever constructions of pathological regions in $\mathbf{R}^2$ for which $-\Delta_N$ has more or less any spectrum one cares to specify.

The following explains the issue of boundary decay more fully for $H_D := -\Delta_D$ acting in $L^2(U)$; see [4, Lemma 6.1.3] for the first part. The second part is a consequence of general elliptic regularity theorems, [2]. We define $C^n(\overline{U})$ to be the set of C^n functions f on U such that f and all derivatives of f of order up to n can be extended continuously to $\overline{U}$.

Theorem 16 *If $f \in C^2(\overline{U})$ and $f|_{\partial U} = 0$ then $f \in W_0^{1,2}(U)$. Conversely if f is an eigenfunction of H_D and ∂U is smooth then $f \in C^\infty(\overline{U})$ and $f|_{\partial U} = 0$.*

If, in two dimensions, one takes the region U to be a square, then the eigenvalues of H_D can be computed in closed form by separation of variables If the square has edge length c then its eigenvalues are given in terms of the positive integers m, n by

$$\lambda_{m,n} = \pi^2(m^2 + n^2)/c^2.$$

If we rearrange the eigenvalues in increasing order and imagine them as lying in the positive quadrant then it may be observed that $\lambda_n \to +\infty$ as $n \to \infty$. Indeed the leading asymptotics of the counting function

$$N(\lambda) := \max\{n : \lambda_n < \lambda\}$$

may be worked out fairly easily (exercise!). A similar argument holds in higher dimensions.

The next theorem provides the crucial idea needed to explain how the variational theorem can lead to spectral information about H_D for a general region U, including the fact that it has discrete spectrum. See [4, Theorem 6.2.3].

Theorem 17 *Let $H_r := -\Delta_D$ acting in $L^2(U_r)$ where $U_1 \subseteq U_2$. Then the eigenvalues of the two operators satisfy $\lambda_{1,n} \geq \lambda_{2,n}$ for all n in an obvious notation.*

The proof is the simple observation that both quadratic forms are given by the same expression but that for H_2 has a larger domain because their definitions imply directly that

$$W_0^{1,2}(U_1) \subseteq W_0^{1,2}(U_2).$$

Therefore when applying the variational formula for the eigenvalues of H_2 one is minimising over a larger set of subspaces, and so gets a smaller answer.

If U_1 is a bounded region then by choosing U_2 to be an N-dimensional cube containing U_1, and using the explicit formula for the eigenvalues of H_2, the above comparison theorem implies that $\lambda_{1,n} \to +\infty$ as $n \to \infty$ for every bounded region U_1, without any regard to how regular its boundary is. This implies that the resolvent operators are compact as in Example 1.

The fact that $W_0^{1,2}(U) \subseteq W^{1,2}(U)$ yields the following theorem by a similar method.

Theorem 18 *Let $H_D := -\Delta_D$ and $H_N := -\Delta_N$ both acting in $L^2(U)$. Then the eigenvalues of the two operators satisfy $\lambda_{D,n} \geq \lambda_{N,n}$ for all n in an obvious notation.*

We give the proof of a variation of this theorem which will be used elsewhere in these lecture courses.

Let $U \subseteq \mathbf{R}^N$ be a bounded region such that $U = U_1 \cup S \cup U_2$, where U_r are subregions separated by the surface S, and all surfaces are piecewise smooth. Let $H, H_D, H_N := -\Delta$ acting in $L^2(U)$ subject to DBC on ∂U, with an extra DBC on S for H_D and an extra NBC on S for H_N.

Theorem 19 *If the eigenvalues of H, H_D, H_N are denoted by $\lambda_n, \lambda_{D,n}, \lambda_{N,n}$ respectively, written in increasing order and repeated according to multiplicity, then*

$$\lambda_{N,n} \leq \lambda_n \leq \lambda_{D,n}$$

for all n.

Proof In each case the quadratic form is

$$Q(f) := \int_U |\nabla f|^2 \mathrm{d}x$$

acting on a dense domain in $L^2(U)$. For H the form domain is $W_0^{1,2}(U)$. For H_D one imposes the extra condition that functions in the domain vanish on S, but for H_N one allows the functions to be discontinuous across S. Hence

$$\mathrm{Dom}(Q_D) \subseteq \mathrm{Dom}(Q) \subseteq \mathrm{Dom}(Q_N)$$

and the theorem follows by an application of the variational theorem.

This theorem has many applications but the most famous is the following, obtained by extending the above theorem to the case in which U is divided into many subregions instead of just two. If U is a general bounded region, say in $\mathbf{R}^2$, one may decompose it into a large number of congruent small squares together with a much smaller number of irregular boundary regions. By evaluating the eigenvalues of the squares and estimating the eigenvalues of the boundary regions, for both Neumann and Dirichlet boundary conditions, one may obtain the leading term in the asymptotic distribution of the counting function $N(\lambda)$ as $\lambda \to \infty$. This method is known as Neumann-Dirichlet bracketing, but it does not yield sharp results concerning the second term in the asymptotic expansion of $N(\lambda)$. The investigation of the second term has occupied the energy of some of the world's leading mathematicians over several decades, and was eventually completed by Ivrii, Safarov and Vassiliev, [12].

Now let us look at more general elliptic operators, with variable second order coefficients. Let $U \subseteq \mathbf{R}^N$ and let H act in $L^2(U)$ according to the formula

$$Hf(x) := -\sum_{i,j=1}^{N} \frac{\partial}{\partial x_i}\left\{a_{i,j}(x)\frac{\partial f}{\partial x_j}\right\}$$

subject to Dirichlet boundary conditions, where the matrix-valued coefficients $a(x)$ are uniformly elliptic in the sense that they satisfy the matrix inequality

$$0 < \alpha I \le a(x) \le \beta I < \infty$$

for all $x \in U$. An integration by parts yields

$$Q(f) := \langle Hf, f\rangle = \int_U \sum_{i,j=1}^{N} a_{i,j}(x) \frac{\partial f}{\partial x_j}\frac{\partial \overline{f}}{\partial x_i} \mathrm{d}x$$

for all $f \in C_c^\infty(U)$. We define H to be the non-negative self-adjoint operator associated with the closure of the quadratic form starting from $C_c^\infty(U)$, or equivalently by this quadratic form with the precise domain $W_0^{1,2}(U)$.

Theorem 20 *Let $\{\lambda_n\}_{n=1}^\infty$ be the eigenvalues of H written in increasing order and repeated according to multiplicity, and let μ_n be the eigenvalues of $H_0 := -\Delta_D$. Then*

$$0 < \alpha\mu_n \le \lambda_n \le \beta\mu_n < \infty$$

for all n, and in particular

$$\lim_{n\to\infty} \lambda_n = +\infty.$$

Proof We observe that the quadratic forms Q and Q_0 have the same domain $W_0^{1,2}(U)$ and that

$$0 \le \alpha Q_0(f) \le Q(f) \le \beta Q_0(f)$$

for all f in this domain. The comparison between the eigenvalues now follows directly from the variational theorem. The divergence of λ_n as $n \to \infty$ follows from the corresponding fact for μ_n, already established.

This coefficient monotonicity theorem still holds for Neumann boundary conditions, since we need only replace $W_0^{1,2}(U)$ by $W^{1,2}(U)$ in the proof.

We now turn to the effect of potentials. The simplest case is as follows.

Theorem 21 *Let $Hf := -\Delta f + Vf$ in $L^2(U)$ subject to Dirichlet boundary conditions, where U is a bounded region and V is a bounded potential. Then the eigenvalues λ_n of H and the eigenvalues μ_n of $H_0 := -\Delta_D$ satisfy*

$$|\lambda_n - \mu_n| \le \|V\|_\infty$$

for all n.

Proof We apply the variational theorem to the quadratic form inequality

$$Q_0(f) - \|V\|_\infty \|f\|_2^2 \le Q(f) \le Q_0(f) + \|V\|_\infty \|f\|_2^2$$

valid for all $f \in W_0^{1,2}(U)$.

If $H_\alpha := -\Delta_D + \alpha V$ where α is a real parameter, then a similar argument (exercise!) establishes that its eigenvalues $\lambda_n(\alpha)$ satisfy

$$|\lambda_n(\alpha) - \lambda_n(\beta)| \le |\alpha - \beta| \, \|V\|_\infty$$

for all real α, β. In fact much more is true: the eigenvalues of H_α are real analytic functions of α under these hypotheses, but this is not an elementary fact; see [8].

The ideas in the above theorems can be extended to perturbations which are not potentials and are not bounded, under suitable hypotheses.

Theorem 22 *Let $H_0 = H_0^* \ge 0$ have eigenvalues $\{\lambda_{0,n}\}_{n=1}^\infty$ written in increasing order and repeated according to multiplicity, and let Q_0 be its associated form. Let V be some symmetric operator such that*

$$|\langle Vf, f\rangle| \le \alpha Q_0(f) + \beta \|f\|^2$$

for some $0 \le \alpha < 1$ and all $f \in \mathrm{Dom}(Q_0)$ and put

$$Q(f) := Q_0(f) + \langle Vf, f\rangle.$$

The self-adjoint operator H associated with this form has eigenvalues λ_n which satisfy

$$(1-\alpha)\lambda_{0,n} - \beta \le \lambda_n \le (1+\alpha)\lambda_{0,n} + \beta$$

for all n.

Proof To prove the upper bound we note that

$$Q(f) \le (1+\alpha) Q_0(f) + \beta \|f\|^2$$

for all f, and then use the variational theorem. The lower bound has a similar proof. If the constants λ_n determined by the variational formula for Q are all equal to some number λ for large n then λ is the bottom of the essential spectrum rather than an eigenvalue. If $\alpha > 1$ then the quadratic form Q defined above need not be semibounded nor lower semicontinuous, and one cannot associate it with a self-adjoint operator H. See [4, p160].

If $H_0 := -\Delta$ acting in $L^2(\mathbf{R}^N)$ and V is a potential then there the precise conclusions which one can draw depend upon the hypotheses on the potential, and the following case is important, [4, p168]. The condition on V in the theorem means roughly that $V(x) \to 0$ as $|x| \to \infty$ and that the local singularities of V are not too extreme.

Theorem 23 *Let $H := -\Delta + V$ acting in $L^2(\mathbf{R}^N)$ where for some $p > N/2$ and every $\varepsilon > 0$ the potential V can be written in the form $V = V_p + V_\infty$ where $V_p \in L^p$ and $\|V_\infty\|_\infty \le \varepsilon$. Then*

$$\mathrm{Spec}(H) = [0, \infty) \cup \{\lambda_n\}_{n=1}^M$$

where M may be 0, ∞ or a finite number. The negative numbers λ_n are eigenvalues of finite multiplicity written in increasing order and the multiplicity of λ_1 is 1. If $M = \infty$ then $\lim_{n\to\infty} \lambda_n = 0$.

3 Computation of Spectra

In my third lecture I present a few topics from computational spectral theory, by which I mean rigorous methods from analysis which have also proved useful in obtaining accurate bounds on the spectrum of certain classes of differential operator.

For the sake of definiteness let us assume that $H := -\Delta$ in $L^2(U)$ where U is a bounded Euclidean region, and we are imposing Dirichlet boundary conditions. There are two important methods of determining eigenvalues numerically, both of which depend upon the choice of a sufficiently large finite-dimensional subspace L. In the first method one depends upon approximation theorems which state that if the subspace is large enough in a certain specified sense then the first k eigenvalues of H restricted to L are within a specified error of the first k eigenvalues of H. The main problem with this approach is that the error guaranteed by the theorems is usually much bigger than the true error. The second method is to use certain theoretical arguments, presented below, to obtain rigorous upper and lower bounds on the eigenvalues of H from calculations carried out within L. Here one has no promise a priori that the upper and lower bounds will be close to each other, but if one chooses L with sufficient expertise one can often obtain very accurate and rigorous upper and lower bounds on the eigenvalues of H.

The upper bound on the eigenvalues of H depends upon the Rayleigh-Ritz theorem, and is quite easy.

Theorem 24 *Let L be an n-dimensional subspace of $W_0^{1,2}(U)$ and let $\phi_1, \ldots, \phi_n$ be an orthonormal basis of L. if we put*

$$B_{r,s} := \int_U \nabla\phi_r \cdot \overline{\nabla\phi_s}\, dx$$

and if λ_r, μ_r are the eigenvalues of H, B respectively then

$$\lambda_r \le \mu_r$$

for all $1 \le r \le n$.

Proof We compare the form Q associated with H with the form Q_L defined by

$$Q_L(f) := \begin{cases} Q(f) & \text{if } f \in L \\ +\infty & \text{otherwise.} \end{cases}$$

Since $Q \le Q_L$ the variational theorem implies that the eigenvalues of H are less than or equal to those associated with Q_L, and these are exactly the eigenvalues of B.

The application of this method to obtain accurate upper bounds depends upon experience to select a suitable subspace L, and standard matrix eigenvalue procedures suitable for large sparse matrices.

One of the standard procedures for choosing the subspace is by the finite elements method. The simplest version is as follows (more complicated versions are needed to treat other boundary conditions). We subdivide the region U into a large enough number of triangles, and consider functions on U which are linear on each triangle, continuous across neighbouring triangles and zero on the boundary of U, assumed to be a polygon. These functions are uniquely determined by their values at the internal vertices, which may be chosen independently, so the dimension of L equals the number of internal vertices. A large number of problems immediately present themselves, and the skill of the numerical analyst is to know how to answer them and how to produce code which implements the answers automatically.

(1) How is the initial triangulation chosen?

(2) Once one has chosen a triangulation, how does one decide which part of the computation is the least accurate and hence which triangles should be subdivided further?

(3) What is the effect of corners on the accuracy of the computation?

(4) How does one modify the method if the region has a piecewise smooth curved boundary? Alternatively what error is caused by approximating the boundary by a polygon?

(5) Is it better to use higher order polynomials when defining the elements, or lower order (e.g. linear) polynomials with a finer triangulation?

(6) How should one define the elements if one is considering higher order differential operators or higher dimensions?

Rather than attempting to answer these questions, I pass to the problem of obtaining lower bounds on the eigenvalues of H to complement the upper bounds. The Temple-Lehmann method depends upon two inputs, a constant ρ and a vector ϕ.

Theorem 25 *Let $\lambda_n < \lambda_{n+1}$ be two consecutive eigenvalues of H, $\phi \in \mathrm{Dom}(H)$ and $\rho \in \mathbf{R}$. If*

$$\lambda_n \leq \frac{\langle H\phi, \phi\rangle}{\langle \phi, \phi\rangle} < \rho \leq \lambda_{n+1}$$

then

$$\lambda_n \geq \frac{\rho\langle H\phi, \phi\rangle - \langle H\phi, H\phi\rangle}{\rho\langle \phi, \phi\rangle - \langle H\phi, \phi\rangle}.$$

Proof Since H has no spectrum in the interval (λ_n, ρ), the operator $(H-\lambda_n)(H-\rho)$ is non-negative; to see this use the spectral theorem. The required lower bound on λ_n is now obtained by rearranging the inequality

$$\langle (H - \lambda_n)(H - \rho)\phi, \phi\rangle \geq 0.$$

The application of the above theorem to obtain accurate lower bounds involves the following issues.

(1) A little thought shows that the theorem above only yields an accurate lower bound on λ_n if ϕ is a good approximation to the corresponding eigenvector: such a vector ϕ can be obtained by the Rayleigh-Ritz method.

(2) If the region U has repeated eigenvalues, i.e. $\lambda_n = \lambda_{n+1}$ then the Temple-Lehmann theorem cannot work in the form given. Moreover if $\lambda_{n+1} - \lambda_n$ is very small then it yields a poor lower bound even if a suitable choice of ρ can be made. This is not an essential difficulty since there are more sophisticated versions of the theorem designed precisely to cope with this possibility.

(3) A much more serious problem is the need for a crude lower bound ρ on the next highest eigenvalue. This would be an insuperable defect of the method if a suitable value of ρ could only be obtained from a lower bound on the eigenvalue λ_{n+2}, leading to an infinite regress.

(4) The method involves using test functions which lie in the operator domain of H, not just in its quadratic form domain. If one is using the finite elements method to construct the space of test functions, then this forces one to use more sophisticated elements than were needed for the Rayleigh-Ritz method.

In simple cases the requisite lower bounds can be obtained by an elementary comparison theorem which we have already discussed.

Example 26 Let $V \subseteq U \subseteq cV \subseteq \mathbf{R}^N$ where $c > 1$, and let λ_n, μ_n be the eigenvalues of U, V respectively. Suppose also that the eigenvalues of V are explicitly known; for example V might be a cube. Then

$$c^{-2}\mu_{n+1} \le \lambda_{n+1} \le \mu_{n+1}$$

for all n by the variational theorem. If c is close enough to 1 then this yields a lower bound $\rho := c^{-2}\mu_{n+1}$ on the eigenvalue λ_{n+1} which might be used in the Temple-Lehmann theorem. However, it turns out in practice that unless n is very small, it is unlikely that one can find a comparison region V for which the lower bound obtained by this method is good enough for the required purpose.

There is a corresponding application to elliptic operators with variable coefficients.

Example 27 Let $U \subseteq \mathbf{R}^N$ and let H act in $L^2(U)$ according to the formula

$$Hf(x) := -\sum_{i,j=1}^{N} \frac{\partial}{\partial x_i}\left\{a_{i,j}(x)\frac{\partial f}{\partial x_j}\right\}$$

subject to Dirichlet boundary conditions, where the matrix-valued coefficients $a(x)$ are uniformly elliptic in the sense that they satisfy the matrix inequality

$$0 < \alpha I \le a(x) \le \beta I < \infty$$

for all $x \in U$. Let λ_n be the eigenvalues of H written in increasing order and repeated according to multiplicity. Suppose also that the eigenvalues μ_n of $H_0 := -\Delta$ acting in $L^2(U)$ subject to DBC are explicitly known.

If the constants α and β are close enough to each other, i.e. the variation in the coefficients of H over U is sufficiently small, then the eigenvalues of H and H_0 are very close since we have already shown that

$$\alpha\mu_n \leq \lambda_n \leq \beta\mu_n.$$

One may then take $\rho := \alpha\mu_{n+1}$ in the Temple-Lehmann theorem provided $\beta\mu_n < \alpha\mu_{n+1}$. This is only likely to be valid for fairly small values of n unless α and β are very close indeed.

4 The Homotopy Method

The homotopy method was introduced fairly recently in the context of spectral bounds, by Goerisch [7] and Plum [9, 10]. We describe the coefficient homotopy method, but there are equally important region homotopy and boundary condition homotopy methods, [9, 10, 6].

As before let $U \subseteq \mathbf{R}^N$ and let H act in $L^2(U)$ according to the formula

$$Hf(x) := -\sum_{i,j=1}^{N} \frac{\partial}{\partial x_i}\left\{a_{i,j}(x)\frac{\partial f}{\partial x_j}\right\}$$

subject to Dirichlet boundary conditions, where the matrix-valued coefficients $a(x)$ are uniformly elliptic in the sense that they satisfy the matrix inequality

$$0 < \alpha I \leq a(x) \leq \beta I < \infty$$

for all $x \in U$. Let λ_n be the eigenvalues of H written in increasing order and repeated according to multiplicity. Suppose that the eigenvalues μ_n of $H_0 := -\alpha\Delta$ acting in $L^2(U)$ subject to DBC are explicitly known.

We do not assume that α and β are close to each other, but introduce a family of intermediate operators

$$H_t := (1-t)H_0 + tH$$

for $0 \leq t \leq 1$. Note that these operators, and therefore their eigenvalues, are monotone increasing functions of t. By using the Rayleigh-Ritz and Temple-Lehmann methods together with the known eigenvalues of H_0, we may compute the first k eigenvalues of $H_{t(1)}$ accurately for some small value $t(1) > 0$ as described in Example 27. Provided with this information we may then compute the first $k-1$ eigenvalues of $H_{t(2)}$ accurately for some value $t(2) > t(1)$ as described in Example 27. Proceeding inductively we obtain $k-m$ eigenvalues of H accurately if the interval $[0,1]$ is covered in m steps. At each stage the method provides rigorous upper and lower bounds, but with no a priori guarantee of how close the upper and lower bounds are to each other.

This method has been implemented in a number of test cases, but some comments about the problems faced in doing so are needed.

(1) The computational cost of the method is heavy, since one has to determine a series of eigenvalues of intermediate operators which are not of interest for their own sake.

(2) It seems from our description that the number of intermediate steps might be infinite, and one needs to prove that this is not so, and to devise a procedure for minimising the number.

(3) As the parameter t increases two eigenvalues may well cross. In this case the version of the Temple-Lehmann method which we have presented fails and we must use a more sophisticated version. This needs to be done automatically by the code.

(4) The starting point is the assumption that the eigenvalues μ_n are already known. This requires a preliminary region homotopy to reduce to the case of a known region, such as a ball or rectangle. If the region is not simply connected then one uses a boundary condition homotopy method involving slicing the region up into smaller parts, [6].

We describe briefly the boundary condition homotopy method for a simple example. Let U be the union of two rectangles U_1 and U_2 together with a separating interval I. Let $H := -\Delta$ acting in $L^2(U)$ subject to NBC, so that its quadratic form Q has domain $W^{1,2}(U)$. Let H_0 be the same operator but with an extra Neumann boundary condition on I. Since H_0 is the sum of the Neumann Laplacian on the subregions U_1 and U_2 its eigenvalues are exactly computable. Now let H_t be the self-adjoint operator associated with the quadratic form

$$Q_t(f) := Q_0(f) + t\int_I |f_+ - f_-|^2$$

where $0 \le t < \infty$ and $f_\pm$ are the boundary values of f on either side of I; the domain of Q_t is by definition equal to that of Q_0 and so allows functions to have different limiting values on either side of I. We proceed as before, noting that the quadratic forms are monotonically increasing functions of t, and therefore the eigenvalues of these operators are also monotonically increasing functions of t, by the variational theorem. Moreover the forms converge to the quadratic form Q as $t \to +\infty$, with the domain decreasing as the limit is achieved, so $\lim_{n\to\infty} \lambda_{t,n} = \lambda_n$ for all n.

By using the Rayleigh-Ritz and Temple-Lehmann methods together with the known eigenvalues of H_0, we may compute the first k eigenvalues of $H_{t(1)}$ accurately for some small value $t(1) > 0$ as described in Example 27. Provided with this information we may then compute the first $k-1$ eigenvalues of $H_{t(2)}$ accurately for some value $t(2) > t(1)$ as described in Example 27. Eventually inductively we obtain accurate bounds on the eigenvalues of H_t for large enough t for these to provide good enough crude lower bounds on the eigenvalues of H itself for the purposes of the Temple-Lehmann theorem.

There is a problem with this method which does not exist for the coefficient homotopy method. The Temple-Lehmann method requires the use of a test function space contained within the domain of the operator being considered. Although the operators H_t have the same form domains their operator domains are all different. The integral over I which appears in the definition of the quadratic form changes to a boundary condition on I for the operator. This boundary condition is of the form

$$\frac{\partial f}{\partial n}(x+) - \frac{\partial f}{\partial n}(x-) = t\{f(x+) - f(x-)\}$$

where the normal derivative is going from the $-$ side of I to the $+$ side. The boundary condition must be incorporated into the definition of the elements in the finite element method.

Lest the reader doubt the practicability of this scheme, we finally mention that it has been implemented for a number of test examples in [6].

References

[1] R A Adams: Sobolev Spaces. Academic Press, New York, 1975.

[2] S Agmon, A Douglis, L Nirenberg: Estimates near the boundary for solutions of elliptic boundary value equations satisfying general boundary conditions. Comm. Pure Appl. Math. 12 (1959) 623-727.

[3] E B Davies: Heat Kernels and Spectral Theory. Camb. Univ Press, Cambridge, 1989.

[4] E B Davies: Spectral Theory and Differential Operators. Camb. Univ Press, Cambridge, 1995.

[5] E B Davies: Corrections to STDO at http://www.mth.kcl.ac.uk/~davies/

[6] E B Davies: A hierarchical method for obtaining eigenvalue enclosures. Preprint 1997.

[7] F Goerisch: Ein Stufenverfahren zur Berechnung von Eigenwertschranken. In 'Numerical Treatment of Eigenvalue Problems' vol. 4 ISNM 83 (Ed. J Albrecht et al) pp 104-114. Birkhauser-Verlag, Basel, 1987.

[8] T Kato: Perturbation Theory of Linear Operators, 1st edition. Springer-Verlag, Berlin, 1966.

[9] M Plum: Eigenvalue inclusions for second order ordinary differential operators by a numerical homotopy method. J. Appl. Math. and Phys. 41 (1990) 205-226.

[10] M Plum: Bounds for eigenvalues of second order elliptic operators. J. Appl. Math. and Phys. 42 (1991) 848-863.

[11] M Reed, B Simon: Methods of Modern Mathematical Physics, vols. 1-4. Academic Press, 1975-1979.

[12] Yu Safarov, D Vassiliev: The asymptotic distribution of eigenvalues of partial differential operators. Transl. of Math. Monographs vol. 155. Amer. Math. Soc., Providence, Rhode Island, 1996.

Department of Mathematics
King's College
Strand
London WC2R 2LS
England
e-mail: E.Brian.Davies@kcl.ac.uk

Isoperimetric and Universal Inequalities for Eigenvalues

Mark S. Ashbaugh*
Department of Mathematics
University of Missouri
Columbia, MO 65211-0001
e-mail: mark@math.missouri.edu

May 7, 1999

1991 Mathematics Subject Classification: Primary 35P15, Secondary 58G25, 49Rxx.

Keywords and phrases: eigenvalues of the Laplacian, isoperimetric inequalities for eigenvalues, Faber-Krahn inequality, Szegő-Weinberger inequality, Payne-Pólya-Weinberger conjecture, Sperner's inequality, biharmonic operator, bi-Laplacian, clamped plate problem, Rayleigh's conjecture, buckling problem, the Pólya-Szegő conjecture, universal inequalities for eigenvalues, Hile-Protter inequality, H.C. Yang's inequality.

Short title: Isoperimetric and Universal Inequalities

Abstract

This paper reviews many of the known inequalities for the eigenvalues of the Laplacian and bi-Laplacian on bounded domains in Euclidean space. In particular, we focus on isoperimetric inequalities for the low eigenvalues of the Dirichlet and Neumann Laplacians and of the vibrating clamped plate problem (i.e., the biharmonic operator with "Dirichlet" boundary conditions). We also discuss the known universal inequalities for the eigenvalues of the Dirichlet Laplacian and the vibrating clamped plate and buckling problems and go on to

*Partially supported by National Science Foundation (USA) grants DMS-9500968 and DMS-9870156.

present some new ones. Some of the names associated with these inequalities are Rayleigh, Faber-Krahn, Szegő-Weinberger, Payne-Pólya-Weinberger, Sperner, Hile-Protter, and H.C. Yang. Occasionally, we will also comment on extensions of some of our inequalities to bounded domains in other spaces, specifically, S^n or H^n.

1 Introduction

1.1 The Eigenvalue Problems

The first eigenvalue problem we shall introduce is that of the *fixed membrane*, or *Dirichlet Laplacian*. We consider the eigenvalues and eigenfunctions of $-\Delta$ on a bounded domain (=connected open set) Ω in Euclidean space $\mathbb{R}^n$, i.e., the problem

$$-\Delta u = \lambda u \quad \text{in } \Omega \subset \mathbb{R}^n, \tag{1.1a}$$
$$u = 0 \quad \text{on } \partial\Omega. \tag{1.1b}$$

It is well-known that this problem has a real and purely discrete spectrum $\{\lambda_i\}_{i=1}^{\infty}$ where

$$0 < \lambda_1 < \lambda_2 \le \lambda_3 \le \ldots \to \infty. \tag{1.2}$$

Here each eigenvalue is repeated according to its multiplicity. An associated orthonormal basis of real eigenfunctions will be denoted u_1, u_2, u_3, ... In fact, throughout this paper we will assume that all functions we consider are real. This entails no loss of generality in the present context.

The next problem we introduce is that of the *free membrane*, or *Neumann Laplacian*. This is the problem

$$-\Delta v = \mu v \quad \text{in } \Omega \subset \mathbb{R}^n, \tag{1.3a}$$
$$\frac{\partial v}{\partial n} = 0 \quad \text{on } \partial\Omega. \tag{1.3b}$$

Here $\partial/\partial n$ denotes the outward normal derivative on $\partial\Omega$, where we now assume that $\partial\Omega$ is sufficiently smooth. With this assumption, problem (1.3) has spectrum $\{\mu_i\}_{i=0}^{\infty}$ where

$$0 = \mu_0 < \mu_1 \le \mu_2 \le \ldots \to \infty, \tag{1.4}$$

with the eigenvalues again repeated according to their multiplicities. A corresponding orthonormal basis of real eigenfunctions will be denoted $\{v_i\}_{i=0}^{\infty}$.

Next we introduce the *clamped plate problem*, or eigenvalue problem for the *Dirichlet biharmonic operator* (for an explanation of this terminology see

[40] or [77]), which describes the characteristic vibrations of a clamped plate. This problem is given by

$$\Delta^2 w = \Gamma w \quad \text{in } \Omega \subset \mathbb{R}^n, \tag{1.5a}$$

$$w = 0 = \frac{\partial w}{\partial n} \quad \text{on } \partial\Omega. \tag{1.5b}$$

We will denote the eigenvalues and an associated orthonormal basis of real eigenfunctions by $\{\Gamma_i\}_{i=1}^{\infty}$ and $\{w_i\}_{i=1}^{\infty}$, respectively. The eigenvalues Γ_i satisfy

$$0 < \Gamma_1 \leq \Gamma_2 \leq \Gamma_3 \leq \ldots \to \infty. \tag{1.6}$$

Lastly, we introduce the *buckling problem*, which determines the critical buckling load of a clamped plate subjected to a uniform compressive force around its boundary (this description applies to the $n = 2$ case of the problem). This problem again involves the biharmonic operator and is formulated as

$$\Delta^2 v = -\Lambda \Delta v \quad \text{in } \Omega \subset \mathbb{R}^n, \tag{1.7a}$$

$$v = 0 = \frac{\partial v}{\partial n} \quad \text{on } \partial\Omega. \tag{1.7b}$$

It also has a discrete spectrum consisting of positive eigenvalues of finite multiplicity with infinity as their only accumulation point. We denote the eigenvalues by $\{\Lambda_i\}_{i=1}^{\infty}$ and a corresponding orthonormal basis of real eigenfunctions by $\{v_i\}_{i=1}^{\infty}$. Thus

$$0 < \Lambda_1 \leq \Lambda_2 \leq \Lambda_3 \leq \ldots \to \infty. \tag{1.8}$$

Good sources of information on many of these problems include the books of Bandle [24], Bérard [26], Courant-Hilbert [38], Davies [39], Kesavan [58], Leis [67], Pólya and Szegő [84], Reed and Simon [87], and Safarov and Vassiliev [88]. The review papers of Davies [40], Payne [79], [80], [81], Protter [85], and Talenti [99], [100] are also quite useful, as is Leissa's monograph [68] on the vibration of plates. Rayleigh's classic *The Theory of Sound* [86] is highly recommended as collateral reading.

1.2 Rearrangement

In this subsection we introduce the notion of *spherically symmetric rearrangement* (or *Schwarz symmetrization*) and some of its properties. Suppose that we have a bounded measurable function f on the bounded measurable set $\Omega \subset \mathbb{R}^n$. We can consider its *distribution function* $\mu_f(t)$ defined by

$$\mu_f(t) = |\{x \in \Omega | |f(x)| > t\}| \tag{1.9}$$

where $|\cdot|$ denotes Lebesgue measure. The distribution function can be viewed as a function from $[0,\infty)$ to $[0,|\Omega|]$. It is clearly a nonincreasing function. The *decreasing rearrangement of* f, denoted f^*, is essentially the inverse of μ_f and is defined by

$$f^*(s) = \inf\{t \geq 0 | \mu_f(t) < s\}. \tag{1.10}$$

It is a nonincreasing function on $[0,|\Omega|]$.

Before defining the spherically decreasing rearrangement of a function we define the *spherical* (or *symmetric*) *rearrangement* of a set. For a bounded measurable set $\Omega \subset \mathbb{R}^n$, we define its spherical rearrangement $\Omega^\star$ as the ball centered at the origin having the same measure as Ω, i.e., $|\Omega^\star| = |\Omega|$. We can now define the *spherically (symmetric) decreasing rearrangement* $f^\star : \Omega^\star \to \mathbb{R}$ by

$$f^\star(x) = f^*(C_n|x|^n) \quad \text{for} \quad x \in \Omega^\star \tag{1.11}$$

where $C_n = \pi^{n/2}/\Gamma\left(\frac{n}{2}+1\right)$ = volume of the unit ball in $\mathbb{R}^n$. A verbal description of $f^\star$ runs as follows: $f^\star$ is that function of $x \in \Omega^\star$ which is spherically symmetric, radially decreasing (in the weak sense of nonincreasing) and equimeasurable with $|f|$, i.e., $f^\star$ and $|f|$ share the same distribution function (see (1.9) above). One way to view this is that for each of the level sets $\Omega_t \equiv \{x \in \Omega | |f(x)| > t\}$ of $|f|$ we take its spherical rearrangement $\Omega_t^\star$ and define this to be the corresponding level set of $f^\star$ (which, we recall, is placed concentrically with $\Omega^\star$). This viewpoint gives an interpretation of $f^\star$ which avoids the intermediate function f^* of a single (=volume) variable. Good sources of further information on rearrangements are [24], [45], [47], [48], [56], [71], [72], [73], [84], [90], [95], [96], [97], [99], [100].

A key fact, which is evident from the equimeasurability of f, f^*, and $f^\star$, is that

$$\int_\Omega f^2 = \int_0^{|\Omega|} f^*(s)^2 ds = \int_{\Omega^\star} (f^\star)^2. \tag{1.12}$$

There is no reason for not putting $|f|$, f^*, and $f^\star$ as integrands here, or even $|f|^p$, $|f^*|^p$, and $|f^\star|^p$, except that (1.12) is all we need later in this paper.

We shall be particularly concerned with how spherical rearrangement affects the first eigenfunction u_1 of the Dirichlet Laplacian. The key property here is that spherical rearrangement typically decreases, and in any case cannot increase, the Dirichlet norm $\int_\Omega |\nabla u_1|^2$. In particular, it is known that for any function f in the Sobolev space $H_0^1(\Omega)$ (for background and notation on Sobolev spaces we recommend [2], [29], [58], and [73]) $f^\star \in H_0^1(\Omega^\star)$ and

$$\int_{\Omega^\star} |\nabla f^\star|^2 \leq \int_\Omega |\nabla f|^2. \tag{1.13}$$

This inequality is crucial to the proof of the Faber-Krahn inequality given in Section 2. Note, too, that by (1.12) the L^2-norm of u_1 does not change when we replace it by $u_1^\star$. For a discussion of (1.13), see Glaser, Martin, Grosse, and Thirring [45], Gunson [47], Kawohl [56], Lieb [71], [72], Lieb and Loss [73], Pólya-Szegő [84], and/or Talenti [95], [96], [99], [100].

We also note a certain elementary property of rearrangement as it affects integrals of products of functions. This is that for two nonnegative measurable functions f and g on Ω

$$\int_\Omega fg \le \int_{\Omega^\star} f^\star g^\star. \tag{1.14}$$

There is a corresponding lower bound if we rearrange f and g in opposite senses. For this we need the notion of *spherically (symmetric) increasing rearrangement*, which we denote by a lower $\star$. The definition is almost identical to that of spherically decreasing rearrangement, except that $g_\star$ should be radially increasing (in the weak sense) on $\Omega^\star$. Then we have, for example,

$$\int_\Omega fg \ge \int_{\Omega^\star} f^\star g_\star. \tag{1.15}$$

In fact, these inequalities take their most elementary form if we use "signed" rearrangements of f and g, that is, we define $f^\star$, $g^\star$, $f_\star$, $g_\star$, etc., in terms of the distribution function as given in (1.9) *except* that we use just $f(x)$ in place of $|f(x)|$ in that formula. With $f^\star$, $g^\star$, etc., defined in this alternative way, (1.14) and (1.15) hold without any need to restrict f and g to be nonnegative functions. All of this basic material is admirably presented in [48], where the essential features of the process (similarly ordered, oppositely ordered) are brought to the fore. In a sense the essence of the whole business is the algebraic identity

$$(a-b)(c-d) \ge 0 \tag{1.16}$$

if (a,b) and (c,d) are similarly ordered (and the reverse if these vectors are oppositely ordered). Thus, if $\vec{v} = (a,b)$ and $\vec{w} = (c,d)$ are similarly ordered

$$\vec{v} \cdot \vec{w} = ac + bd \ge ad + bc = \vec{v} \cdot (d,c). \tag{1.17}$$

The relevance of this simple vector identity for (1.14) (in its general "signed" version) is that we can approach (1.14) by approximating f and g by simple functions decomposed over sets of equal measures. In this setting the relevance of (1.17) is apparent (as a first approximation in the single-variable setting one might think of passing to Riemann sums over subintervals of equal length). See [48], and the more recent article by Baernstein [25], for further development of these ideas.

We come finally to a result on how rearrangement affects the solution to the Poisson equation on a bounded domain Ω with homogeneous Dirichlet boundary conditions

$$-\Delta u = f \quad \text{in } \Omega \subset \mathbb{R}^n, \tag{1.18a}$$
$$u = 0 \quad \text{on } \partial\Omega. \tag{1.18b}$$

Suppose we solve this problem for u and then take its spherically decreasing rearrangement $u^\star$. We want to compare $u^\star$ to the solution v of the symmetrized (=spherically rearranged) problem

$$-\Delta v = f^\star \quad \text{in } \Omega^\star, \tag{1.19a}$$
$$v = 0 \quad \text{on } \partial\Omega^\star. \tag{1.19b}$$

It turns out that if $u \geq 0$ on Ω, then

$$0 \leq u^\star \leq v. \tag{1.20}$$

For $f^\star$ in (1.19a) we can even use a signed rearrangement of f (still under the assumption that f is sufficiently positive that $u \geq 0$ on Ω, which implies, in particular, that $\int_\Omega f \geq 0$). For example, if f is bounded below by c we can take $f^\star = (f - c)^\star + c$; this gives a radially decreasing function that passes into negative values out near $\partial\Omega$ if f itself is not always nonnegative.

The key to proving (1.20) is a differential inequality for $u^\star$ (or equivalently u^*) and the fact that (1.19) reduces to a one-dimensional problem that can be solved explicitly. In particular, we have (in the radial variable $r = |x|$)

$$-\frac{1}{r^{n-1}}(r^{n-1}v')' = f^\star \tag{1.21}$$

and hence

$$r^{n-1}v'(r) = -\int_0^r \tau^{n-1} f^\star(\tau)d\tau \tag{1.22}$$

and finally

$$v(r) = \int_r^R t^{1-n} \int_0^t \tau^{n-1} f^\star(\tau)d\tau dt, \tag{1.23}$$

where we have incorporated the conditions $v'(0) = 0$ (necessary for v to be smooth at the origin in $\Omega^\star$) and $v(R) = 0$, with R denoting the radius of $\Omega^\star$. Recalling the condition $\int_\Omega f \geq 0$, we see that this is precisely the condition needed to keep v nonnegative, since this guarantees that $\int_0^t \tau^{n-1} f^\star(\tau)d\tau \geq 0$ for all $t \in [0, R]$ (recall, too, that up to the constant factor nC_n, $\int_0^t \tau^{n-1} f^\star(\tau)d\tau$ is $\int_0^s f^\star(\sigma)d\sigma$ where σ is now the volume variable $\sigma = C_n\tau^n$ and similarly s is related to t by $s = C_n t^n$; in general, we shall reserve s and σ as "volume variables", while t, τ, r, and ρ will be used as radial variables).

The basic idea used to get (1.20) goes back at least to Weinberger [103]. The method is also given by Bandle in her book [24] and figures prominently in works of Talenti [95], [96], [97], and Chiti [34], [35], [36], [37]. The extended form presented above (allowing a signed rearrangement of f) occurs in Talenti [98], where it is instrumental in his treatment of the first eigenvalue Γ_1 of the clamped plate problem. As such, it also figures in the later papers on the subject by Nadirashvili [74], [75], [76] and Ashbaugh and Benguria [18] (see also [22]). A useful discussion of the extended form is also found in the papers of Kesavan [57], [59], [60], [61]. The interaction between rearrangements and partial differential equations encompasses a variety of topics by a wide range of authors. A small sampling of other works in this field includes [3], [4], [5], [6], [7], and [25].

That one can allow the function f to have variable sign while using a signed rearrangement of it so long as the condition $u \geq 0$ holds was perhaps first observed by Talenti. Earlier authors had assumed $f \geq 0$, apparently because this is a clean and easily checked condition from which $u \geq 0$ follows immediately via the maximum principle. At any rate, throughout this paper when we invoke the result (1.20) under the condition $u \geq 0$ while employing a signed rearrangement of f, we shall refer to it as Talenti's theorem.

To show (1.20), one applies a fairly standard procedure to (1.18a) integrated over level sets of u. One then uses the classical isoperimetric inequality and arrives at an integral-differential inequality for u^* in which v' (for v as given by (1.23)) appears on one side. Inequality (1.20) then follows by integration. For details, one might consult the papers of Talenti or Kesavan mentioned above, or books such as [24] or [56].

1.3 The Rayleigh-Ritz Inequality

Throughout this paper we shall have many occasions to use the Rayleigh-Ritz inequality, which gives a simple way to bound eigenvalues from above based on trial functions. For example, for the Dirichlet Laplacian on the bounded domain Ω one has

$$\lambda_1(\Omega) = \inf_{\substack{\varphi \in D(-\Delta) \\ \varphi \not\equiv 0}} \frac{\int_\Omega \varphi(-\Delta\varphi)}{\int_\Omega \varphi^2} \tag{1.24}$$

where φ is a real trial function in the domain of $-\Delta$ (denoted $D(-\Delta)$). One can also get at the higher eigenvalues by imposing orthogonality conditions on the class of trial functions used. For example,

$$\lambda_{k+1}(\Omega) = \inf_{\substack{\varphi \in D(-\Delta) \\ 0 \not\equiv \varphi \perp u_1, \ldots, u_k}} \frac{\int_\Omega \varphi(-\Delta\varphi)}{\int_\Omega \varphi^2} \tag{1.25}$$

where $u_1, \ldots, u_k$ denote the first k eigenfunctions of $-\Delta$ on Ω. Beyond this, and somewhat more useful for our purposes below, there is a quadratic form formulation of the Rayleigh-Ritz inequality. For the Dirichlet Laplacian it reads as follows:

$$\lambda_{k+1}(\Omega) = \inf_{\substack{\varphi \in H_0^1(\Omega) \\ 0 \not\equiv \varphi \perp u_1, \ldots, u_k}} \frac{\int_\Omega |\nabla\varphi|^2}{\int_\Omega \varphi^2}. \tag{1.26}$$

We note that the Sobolev space $H_0^1(\Omega) = Q(-\Delta) =$ the form domain of $-\Delta$ in this case (see, for example [39], [58], and [87]). This formulation has the advantage over (1.25) that the trial function φ can be chosen from a larger class of functions (and, in particular, it need have essentially only one square-integrable derivative, not two).

The Rayleigh-Ritz inequality (and the closely related Min-Max Principle) applies to any semi-bounded (from below) self-adjoint operator on a Hilbert space. For more details and discussion, the reader might consult Bandle [24], Bérard [26], Chavel [30], [31], Davies [39], Kesavan [58], and/or Reed and Simon, vol. 4 [87].

2 Isoperimetric Inequalities for Eigenvalues

2.1 The Faber-Krahn Inequality

One of the earliest isoperimetric inequalities for an eigenvalue is certainly that for the first eigenvalue of the Dirichlet Laplacian (= fixed membrane problem) conjectured by Rayleigh [86] in 1877:

$$\lambda_1(\Omega) \geq \lambda_1(\Omega^\star) \quad \text{for } \Omega \subset \mathbb{R}^n \tag{2.1}$$

with equality if and only if Ω is a ball, i.e., $\Omega = \Omega^\star$. This result was subsequently proved (independently) by Faber [43] and Krahn [63], [64] in the 1920's using symmetrization. In terms of spherical symmetrization (=spherical rearrangement) the proof can now be reduced to a few lines. One uses the first eigenfunction u_1 for Ω and (1.12), (1.13) to conclude

$$\begin{aligned} \lambda_1(\Omega) &= \int_\Omega |\nabla u_1|^2 \\ &\geq \int_{\Omega^\star} |\nabla u_1^\star|^2 \\ &\geq \lambda_1(\Omega^\star), \end{aligned} \tag{2.2}$$

where the last line follows from the Rayleigh-Ritz inequality for λ_1 of $-\Delta$ on $\Omega^\star$ and the fact (mentioned in connection with (1.13)) that $u_1 \in H_0^1(\Omega)$ implies that $u_1^\star \in H_0^1(\Omega^\star)$, and thus that $u_1^\star$ is a valid trial function in the Rayleigh-Ritz inequality for $\lambda_1(\Omega^\star)$. The characterization of the case of equality is somewhat technical, so we refer the reader to the literature for it. Good sources are Kawohl's book [56], or Kesavan's articles [59], [60].

2.2 The Szegő-Weinberger Inequality

We next turn to the isoperimetric result for $\mu_1(\Omega)$, the first nonzero Neumann eigenvalue, due to Szegő [93] (for $n = 2$ and Ω simply connected) and Weinberger [102] (in full generality). This is the next simplest (or even perhaps the simplest) isoperimetric result for eigenvalues. The result was first suggested by Kornhauser and Stakgold [62], who also obtained some results in support of it. The Szegő-Weinberger result states that

$$\mu_1(\Omega) \leq \mu_1(\Omega^\star) \quad \text{for } \Omega \subset \mathbb{R}^n, \tag{2.3}$$

with equality if and only if Ω is a ball.

We follow Weinberger's method of proof, since it is very natural and lends itself to the problem we treat next, that of finding the optimal upper bound for λ_2/λ_1. First we recall that $\mu_1(\Omega)$ may be characterized through the variational principle

$$\mu_1(\Omega) = \min_{\substack{\varphi \in H^1(\Omega) \\ 0 \not\equiv \varphi \perp 1}} \frac{\int_\Omega |\nabla \varphi|^2}{\int_\Omega \varphi^2}. \tag{2.4}$$

This is just the (quadratic form version of the) Rayleigh-Ritz inequality for μ_1, since v_0 is a constant. Following Weinberger, we take as trial functions $\varphi = P_i$, $i = 1, \ldots, n$, such that $\int_\Omega P_i = 0$ for $i = 1, \ldots, n$ (this is proved via a topological argument, but given (2.5) below all it says is that we can choose to place our origin of coordinates at an appropriate generalized center of mass) with

$$P_i(x) = g(r)\frac{x_i}{r}, \tag{2.5}$$

where the x_i's are Cartesian coordinates, $x = (x_1, \ldots, x_n) \in \mathbb{R}^n$, $r = |x|$, and

$$g(r) = \begin{cases} w(r) = & [\text{“right” radial function for a ball } B_R \text{ of radius } R \\ & \quad \text{where } B_R = \Omega^\star] \qquad \text{for } 0 \leq r \leq R, \\ w(R) & \qquad\qquad\qquad\qquad\quad \text{for } r \geq R. \end{cases} \tag{2.6}$$

The function w arises as a solution of the radial equation when one separates variables on the ball B_R and is therefore basically a Bessel function. We use only the facts that $w(0) = 0$ and that w satisfies

$$w'' + \frac{n-1}{r}w' - \frac{n-1}{r^2}w + \mu_1(B_R)w = 0 \quad \text{for} \quad 0 < r < R. \tag{2.7}$$

Since $w'(R) = 0$ and $\mu_1(B_R)$ is defined as the first eigenvalue of this boundary value problem, it follows (by an appropriate choice of sign) that $w(r)$ is increasing on $[0, R]$ and hence that g is everywhere nondecreasing for $r \geq 0$. By substituting our trial functions P_i into the Rayleigh-Ritz inequality for μ_1, we find

$$\mu_1(\Omega) \int_\Omega P_i^2 \leq \int_\Omega |\nabla P_i|^2. \tag{2.8}$$

Summing this in i for $1 \leq i \leq n$, we arrive at

$$\begin{aligned} \mu_1(\Omega) &\leq \frac{\int_\Omega \sum_{i=1}^n |\nabla P_i|^2}{\int_\Omega \sum_{i=1}^n P_i^2} = \frac{\int_\Omega \left[g'(r)^2 + \frac{n-1}{r^2} g(r)^2\right]}{\int_\Omega g(r)^2} \\ &= \frac{\int_\Omega B(r)}{\int_\Omega g(r)^2} \end{aligned} \tag{2.9}$$

where we have defined

$$B(r) \equiv g'(r)^2 + \frac{n-1}{r^2} g(r)^2. \tag{2.10}$$

Now $B(r)$ is easily seen to be decreasing for $0 \leq r \leq R$ by differentiating and using the differential equation (2.7). One finds

$$B'(r) = -2[\mu_1(B_R)gg' + (n-1)(rg' - g)^2/r^3] < 0 \quad \text{for} \quad 0 < r < R. \tag{2.11}$$

In addition, $B(r) = (n-1)w(R)^2/r^2$ for $r \geq R$ shows that B is decreasing for $r \geq R$. Since B is continuous for all $r \geq 0$, it is clearly also decreasing there. Now one has only to observe that

$$\int_\Omega B(r) \leq \int_{\Omega^\star} B(r) \tag{2.12}$$

since the volumes integrated over are the same in both cases, while in passing from the left- to right-hand sides you are exchanging integrating over $\Omega \backslash \Omega^\star$ for integrating over $\Omega^\star \backslash \Omega$ (which are sets of equal volume). Since B is (strictly) decreasing this clearly increases the value of the integral unless $\Omega = \Omega^\star$, when equality obtains. Similarly we find that

$$\int_\Omega g(r)^2 \geq \int_{\Omega^\star} g(r)^2 \tag{2.13}$$

since g is nondecreasing. Thus we arrive at

$$\mu_1(\Omega) \le \frac{\int_{\Omega^*} B(r)}{\int_{\Omega^*} g(r)^2} = \mu_1(\Omega^*), \tag{2.14}$$

since each P_i is precisely an eigenfunction of $-\Delta$ with eigenvalue $\mu_1(B_R)$ for the domain $B_R = \Omega^*$. This completes the proof of the Szegő-Weinberger inequality, including the characterization of the case of equality.

2.3 The Payne-Pólya-Weinberger Conjecture

The next isoperimetric result that we consider is that for λ_2/λ_1 for the fixed membrane problem. In 1955 and 1956, Payne, Pólya, and Weinberger [82], [83] proved that

$$\frac{\lambda_2}{\lambda_1} \le 3 \quad \text{for} \quad \Omega \subset \mathbb{R}^2 \tag{2.15}$$

and conjectured that

$$\frac{\lambda_2}{\lambda_1} \le \left.\frac{\lambda_2}{\lambda_1}\right|_{\text{disk}} = \frac{j_{1,1}^2}{j_{0,1}^2} \approx 2.5387 \tag{2.16}$$

with equality if and only if Ω is a disk (i.e., $\Omega = \Omega^*$) and where $j_{p,k}$ denotes the kth positive zero of the Bessel function $J_p(t)$ (we follow the notation of Abramowitz and Stegun [1] here). For general dimension $n \ge 2$ the analogous statements are

$$\frac{\lambda_2}{\lambda_1} \le 1 + \frac{4}{n} \quad \text{for } \Omega \in \mathbb{R}^n, \tag{2.17}$$

and the *PPW conjecture*

$$\frac{\lambda_2}{\lambda_1} \le \left.\frac{\lambda_2}{\lambda_1}\right|_{n-\text{ball}} = \frac{j_{n/2,1}^2}{j_{n/2-1,1}^2}, \tag{2.18}$$

with equality if and only if Ω is an n-ball. This PPW conjecture was proved in 1990 by Rafael Benguria and the author (see [10], [11], [13]).

We proceed now with the proof of (2.18). This proof follows the main outline of Weinberger's proof of the μ_1 result given previously, but it is substantially more complicated due mainly to the fact that, unlike $v_0 = 1/\sqrt{|\Omega|}$, u_1 is not constant. We start from the variational principle for λ_2

$$\lambda_2(\Omega) = \min_{\substack{\varphi \in H_0^1(\Omega) \\ 0 \not\equiv \varphi \perp u_1}} \frac{\int_\Omega |\nabla\varphi|^2}{\int_\Omega \varphi^2}, \tag{2.19}$$

or, better for our purposes here,

$$\lambda_2(\Omega) - \lambda_1(\Omega) \leq \min_{\substack{Pu_1 \in H_0^1(\Omega) \\ \int_\Omega Pu_1^2 = 0 \\ P \not\equiv 0}} \frac{\int_\Omega |\nabla P|^2 u_1^2}{\int_\Omega P^2 u_1^2}, \tag{2.20}$$

which follows from (2.19) via integration by parts. (Note that with $\varphi = Pu_1$,

$$\begin{aligned} \int_\Omega |\nabla\varphi|^2 &= \int_\Omega [|\nabla P|^2 u_1^2 + 2Pu_1 \nabla P \cdot \nabla u_1 + P^2 |\nabla u_1|^2] \\ &= \int_\Omega |\nabla P|^2 u_1^2 + \int_\Omega [\nabla(P^2 u_1)] \cdot \nabla u_1 \\ &= \int_\Omega |\nabla P|^2 u_1^2 + \int_\Omega P^2 u_1(-\Delta u_1) \\ &= \int_\Omega |\nabla P|^2 u_1^2 + \lambda_1(\Omega) \int_\Omega P^2 u_1^2, \end{aligned}$$

where we integrated by parts in the second-to-last step.) In (2.20) we shall use n trial functions $P = P_i$, $i = 1, \ldots, n$, such that $\int_\Omega P_i u_1^2 = 0$ for $i = 1, \ldots, n$ (again proved by a topological argument and interpretable as a generalized center of mass result) where

$$P_i = g(r)\frac{x_i}{r} \tag{2.21}$$

and

$$g(r) = \begin{cases} w(r) = & [\text{“right” radial function for a ball } B_R \\ & \quad \text{of radius } R] \qquad \text{for } 0 \leq r \leq R, \\ w(R) & \qquad\qquad\qquad\qquad \text{for } r \geq R. \end{cases} \tag{2.22}$$

The right R in this case turns out to be the unique R such that $\lambda_1(B_R) = \lambda_1(\Omega)$. This is a key fact and explains why at bottom our proof of the PPW conjecture is a "fixed-λ_1 result". A major motivation for this is the comparison result of Chiti, to be explained shortly. The right function $w(r)$ for $0 \leq r \leq R$ is fairly complicated, being a ratio of Bessel functions.

From here we can proceed much as before finding

$$\lambda_2(\Omega) - \lambda_1(\Omega) \leq \frac{\int_\Omega B(r) u_1^2}{\int_\Omega g(r)^2 u_1^2} \tag{2.23}$$

where

$$B(r) \equiv g'(r)^2 + \frac{n-1}{r^2} g(r)^2. \tag{2.24}$$

Again it can be confirmed (though it is harder this time around) that B is decreasing (and positive) and g is increasing (and positive). For the proof of these facts, see [10], [11], [13], or [16]. The proof found in [10] (for $n = 2$ and 3) and [11] (for all $n \geq 2$) is based on the product representation for the Bessel functions involved and ultimately comes down to certain inequalities between Bessel function zeros (see also Section 4 of [13] for a brief further discussion of these issues). On the other hand, the proof found in [13] (see also [16]) is somewhat simpler and lends itself to generalization to a version of the PPW conjecture for domains in S^n. These topics are dealt with further in [20], [21], and also to some extent below.

To continue from (2.23) we use rearrangement as follows:

$$\lambda_2(\Omega) - \lambda_1(\Omega) \leq \frac{\int_{\Omega^\star} B(r)^\star u_1^{\star 2}}{\int_{\Omega^\star} g(r)_\star^2 u_1^{\star 2}} \leq \frac{\int_{\Omega^\star} B(r) u_1^{\star 2}}{\int_{\Omega^\star} g(r)^2 u_1^{\star 2}}. \tag{2.25}$$

The basic ideas here are that the integral of a product is increased if we similarly rearrange both functions (see (1.14)), while it is decreased if we oppositely rearrange them (see (1.15)). Note that in (2.25) we similarly rearrange in the numerator, while we oppositely rearrange in the denominator. The removal of the $\star$'s from $g(r)^2$ and $B(r)$ in the last step is allowed due to their respective monotonicity properties.

Finally, we need to invoke a comparison result due to Chiti [34], [35], [36], [37] to replace the $u_1^\star$'s in (2.25) by something more tractable. This result says that if we take $z(r)$ as the normalized first eigenfunction of B_R (recall that R is so chosen that $\lambda_1(\Omega) = \lambda_1(B_R)$) then $B_R \subset \Omega^\star$ and z will be larger than $u_1^\star$ (thought of as a function of r) for r near 0, say for $0 < r < r_1$, and smaller farther out, in this case for $r > r_1$. Under this condition one finds that if $u_1^\star$ is replaced by z in the final member of (2.25) the numerator can only go up and the denominator can only go down. Thus

$$\lambda_2(\Omega) - \lambda_1(\Omega) \leq \frac{\int_{B_R} B(r) z^2}{\int_{B_R} g(r)^2 z^2} = \lambda_2(B_R) - \lambda_1(B_R), \tag{2.26}$$

the final equality holding because our choices of g and w were made precisely to make things come back together in this way. Since $\lambda_1(B_R) = \lambda_1(\Omega)$, (2.26) implies

$$\lambda_2(\Omega) \leq \lambda_2(B_R) \tag{2.27}$$

and hence also

$$\frac{\lambda_2(\Omega)}{\lambda_1(\Omega)} \leq \frac{\lambda_2(B_R)}{\lambda_1(B_R)} = \left.\frac{\lambda_2}{\lambda_1}\right|_{\text{any } n\text{-ball}} = \frac{\lambda_2(\Omega^\star)}{\lambda_1(\Omega^\star)}. \tag{2.28}$$

We note that the PPW inequality (2.28) was really proved as a subsidiary result to (2.27), which should be regarded as a "fixed-λ_1 result". It is this result that is in a certain sense more fundamental and which is most easily extended to other settings, for example, domains contained in a hemisphere of S^n.

Indeed, in subsequent work [16], [20], [21] we have established the fixed-λ_1 result (2.27) for domains contained in a hemisphere of S^n. In addition, from this result we can then derive the result

$$\frac{\lambda_2(\Omega)}{\lambda_1(\Omega)} \leq \frac{\lambda_2(\Omega^\star)}{\lambda_1(\Omega^\star)} \tag{2.29}$$

for a domain Ω contained in a hemisphere of S^n. Here $\Omega^\star$ denotes the geodesic ball (=polar cap) having the same volume as Ω and B_R must be taken as the (unique) polar cap having the same λ_1 as Ω. Furthermore, there is a corresponding result for μ_1 (that is, $\mu_1(\Omega) \leq \mu_1(\Omega^\star)$) for a domain Ω contained in a hemisphere of S^n and for Ω a bounded domain in H^n. See [16], [17] for more on this aspect of the μ_1 problem (in particular, the result for H^n is due to Chavel, while both Chavel and Bandle had earlier variants of the S^n result). The Faber-Krahn inequality also extends in sharp form to the spaces S^n and H^n. See, for example, Sperner [91] or Friedland and Hayman [44]. In particular, Sperner's bound $\lambda_1(\Omega) \geq \lambda_1(\Omega^\star)$ for $\Omega \subset S^n$ is one of the elements necessary to our proof of (2.27) and (2.29) for domains contained in a hemisphere of S^n.

2.4 Rayleigh's Conjecture for the Vibrating Clamped Plate

We turn now to Rayleigh's conjecture for the vibration of a clamped plate. Rayleigh made this conjecture in 1877 in the first edition of his book **The Theory of Sound** [86] (see p. 382 of volume 1 of the second edition). In terms of the notation introduced in Section 1, Rayleigh's conjecture states that

$$\Gamma_1(\Omega) \geq \Gamma_1(\Omega^\star) \quad \text{for } \Omega \subset \mathbb{R}^2, \tag{2.30}$$

with equality if and only if Ω is itself a disk. It is natural to conjecture that this inequality might apply equally well in $\mathbb{R}^n$.

This conjecture seems to have lain dormant until around 1950, when Szegő made some progress on it (see [92], [94], and also the treatment in Pólya and Szegő's book [84]). In fact, Szegő was able to prove (2.30) for simply connected domains Ω having a nonnegative first eigenfunction, $w_1 \geq 0$. At the time it seems to have been thought possible that $w_1 \geq 0$ for all domains.

However, it soon developed that this cannot be expected to hold in general. Results of Duffin [41] and Duffin and Shaffer [42] were enough to disabuse people of the notion that $w_1 \geq 0$ should always hold (see also the much more recent article of Kozlov, Kondrat'ev, and Maz'ya [65], as well as further references given in [22] and [23]).

The next advance toward the proof of Rayleigh's conjecture (2.30) came in 1981 when Talenti [98] developed an approach to the sign-indeterminate case using separate rearrangements on the sets $\Omega_+ = \{x \in \Omega | w_1 > 0\}$ and $\Omega_- = \{x \in \Omega | w < 0\}$. This procedure led Talenti to two radial subproblems (tied to the two separate balls $(\Omega_+)^\star$ and $(\Omega_-)^\star$ and also to the full ball $\Omega^\star$) which he decoupled and considered as variational problems in their own right. By this means, he obtained a lower bound to $\Gamma_1(\Omega)$ depending on the parameter $t \equiv |\Omega_+|/|\Omega|$ for $t \in [0,1]$. If this problem had been minimized at $t = 0$ (and therefore also at $t = 1$, by symmetry) then Talenti's approach would have proved Rayleigh's conjecture. But unfortunately Talenti's minimum occurred at $t = 1/2$ (for each $n = 2, 3, \ldots$) and he was only able to obtain inequalities

$$\Gamma_1(\Omega) \geq d_n' \Gamma_1(\Omega^\star) \quad \text{for } \Omega \subset \mathbb{R}^n. \tag{2.31}$$

These bounds fall short of Rayleigh's conjecture, since the constants d_n' are less than 1. However, Talenti was able to show that $d_n' \in (1/2, 1)$, and, in particular, he found $d_2' \approx 0.9777$, $d_2' \approx 0.7391$, $d_4' \approx 0.6524$, ... In fact, the d_n''s seem to decrease monotonically to 1/2 as n goes to infinity (that $\lim_{n\to\infty} d_n' = 1/2$ is proved in [23]).

In 1992 Nadirashvili (see [74], [75], [76]) saw how to improve upon the approach of Talenti by using slightly different radial subproblems, each living only on $(\Omega_+)^\star$ or $(\Omega_-)^\star$, but coupled together via boundary conditions. With his approach and a geometric rearrangement argument, Nadirashvili was able to give a proof of (2.30) for $n = 2$. However, it was not clear that the same approach could handle cases with $n \geq 3$.

To that end, Rafael Benguria and the author introduced a more analytical variant of Nadirashvili's approach, wherein the final analysis comes down to understanding a certain function defined explicitly in terms of Bessel functions. (Bessel functions, while certainly in the background in the clamped plate problem, were not in evidence in any part of Nadirashvili's proof.) By studying the behavior of this function we were able to show [18] that for $n = 2$ and 3 the minimizer of our parametrized two-ball variational problem occurs when $t = |\Omega_+|/|\Omega|$ is 0 (or 1), thereby proving Rayleigh's conjecture in dimensions 2 and 3. In subsequent work with Richard Laugesen [23] (see also [22]) it was shown that for all $n \geq 4$ things go the other way, i.e., the parametrized two-ball minimizer occurs at $t = 1/2$. This yields bounds of the form

$$\Gamma_1(\Omega) > d_n \Gamma_1(\Omega^\star) \quad \text{for } \Omega \subset \mathbb{R}^n, n \geq 4 \tag{2.32}$$

where our dimension-dependent constants d_n turn out always to be better than the corresponding d_n''s of Talenti, and, in fact, go to 1 as n goes to infinity. For example, we have $d_4 \approx 0.9537$, $d_5 \approx 0.9218$, $d_6 \approx 0.9077$, ... ($d_8 \approx 0.8998$ appears to be the minimum over all dimensions n).

We now give the proof of Rayleigh's conjecture (2.30) for dimensions 2 and 3. As a warm-up, we first treat the case where $w_1 \geq 0$ on Ω. This is the case first successfully handled by Szegő [92], [94] (see also Pólya-Szegő [84] and Talenti [98]). We recall the variational principle for Γ_1

$$\Gamma_1(\Omega) = \min_{\substack{\varphi \in H_0^2(\Omega) \\ \varphi \not\equiv 0}} \frac{\int_\Omega (\Delta\varphi)^2}{\int_\Omega \varphi^2}. \tag{2.33}$$

In particular, this gives equality if we take $\varphi = w_1$. If we now decreasing rearrange $f = -\Delta w_1$ to $f^\star = (-\Delta w_1)^\star$ where our rearrangement respects signs, we can invoke Talenti's rearrangement result (1.20) for the problem $-\Delta w_1 = f$ in Ω, $w_1 = 0$ on $\partial\Omega$, to get

$$\Gamma_1(\Omega) = \frac{\int_\Omega (\Delta w_1)^2}{\int_\Omega w_1^2} \geq \frac{\int_{\Omega^\star} [(-\Delta w_1)^\star]^2}{\int_{\Omega^\star} v^2} = \frac{\int_{\Omega^\star} (\Delta v)^2}{\int_{\Omega^\star} v^2} \geq \Gamma_1(\Omega^\star), \tag{2.34}$$

which is the result we want, assuming that v is truly an admissible trial function for the Rayleigh-Ritz inequality (= variational principle) for Γ_1 on $\Omega^\star$. To verify this, we note that v is a radial C^2 function on the ball $\Omega^\star$ and satisfies $v(R) = 0$ where R is the radius of $\Omega^\star$. To see that v also satisfies $\frac{\partial v}{\partial n} = 0$ on $\partial\Omega^\star$ we compute (since $\frac{\partial w_1}{\partial n} = 0$ on $\partial\Omega$)

$$0 = \int_{\partial\Omega} \frac{\partial w_1}{\partial n} = -\int_\Omega (-\Delta w_1) = -\int_{\Omega^\star} f^\star = -\int_{\Omega^\star} (-\Delta v) = \int_{\partial\Omega^\star} \frac{\partial v}{\partial n} \tag{2.35}$$

and this shows that $\frac{\partial v}{\partial n} = 0$ on $\partial\Omega^\star$ since v is radial and therefore $\int_{\partial\Omega^\star} \frac{\partial v}{\partial n} = |\partial\Omega^\star| v'(R) = nC_n R^{n-1} v'(R)$. This proves Szegő's clamped plate result, i.e., Rayleigh's conjecture for the clamped plate in the case of a first eigenfunction of fixed sign. This proof holds for all dimensions n.

We now proceed to the proof of the general case for $n = 2$ and 3. It turns out that the proof does not work for $n \geq 4$ (though Rayleigh's conjecture may well still be true there). As mentioned earlier, in the general case we proceed by decomposing the problem into two (coupled) subproblems, each on a separate ball (on $(\Omega_+)^\star$ and $(\Omega_-)^\star$ in the notation introduced above). Following this decomposition procedure, we have

$$\begin{aligned}\Gamma_1(\Omega) = \frac{\int_\Omega (\Delta w_1)^2}{\int_\Omega w_1^2} &= \frac{\int_{\Omega^+} (\Delta w_1)^2 + \int_{\Omega^-} (\Delta w_1)^2}{\int_{\Omega^+} w_1^2 + \int_{\Omega^-} w_1^2} \\ &\geq \frac{\int_{B_a} (\Delta u)^2 + \int_{B_b} (\Delta v)^2}{\int_{B_a} u^2 + \int_{B_b} v^2}\end{aligned} \tag{2.36}$$

where B_a = ball of radius $a = (\Omega_+)^\star$ and B_b =ball of radius $b = (\Omega_-)^\star$ and u and v satisfy symmetrized Dirichlet problems on B_a and B_b, respectively. Note that the numerator in (2.36) is unchanged by the introduction of u and v, while the denominator is (typically) increased (and certainly does not decrease) by virtue of inequality (1.20) as applied to u and v (and $(w_1^+)^\star$ and $(w_1^-)^\star$), respectively. Now, not only do u and v satisfy

$$u(a) = 0 = v(b) \tag{2.37}$$

but $u'(a)$ and $v'(b)$ are coupled by the equation

$$a^{n-1} u'(a) = b^{n-1} v'(b). \tag{2.38}$$

This comes about by an argument similar to that in (2.35). We have

$$\begin{aligned}0 &= \int_{\partial\Omega} \frac{\partial w_1}{\partial n} = \int_\Omega \Delta w_1 = -\int_{B_a} (-\Delta w_1)^\star + \int_{B_b} (\Delta w_1)^\star \\ &= \int_{B_a} \Delta u - \int_{B_b} \Delta v = \int_{\partial B_a} \frac{\partial u}{\partial n} - \int_{\partial B_b} \frac{\partial v}{\partial n} \\ &= nC_n [a^{n-1} u'(a) - b^{n-1} v'(b)].\end{aligned} \tag{2.39}$$

The seemingly odd changes of sign in certain terms come about because we introduce a sign change for Ω_- so that we will still be comparing positive functions (and thus u and v are both positive). As a curiosity we mention that in fact (2.39) (and also (2.35)) uses only the fact that $\partial w_1/\partial n$ has average value 0 on the boundary of Ω. Thus our results all hold under the weaker assumption that our plate has its edge fixed and "clamped on average", a somewhat curious result first observed by Richard Laugesen.

To finish the argument we view the final member of (2.36) as a variational problem in its own right where we now treat u and v as trial functions subject to (2.37), (2.38), and the condition that they be radial functions on B_a and B_b, respectively. Thus we consider the minimization problem

$$J(a) \equiv \inf_{\varphi,\psi} \frac{\int_{B_a} (\Delta\varphi)^2 + \int_{B_b} (\Delta\psi)^2}{\int_{B_a} \varphi^2 + \int_{B_b} \psi^2} \tag{2.40}$$

with φ and ψ radial and satisfying the boundary conditions (2.37) and (2.38) (with φ and ψ replacing u and v respectively). This is what we refer to as our parametrized two-ball variational problem. Note that a now appears as a final parameter (which we will eventually need to minimize over, since we have no control on it). We observe that, due to scaling freedom, we can assume that $a^n + b^n = 1$, which is equivalent to normalizing $|\Omega|$ at C_n. In (2.40), and henceforth, we regard b as defined implicitly in terms of a by this relation. We also remark that in terms of our earlier parameter $t = |\Omega_+|/|\Omega|$ we have $t = a^n$. In a sense t is the preferred variable here, since as a function of $t \in [0,1]$ J is symmetric about 1/2. We compromise by speaking in terms of a^n at times as we proceed.

Leaving $a \in (0,1)$ fixed for the time being, standard variational theory can be applied to (2.40) with the result that we find the Euler equations

$$\Delta^2\varphi = \mu\varphi \quad \text{in } B_a, \tag{2.41a}$$
$$\Delta^2\psi = \mu\psi \quad \text{in } B_b, \tag{2.41b}$$

and boundary conditions

$$\varphi(a) = 0 = \psi(b), \tag{2.41c}$$
$$a^{n-1}\varphi'(a) = b^{n-1}\psi'(b), \tag{2.41d}$$

and

$$\Delta\varphi(a) + \Delta\psi(b) = 0. \tag{2.41e}$$

The expressions in the last formula make sense because φ and ψ are radial. This formula comes about as a product of the variational theory, and is what is known as a "natural" boundary condition (see, for example, [46] or [104]). See [18] for more on its derivation and for more details on this material in general (see also [22] for a general overview of the topic).

The boundary value problem (2.41) can now be solved explicitly in terms of Bessel functions and modified Bessel functions. The result is an implicit relation for the eigenvalues (and in particular the first eigenvalue $\mu_1 = J(a)$) which we can analyze to determine how $J(a)$ behaves for $0 \le a \le 1$. Since

$$\Gamma_1(\Omega) \ge \min_a J(a) \tag{2.42}$$

and $J(0) = J(0^+) = J(1) = J(1^-) = \Gamma(\Omega^\star)$ (this requires some work to see, though it certainly should already appear plausible; essentially one wants to confirm that taking $a = 0$ gives the same result as the limit $a \to 0^+$ and similarly for $a = 1$), we will be done if we can show that $\min J(a)$ occurs at $a = 0$ (and $a = 1$). For dimensions 2 and 3 this can be shown (see [18]) and this completes the proof of the general Rayleigh conjecture for the clamped plate in these cases.

For $n \geq 4$, the analysis given above holds all the way until the final step, where it is found that $\min_a J(a)$ occurs not at $a = 0$ and 1, but at $a^n = 1/2$. Following this outcome to its logical conclusion leads to the nonoptimal bounds (2.32).

2.5 The Pólya-Szegő Conjecture for the Buckling of a Clamped Plate

Much the same strategy can be used on the first eigenvalue $\Lambda_1(\Omega)$ of the buckling problem. For $n = 2$, this eigenvalue determines the critical buckling load of the clamped plate of the shape of Ω under uniform compressive loading around its boundary. The analog of the Rayleigh conjecture in this setting,

$$\Lambda_1(\Omega) \geq \Lambda_1(\Omega^\star) \quad \text{for } \Omega \subset \mathbb{R}^n, \tag{2.43}$$

with equality if and only if Ω is a ball, was conjectured by Pólya and Szegő around 1950. It is discussed in their book [84]. At around the same time Szegő proved this conjecture [92], [94] under the assumption that $\Lambda_1(\Omega)$ has a first eigenfunction of fixed sign (which we can take to be nonnegative, i.e., $v_1 \geq 0$ on Ω). Again this assumption turns out not to hold in all cases (see the article by Kozlov, Kondrat'ev, and Maz'ya [65]).

The proof of the fixed-sign case (we assume $v_1 \geq 0$) can be accomplished much as was the corresponding case of the vibrating clamped plate above. One simply starts from the variational principle

$$\Lambda_1(\Omega) = \min_{\substack{\varphi \in H_0^2(\Omega) \\ \varphi \not\equiv 0}} \frac{\int_\Omega (\Delta\varphi)^2}{\int_\Omega |\nabla\varphi|^2} = \frac{\int_\Omega (\Delta v_1)^2}{\int_\Omega |\nabla v_1|^2} \tag{2.44}$$

and proceeds as before. One rearranges $-\Delta v_1$ to obtain a symmetrized comparison problem. Since Talenti's version of the comparison argument (for (1.18) and (1.19)) can also be used to compare derivatives (indeed, this can be done directly from the integral-differential inequality for $u^\star$) we find easily that

$$\Lambda_1(\Omega) \geq \frac{\int_{\Omega^\star} (\Delta\varphi)^2}{\int_{\Omega^\star} |\nabla\varphi|^2} \tag{2.45}$$

where φ is now a radial function on $\Omega^\star$ satisfying $\varphi(R) = 0 = \varphi'(R)$ (the proof that $\varphi'(R) = 0$ is exactly the same as that given for v in (2.35)). Since this means that φ is a valid trial function for $\Lambda_1(\Omega^\star)$, the proof of $\Lambda_1(\Omega) \geq \Lambda_1(\Omega^\star)$ is complete.

For the general case (where v_1 is not necessarily of fixed sign) we can also proceed as in the vibrating clamped plate problem. That is, we break Ω into the two parts $\Omega_\pm$ and rearrange $\mp\Delta v_1$ on each part separately to obtain two symmetrized subproblems. The hitch again comes at the very last step, where the minimum over $a \in [0,1]$ turns out to occur at $a^n = 1/2$ for all dimensions $n \geq 2$. Thus we do not obtain the general result (2.43) for any $n \geq 2$, but we can again get nonoptimal lower bounds for $\Lambda_1(\Omega)$ in the form

$$\Lambda_1(\Omega) > c_n \Lambda_1(\Omega^\star) \quad \text{for } \Omega \subset \mathbb{R}^n, \tag{2.46}$$

where the dimension-dependent constants c_n are found to tend to 1 as n goes to infinity. Some values of c_n for low dimensions are $c_2 \approx 0.7877$, $c_3 \approx 0.7759$, $c_4 \approx 0.7872$, $c_5 \approx 0.8020$, $c_6 \approx 0.8163$.

It is a rather remarkable fact that the bounds (2.46) with precisely the same constants c_n can also be obtained by combining two classical eigenvalue inequalities. These are the inequality of Payne [78]

$$\Lambda_1(\Omega) \geq \lambda_2(\Omega) \quad \text{for } \Omega \subset \mathbb{R}^n, \tag{2.47}$$

with equality if and only if Ω is a ball, and Krahn's λ_2 inequality [64]

$$\lambda_2(\Omega) > 2^{2/n} \lambda_1(\Omega^\star) \quad \text{for } \Omega \subset \mathbb{R}^n \tag{2.48}$$

(this saturates as Ω disconnects into two equal disjoint balls). Together these yield

$$\begin{aligned} \Lambda_1(\Omega) > 2^{2/n}\lambda_1(\Omega^\star) &= 2^{2/n}(C_n/|\Omega|)^{2/n} j_{n/2-1,1}^2 \\ &= 2^{2/n}\left(\frac{j_{n/2-1,1}}{j_{n/2,1}}\right)^2 \left(\frac{C_n}{|\Omega|}\right)^{2/n} j_{n/2,1}^2 \\ &= c_n \Lambda_1(\Omega^\star) \end{aligned} \tag{2.49}$$

since $\Lambda_1(\Omega^\star) = (C_n/|\Omega|)^{2/n} j_{n/2,1}^2$ and hence the constants c_n are given by

$$c_n = 2^{2/n}\left(\frac{j_{n/2-1,1}}{j_{n/2,1}}\right)^2. \tag{2.50}$$

The basic observation here was made by Bramble and Payne [27], though they only gave the inequality for $n = 2$ and hence did not investigate the behavior of the c_n's with varying n. Neither did they mention the connection between (2.49) and the conjecture (2.43) of Pólya and Szegő.

3 Universal Inequalities for Eigenvalues

3.1 The General Inequalities of Payne-Pólya-Weinberger for the Fixed Membrane Eigenvalues and their Extensions

In this section we turn our attention to universal eigenvalue inequalities. We begin with a study of the Dirichlet eigenvalues of $-\Delta$ on a bounded domain $\Omega \subset \mathbb{R}^n$. This subject began in 1955 with the work of Payne, Pólya, and Weinberger [82], [83], who proved (among other things) the bound

$$\lambda_{m+1} - \lambda_m \leq \frac{2}{m}\sum_{i=1}^{m}\lambda_i, \quad m = 1, 2, \ldots \tag{3.1}$$

for $\Omega \subset \mathbb{R}^2$. This result easily extends to $\Omega \subset \mathbb{R}^n$ as

$$\lambda_{m+1} - \lambda_m \leq \frac{4}{mn}\sum_{i=1}^{m}\lambda_i, \quad m = 1, 2, \ldots. \tag{3.2}$$

Since Payne, Pólya, and Weinberger's paper [83], (3.2) has been extended in several ways by a number of authors. For the Euclidean case there have been two main advances: that of Hile and Protter [53] in 1980 and that of H.C. Yang [105] in 1991 (this paper is yet to be published, as far as this author knows). It turns out that, even though the proofs given by Hile-Protter and Yang are dissimilar in several respects and are rather involved, the main approach is still that of Payne-Pólya-Weinberger. Moreover, we have succeeded in streamlining these proofs to the point where, with the addition of one key idea, they can be unified into a single overall approach that is no harder than the original proof of (3.1) by Payne, Pólya, and Weinberger. The process of reducing these proofs to their essence was begun in [16],[19] and is concluded in [9]. Since our approach is unified we head directly for the result of H.C. Yang, remarking only at the end on how the results of PPW and Hile-Protter follow "by simplification".

To give an idea of Payne, Pólya, and Weinberger's basic method, we begin by proving their bound

$$\frac{\lambda_2}{\lambda_1} \leq 1 + \frac{4}{n} \quad \text{for } \Omega \subset \mathbb{R}^n \tag{3.3}$$

as a warm-up exercise (note that this is the $m = 1$ case of (3.2)). The n-dimensional inequality (3.3) was first given explicitly by Thompson [101] in 1969, but certainly it (and more) is implicit in the work of Payne, Pólya, and Weinberger, as the generalization of their results to $\Omega \subset \mathbb{R}^n$ is entirely straightforward. To prove (3.3), we introduce the trial function

$$\varphi = x u_1 \tag{3.4}$$

where x represents any Cartesian coordinate $x_\ell (1 \le \ell \le n)$ for $\mathbb{R}^n$. By an appropriate choice of origin we can arrange that $\int_\Omega x_\ell u_1^2 = 0$ for $1 \le \ell \le n$ (i.e., we put the origin at the center of mass of Ω for the mass distribution defined by u_1^2). This guarantees $\varphi \perp u_1$ (also $\varphi^{(\ell)} \perp u_1$ for $1 \le \ell \le n$ where $\varphi^{(\ell)} \equiv x_\ell u_1$) and hence, by the Rayleigh-Ritz inequality (for λ_2),

$$\lambda_2 \le \frac{\int_\Omega \varphi(-\Delta\varphi)}{\int_\Omega \varphi^2}. \tag{3.5}$$

Since

$$-\Delta\varphi = x(-\Delta u_1) - 2u_{1x} = \lambda_1 x u_1 - 2u_{1x} = \lambda_1 \varphi - 2u_{1x}, \tag{3.6}$$

(3.5) gives

$$\lambda_2 - \lambda_1 \le -\frac{2\int_\Omega \varphi u_{1x}}{\int_\Omega \varphi^2}. \tag{3.7}$$

Now

$$0 \le -2\int_\Omega \varphi u_{1x} = -2\int_\Omega x u_1 u_{1x} = -\int_\Omega x(u_1^2)_x = \int_\Omega u_1^2 = 1 \tag{3.8}$$

where in the second-to-last step we integrated by parts, and also by the Cauchy-Schwarz inequality,

$$(-2\int_\Omega \varphi u_{1x})^2 \le 4(\int_\Omega \varphi^2)(\int_\Omega u_{1x}^2), \tag{3.9}$$

implying (since $-2\int_\Omega \varphi u_{1x} > 0$)

$$\lambda_2 - \lambda_1 \le \frac{-2\int_\Omega \varphi u_{1x}}{\int_\Omega \varphi^2} \le \frac{4\int_\Omega u_{1x}^2}{-2\int_\Omega \varphi u_{1x}} = 4\int_\Omega u_{1x}^2. \tag{3.10}$$

Obviously this same argument applies to $\varphi^{(\ell)} \equiv x_\ell u_1$ for $1 \le \ell \le n$, allowing us to arrive at

$$\lambda_2 - \lambda_1 \le 4\int_\Omega u_{1x_\ell}^2 \quad \text{for } 1 \le \ell \le n. \tag{3.11}$$

If we now average these inequalities over ℓ we find

$$\lambda_2 - \lambda_1 \le \frac{4}{n}\int_\Omega |\nabla u_1|^2 = \frac{4}{n}\lambda_1 \tag{3.12}$$

and hence (3.3) follows.

Next we address the general results for λ_{m+1}. The basic strategy is the same as above in that we base our trial functions φ on xu_i where x is a Cartesian coordinate and u_i is a lower eigenfunction (i.e., $1 \le i \le m$), but now to enforce orthogonality to $u_1, \ldots, u_m$ we no longer rely on the device of locating the origin at the center of mass but rather we subtract away counterterms which are just the projections of xu_i along the eigenfunctions u_j for $1 \le j \le m$ (when $m = 1$, this amounts to shifting to the center of mass as before). Thus as our trial functions we take

$$\varphi_i = xu_i - \sum_{j=1}^{m} a_{ij} u_j \quad \text{for } 1 \le i \le m \tag{3.13}$$

where

$$a_{ij} \equiv \int_\Omega xu_i u_j = a_{ji} \tag{3.14}$$

are the components of xu_i along u_j for $1 \le j \le m$ and thus clearly $\varphi_i \perp u_j$ for $1 \le i, j \le m$. Also, it is straightforward to compute

$$\int_\Omega \varphi_i^2 = \int_\Omega \varphi_i x u_i = \int_\Omega x^2 u_i^2 - \sum_{j=1}^{m} a_{ij}^2 \tag{3.15}$$

and

$$-\Delta\varphi_i = \lambda_i x u_i - 2u_{ix} - \sum_{j=1}^{m} a_{ij}\lambda_j u_j \tag{3.16}$$

so that (since $\varphi_i \perp u_j$ for $1 \le j \le m$)

$$\int_\Omega \varphi_i(-\Delta\varphi_i) = \lambda_i \int_\Omega \varphi_i^2 - 2\int_\Omega \varphi_i u_{ix}. \tag{3.17}$$

It therefore follows from the Rayleigh-Ritz inequality for λ_{m+1} that

$$(\lambda_{m+1} - \lambda_i)\int_\Omega \varphi_i^2 \le -2\int_\Omega \varphi_i u_{ix}. \tag{3.18}$$

Proceeding much as before, we find

$$\begin{aligned} 0 \le -2\int_\Omega \varphi_i u_{ix} &= -2\int_\Omega u_{ix}\left[xu_i - \sum_{j=1}^{m} a_{ij}u_j\right] \\ &= -\int_\Omega x(u_1^2)_x + 2\sum_{j=1}^{m} a_{ij}\int_\Omega u_{ix}u_j \\ &= 1 + 2\sum_{i=1}^{m} a_{ij}b_{ij} \end{aligned} \tag{3.19}$$

where we have defined b_{ij} by

$$b_{ij} = \int_\Omega u_{ix} u_j. \tag{3.20}$$

At this point, we can both recognize the b_{ij}'s as the components of the u_{ix}'s along the u_j's and compute them explicitly in terms of the a_{ij}'s. It turns out that these observations are both important for us. (Also, b_{ij} is antisymmetric in i and j, as can be recognized from (3.20) immediately via integration by parts.) We first relate b_{ij} to a_{ij}:

$$\begin{aligned}
\lambda_i a_{ij} &= \int_\Omega (-\Delta u_i) x u_j \\
&= \int_\Omega u_i[-\Delta(x u_j)] \\
&= \int_\Omega u_i[\lambda_j x u_j - 2u_{jx}] \\
&= \lambda_j a_{ij} + 2\int_\Omega u_{ix} u_j \\
&= \lambda_j a_{ij} + 2b_{ij}
\end{aligned} \tag{3.21}$$

or

$$2b_{ij} = (\lambda_i - \lambda_j) a_{ij}. \tag{3.22}$$

Thus, from (3.19),

$$0 \le -2\int_\Omega \varphi_i u_{ix} = 1 + \sum_{j=1}^m (\lambda_i - \lambda_j) a_{ij}^2. \tag{3.23}$$

Furthermore, by the Cauchy-Schwarz inequality,

$$\begin{aligned}
(-2\int_\Omega \varphi_i u_{ix})^2 &= (-2\int_\Omega \varphi_i[u_{ix} - \sum_{j=1}^m b_{ij} u_j])^2 \\
&\le 4(\int_\Omega \varphi_i^2)(\int_\Omega [u_{ix} - \sum_{j=1}^m b_{ij} u_j]^2) \\
&= 4(\int_\Omega \varphi_i^2)[\int_\Omega u_{ix}^2 - \sum_{j=1}^m b_{ij}^2].
\end{aligned} \tag{3.24}$$

From (3.18) we now find

$$\begin{aligned}
(\lambda_{m+1} - \lambda_i)(\int_\Omega \varphi_i^2)(-2\int_\Omega \varphi_i u_{ix}) &\le (-2\int_\Omega \varphi_i u_{ix})^2 \\
&\le 4(\int_\Omega \varphi_i^2)[\int_\Omega u_{ix}^2 - \sum_{j=1}^m b_{ij}^2]
\end{aligned} \tag{3.25}$$

and hence, dividing by $\int_\Omega \varphi_i^2$,

$$(\lambda_{m+1} - \lambda_i)(-2\int_\Omega \varphi_i u_{ix}) \leq 4[\int_\Omega u_{ix}^2 - \sum_{j=1}^m b_{ij}^2]. \tag{3.26}$$

Inequality (3.26) holds even in the (unlikely) event that $\varphi_i \equiv 0$, since in that case clearly its left-hand side is 0 while its right-hand side is nonnegative (it is 4 times the square of the norm of $u_{ix} - \sum_{j=1}^m b_{ij}u_j$). Now, using (3.22) and (3.23), (3.26) can be put in the form

$$(\lambda_{m+1} - \lambda_i)[1 + \sum_{j=1}^m (\lambda_i - \lambda_j)a_{ij}^2] \leq 4\int_\Omega u_{ix}^2 - \sum_{j=1}^m (\lambda_i - \lambda_j)^2 a_{ij}^2 \tag{3.27}$$

or

$$\lambda_{m+1} - \lambda_i + \sum_{j=1}^m (\lambda_{m+1} - \lambda_j)(\lambda_i - \lambda_j)a_{ij}^2 \leq 4\int_\Omega u_{ix}^2. \tag{3.28}$$

From here it is clear how to finish our argument: we want to introduce a factor of $(\lambda_{m+1} - \lambda_i)$ to make the term involving a_{ij}^2 antisymmetric in i and j, and then sum on i from 1 to m to make this term drop out. We thus arrive at

$$\sum_{i=1}^m (\lambda_{m+1} - \lambda_i)^2 \leq 4\sum_{i=1}^m (\lambda_{m+1} - \lambda_i)\int_\Omega u_{ix}^2. \tag{3.29}$$

If we now recall that x here could be any $x_\ell (1 \leq \ell \leq n)$ and sum over ℓ from 1 to n we obtain

$$n\sum_{i=1}^m (\lambda_{m+1} - \lambda_i)^2 \leq 4\sum_{i=1}^m \lambda_i(\lambda_{m+1} - \lambda_i) \tag{3.30}$$

or

$$\sum_{i=1}^m (\lambda_{m+1} - \lambda_i)\left(\lambda_{m+1} - \left(1 + \frac{4}{n}\right)\lambda_i\right) \leq 0, \tag{3.31}$$

which is the main (or "first") inequality of Hong Cang Yang [105]. This completes our derivation of *Hong Cang Yang's first inequality*.

We now make a variety of comments about this inequality and the inequalities of Payne-Pólya-Weinberger and Hile-Protter. First, noting that the left-hand side of (3.31) is a quadratic in λ_{m+1}, we write it as

$$m\lambda_{m+1}^2 - 2\left(1 + \frac{2}{n}\right)\left(\sum_{i=1}^m \lambda_i\right)\lambda_{m+1} + \left(1 + \frac{4}{n}\right)\sum_{i=1}^m \lambda_i^2 \leq 0 \tag{3.32}$$

and derive the explicit upper bound

$$\lambda_{m+1} \leq [\text{ larger root of the quadratic }]$$
$$= \frac{1}{m}\left[\left(1+\frac{2}{n}\right)\sum_{i=1}^{m}\lambda_i + \left\{\left(1+\frac{2}{n}\right)^2\left(\sum_{i=1}^{m}\lambda_i\right)^2 - m\left(1+\frac{4}{n}\right)\sum_{i=1}^{m}\lambda_i^2\right\}^{1/2}\right]. \tag{3.33}$$

This bound is the best general upper bound yet derived by the methods of Payne, Pólya, and Weinberger (or any other methods, for that matter). As was already observed by H.C. Yang [105], the bound (3.33) is much sharper than previously known bounds for large m, since it comes much closer to incorporating the Weyl asymptotic behavior of the eigenvalues λ_i. For further comments and observations in this regard, see [52], [19], [9].

A simpler inequality due to H.C. Yang, *Yang's second inequality*, follows readily if we use the Cauchy-Schwarz inequality to replace $m\sum_{i=1}^{m}\lambda_i^2$ by $(\sum_{i=1}^{m}\lambda_i)^2$ on the right-hand side of (3.33):

$$\lambda_{m+1} \leq \left(1+\frac{4}{n}\right)\frac{1}{m}\sum_{i=1}^{m}\lambda_i \quad \text{for } m = 1, 2, \ldots. \tag{3.34}$$

This weaker inequality already implies the *Payne-Pólya-Weinberger inequality*, (3.2), since we can obtain (3.34) from (3.2) by replacing the λ_m occurring on the left-hand side of (3.2) by the average $\frac{1}{m}\sum_{i=1}^{m}\lambda_i$, which is clearly less than λ_m (and is, in fact, strictly less for all $m \geq 2$ since $\lambda_1 < \lambda_i$ for all $i \geq 2$).

To obtain *Hile and Protter's inequality* we go back into our proof of H.C. Yang's inequality and make one modification. In inequality (3.24), where we made use of the Cauchy-Schwarz inequality, we use it, but without incorporating the counterterms involving b_{ij}'s, arriving at

$$(-2\int_\Omega \varphi_i u_{ix})^2 \leq 4(\int_\Omega \varphi_i^2)(\int_\Omega u_{ix}^2). \tag{3.35}$$

This is how everyone had proceeded prior to H.C. Yang. No one had realized previously that one could make better (="optimal") use of the Cauchy-Schwarz inequality in this way (taking advantage of the known orthogonalities $\varphi_i \perp u_j$ for $1 \leq i, j \leq m$ and the fact that the b_{ij}'s have a simple relation to the a_{ij}'s). If our argument above is now carried through with (3.35) replacing (3.24), we arrive at

$$(\lambda_{m+1} - \lambda_i)\left[1 + \sum_{j=1}^{m}(\lambda_i - \lambda_j)a_{ij}^2\right] \leq 4\int_\Omega u_{ix}^2 \tag{3.36}$$

in place of (3.28).

From here the clear thing to do to eliminate the unwanted (because they're not easily controlled) terms in the a_{ij}^2's is to divide through by $\lambda_{m+1} - \lambda_i$ and sum on i from 1 to m, yielding

$$\sum_{i=1}^{m} \frac{\int_\Omega u_{ix}^2}{\lambda_{m+1} - \lambda_i} \geq \frac{m}{4}. \tag{3.37}$$

All the terms in a_{ij}^2 have dropped out due to antisymmetry. Finally, recalling that x could be any x_ℓ, $1 \leq \ell \leq n$, and summing on ℓ we arrive at

$$\sum_{i=1}^{m} \frac{\lambda_i}{\lambda_{m+1} - \lambda_i} \geq \frac{mn}{4}, \tag{3.38}$$

which is the *Hile-Protter inequality*. This inequality is stronger than the Payne-Pólya-Weinberger inequality (3.2), since if we replace the λ_i appearing in the denominator of the left-hand side of (3.38) by λ_m we obtain (3.2). It is also weaker than either of Yang's inequalities, (3.33) or (3.34). It therefore follows that (for each $m = 1, 2, \ldots$)

$$\text{Yang 1} \Rightarrow \text{Yang 2} \Rightarrow \text{Hile-Protter} \Rightarrow \text{PPW}. \tag{3.39}$$

While our derivation above of the Hile-Protter inequality certainly suggests that Yang's inequalities are stronger, it is not altogether straightforward to show the middle implication in (3.39). This is done in our longer paper [9], which is devoted to the topic of universal inequalities for the eigenvalues of the Dirichlet Laplacian. That paper also contains a discussion of when the various inequalities given above are known to hold strictly, as well as various other extensions and generalizations. A more complete set of references and some further comments on them will be found there as well.

For anyone worried by our statement following inequality (3.26) admitting the possibility that our trial functions φ_i might vanish identically and hence worried that our derivation might result in a triviality, we make two remarks to allay such fears. The first is that our proof as given takes account of these matters, and does indeed lead to a nontrivial inequality. The second, and probably more useful, observation is that it can never happen that all the φ_i's vanish identically (if indeed even one of them can vanish). To see this directly, it suffices to consider (3.23) and sum it on i from 1 to m. By antisymmetry the sum in a_{ij}^2 drops out, yielding $m = -2\sum_{i=1}^{m} \int_\Omega \varphi_i u_{ix}$ and showing that not all the φ_i's can vanish identically (or even each be orthogonal to u_{ix}). Thus at least some of the inequalities represented by (3.26) will be nontrivial and will lead to a nontrivial inequality (3.30) as the outcome of our final summations.

3.2 Universal Inequalities for Low Fixed Membrane Eigenvalues

We next turn to some universal inequalities for low Dirichlet eigenvalues, which again stem from the original work of Payne, Pólya, and Weinberger [83]. In 1956 they proved that for $\Omega \subset \mathbb{R}^2$

$$\frac{\lambda_2 + \lambda_3}{\lambda_1} \leq 6. \tag{3.40}$$

This easily extends to $\Omega \subset \mathbb{R}^n$ as

$$\frac{\lambda_2 + \lambda_3 + \ldots + \lambda_{n+1}}{\lambda_1} \leq n\left(1 + \frac{4}{n}\right) = n + 4. \tag{3.41}$$

These inequalities are proved by using the Rayleigh-Ritz inequality in "trace form" or as usually applied, using rotations and translations to enforce the further orthogonalities needed. No a_{ij}'s appear (or can be tolerated) in these arguments. There are also a variety of extensions of results of this type, the simplest being that of Brands [28] from 1964

$$\frac{\lambda_2 + \lambda_3}{\lambda_1} \leq 5 + \frac{\lambda_1}{\lambda_2} \quad \text{for } \Omega \subset \mathbb{R}^2 \tag{3.42}$$

and its extension to $\mathbb{R}^n$

$$\frac{\lambda_2 + \lambda_3 + \ldots + \lambda_{n+1}}{\lambda_1} \leq n + 3 + \frac{\lambda_2}{\lambda_1} \quad \text{for } \Omega \subset \mathbb{R}^n \tag{3.43}$$

(see [12] for a derivation, though this extension was certainly known to Hile and Protter earlier [53]).

Beyond this, much work has been done toward bounding the range of values of $(\lambda_2/\lambda_1, \lambda_3/\lambda_1)$ for an arbitrary domain $\Omega \subset \mathbb{R}^2$ (with corresponding, but less extensive, work for $\Omega \subset \mathbb{R}^n$). In particular, it would be very interesting to know the best bounds for λ_3/λ_1 and $(\lambda_2 + \lambda_3)/\lambda_1$ (the best bound for λ_2/λ_1 is, of course, its value for a disk, given by (2.16)). These quantities can be gotten at by looking at the range of possible values of $(\lambda_2/\lambda_1, \lambda_3/\lambda_1)$, which is one motivation for its study.

The current state of knowledge regarding the range of values of $(\lambda_2/\lambda_1, \lambda_3/\lambda_1)$ is summarized in the paper of Ashbaugh and Benguria [19] (see also its precursors [12] and [15]). In particular, one should consult the graph given as Figure 1 on p. 38 of [19]. It is shown in [19] that

$$5.077^{+} \leq \sup_{\Omega} \frac{\lambda_2 + \lambda_3}{\lambda_1} \leq 5.50661^{-} \tag{3.44}$$

and that

$$3.1818^+ \le \sup_\Omega \frac{\lambda_3}{\lambda_1} \le 3.83103^-. \tag{3.45}$$

The lower bound in (3.44) is simply the value for a disk. It is a conjecture of Payne, Pólya, and Weinberger that this is the actual maximum value of $(\lambda_2 + \lambda_3)/\lambda_1$ as well. The lower bound in (3.45) is 35/11, the value taken by λ_3/λ_1 for a $\sqrt{8}$ by $\sqrt{3}$ rectangle (and is certainly the maximum of λ_3/λ_1 among rectangles). However, there is no current guess as to the precise shape of domain that will maximize λ_3/λ_1 (if a maximizer even exists). The best current thinking would have it be an elongated convex figure, roughly in the shape of an oval or ellipse. (Among ellipses the largest value of λ_3/λ_1 seems to be very near to, but slightly less than, 3.1818.)

3.3 Universal Eigenvalue Inequalities in Other Spaces

There are also versions of the Payne-Pólya-Weinberger inequality (3.2) and its extensions for bounded domains in the constant curvature spaces S^n and H^n. For example, for $\Omega \subset S^2$ in 1975 Cheng [33] derived the bound

$$\frac{\lambda_2}{\lambda_1} \le 1 + 2\left(\frac{2}{1+\cos\Theta}\right)^4 \quad \text{for } 0 < \Theta < \pi \tag{3.46}$$

where Θ is defined as the outradius (=geodesic radius of the circumscribing circle) of $\Omega \subset S^2$. This was improved to

$$\frac{\lambda_2}{\lambda_1} \le 1 + 2\left(\frac{2}{1+\cos\Theta}\right)^2 \quad \text{for } 0 < \Theta < \pi \tag{3.47}$$

by Harrell [49] in 1993. Both of these results have extensions to higher eigenvalues and to S^n.

On a slightly different front, by an extension of the method they used to prove the Payne-Pólya-Weinberger conjecture for λ_2/λ_1, Ashbaugh and Benguria proved [11], [16] (see Section 5) that for $\Omega \subset S^2$

$$\frac{\lambda_2}{\lambda_1} \le 1 + 1.5387\left(\frac{\cdot 2}{1+\cos\Theta}\right)^2 \quad \text{for } 0 < \Theta < \pi. \tag{3.48}$$

The constant here is $j_{1,1}^2/j_{0,1}^2 - 1$ and comes from the proof of the Euclidean PPW conjecture as extended to general second-order elliptic operators in Section 4 of [11].

Beyond this, there are the sharp PPW results for Ω contained in a hemisphere of S^n,

$$\lambda_2(\Omega) \le \lambda_2(B_{\lambda_1}) \tag{3.49}$$

where B_{λ_1} is the geodesic ball in S^n having the same value of λ_1 as Ω, and

$$\frac{\lambda_2(\Omega)}{\lambda_1(\Omega)} \leq \frac{\lambda_2(\Omega^\star)}{\lambda_1(\Omega^\star)} \tag{3.50}$$

where $\Omega^\star$ is the geodesic ball in S^n having the same measure as Ω. These last two results are due to Ashbaugh and Benguria (see [16], [20], [21]). Note that (3.49) and (3.50) are much sharper than the bounds in terms of Θ listed previously, since the outradius of Ω is a much cruder measure of the size of Ω than either $\lambda_1(\Omega)$ or $|\Omega|$. On the other hand, the bounds in terms of Θ apply to all domains Ω, not just to those contained in a hemisphere. While it may be possible that (3.49) and/or (3.50) hold beyond the hemisphere, this is not proved as yet. It might also be mentioned that (3.50) follows easily from (3.49) since it can be shown that λ_2/λ_1 for a geodesic ball is an increasing function of its radius. See [16], [20], and [21] for more details and discussion.

Further results for domains Ω contained in S^n, H^n, or other more general manifolds are due to Cheng [33] in 1975, Li [70] in 1980, P.C.Yang and S.-T. Yau [106] in 1980, Lee [66] and Leung [69] in 1991, Harrell [49] in 1993, and Harrell and Michel [50],[51] in 1994 and 1995. Some of these papers establish analogs of the Hile-Protter inequality in a more general setting. More discussion of this literature will be found in [9]. We mention, in particular, the Hile-Protter-type bound for a domain in S^n

$$\sum_{i=1}^{m} \frac{4\lambda_i + n^2}{\lambda_{m+1} - \lambda_i} \geq mn. \tag{3.51}$$

For $m = 1$ this bound gives us

$$\lambda_2 \leq (1 + \frac{4}{n})\lambda_1 + n. \tag{3.52}$$

(Indeed, one has, more generally, the PPW-type bound

$$\lambda_{m+1} \leq \lambda_m + \frac{4}{mn}\sum_{i=1}^{m} \lambda_i + n \leq (1 + \frac{4}{n})\lambda_m + n.) \tag{3.53}$$

This bound can be regarded as an alternative and quite natural bound for λ_2 in terms of λ_1 for domains $\Omega \subset S^n$ (to be compared with, say, (3.48), at least when $n = 2$). Note that this bound "takes account of" the blow-up of λ_2/λ_1 that we expect when we approach the whole sphere S^n by the presence of the constant term n on its right-hand side. Thus, even though $\lambda_1 \to 0$ as we approach the whole sphere while $\lambda_2 \to n$, this new bound gives us control of λ_2 without the need of a coefficient that blows up in this limit. Moreover, the new bound $\lambda_2 \leq (1 + \frac{4}{n})\lambda_1 + n$ is seen to be sharp in this limit (viz., for the full sphere $\lambda_2 = n$, while the right-hand side becomes n since $\lambda_1 = 0$). The bound (3.51) is actually the "HP-weakening" of a stronger Yang-style bound for $\Omega \subset S^n$ first given in this context by H. C. Yang [105] (in the 1995 version of this paper). For a full discussion of Yang's bounds and related inequalities (such as (3.51)) in this context, see [9]. Weaker, but generally more complex, precursors may be found in [33], [50], [69], and [106] (most of these are directed at compact minimal hypersurfaces in S^n, but the ideas and calculations are much the same).

3.4 Universal Eigenvalue Inequalities for the Vibrating Clamped Plate

In their 1956 paper [83], Payne, Pólya, and Weinberger also established the bound

$$\Gamma_{m+1} - \Gamma_m \leq \frac{8}{m}(\Gamma_1 + \ldots + \Gamma_m), \quad m = 1, 2, \ldots, \tag{3.54}$$

for the eigenvalues Γ_i of the clamped plate problem (1.5) on a domain $\Omega \subset \mathbb{R}^2$. This is the analog for the clamped plate of their bound (3.1) for the fixed membrane. For $\Omega \subset \mathbb{R}^n$ this bound becomes

$$\Gamma_{m+1} - \Gamma_m \leq \frac{8(n+2)}{n^2 m}(\Gamma_1 + \ldots + \Gamma_m), \quad m = 1, 2, \ldots, \tag{3.55}$$

and, even better,

$$\Gamma_{m+1} - \Gamma_m \leq \frac{8(n+2)}{n^2 m^2}(\Gamma_1^{1/2} + \ldots + \Gamma_m^{1/2})^2, \quad m = 0, 1, \ldots. \tag{3.56}$$

This latter inequality was not given by Payne, Pólya, and Weinberger, even for $n = 2$, but it certainly could have been, as it is in a sense implicit in their work. Beyond this, in 1984 Hile and Yeh [54], extending the approach of Hile and Protter to the clamped plate problem, established the bound

$$\sum_{i=1}^{m} \frac{\Gamma_i^{1/2}}{\Gamma_{m+1} - \Gamma_i} \geq \frac{n^2 m^{3/2}}{8(n+2)} \left(\sum_{i=1}^{m} \Gamma_i \right)^{-1/2}, \quad m = 1, 2, \ldots. \tag{3.57}$$

Again, implicit in their work is the better bound

$$\frac{n^2 m^2}{8(n+2)} \leq \left(\sum_{i=1}^{m} \frac{\Gamma_i^{1/2}}{\Gamma_{m+1} - \Gamma_i} \right) \left(\sum_{i=1}^{m} \Gamma_i^{1/2} \right), \quad m = 1, 2, \ldots, \tag{3.58}$$

which was exhibited explicitly only later. In fact Hook [55] in 1990 established (3.58) as a strict inequality. Also in 1990, Chen and Qian [32] independently stated and proved (3.58).

Possibly (3.58) could be improved to

$$\left(\sum_{i=1}^{m}\sqrt{\frac{\Gamma_i}{\Gamma_{m+1}-\Gamma_i}}\right)^2 \geq \frac{n^2m^2}{8(n+2)}, \quad m=1,2,\ldots. \tag{3.59}$$

This inequality would imply all the previous ones in this subsection ((3.58), for example, would follow easily using the Cauchy-Schwarz inequality).

It might be noted that the quantity $8(n+2)/n^2$ that appears in the foregoing inequalities is indeed natural from the PPW/HP point of view, just as $4/n$ is in the case of the fixed membrane. This is because $8(n+2)/n^2$ arises as $(1+4/n)^2-1$ just as $4/n$ arises as $(1+4/n)-1$.

Yet another inequality in this vein is

$$\sum_{i=1}^{m}\frac{\Gamma_i}{\Gamma_{m+1}-\Gamma_i} \geq \frac{n^2m}{8(n+2)}, \quad m=1,2,\ldots. \tag{3.60}$$

This inequality is an easy consequence of (3.58) via Chebyshev's inequality (it would also follow directly from (3.59) via the Cauchy-Schwarz inequality). Its chief appeal is its simplicity. It should be mentioned that as yet no one has established an analog of H.C. Yang's bound (3.31) (or equivalently (3.33)) in this setting, i.e., for the eigenvalues of the vibrating clamped plate. So far (3.58), which is best regarded as an analog of the Hile-Protter inequality (3.38) in this setting, is the closest anyone has come.

3.5 Universal Eigenvalue Inequalities for the Buckling Problem

For the buckling problem for a clamped plate (problem (1.7)) even less is known, largely owing to the fact that the inner product one must employ for this problem, $\langle f,g\rangle \equiv \int_\Omega \nabla f\cdot\nabla g$, does not induce a symmetric a_{ij} matrix when one attempts the usual PPW approach and sets $a_{ij}=\langle xv_i,v_j\rangle$. This leads to all sorts of extra complications, and thus far no one has been able to bring the general case (for $\Lambda_{m+1}-\Lambda_i$ or even just $\Lambda_{m+1}-\Lambda_m$) to a satisfactory conclusion. Payne, Pólya, and Weinberger [83] were able to establish the low eigenvalue result

$$\Lambda_2/\Lambda_1 < 3 \quad \text{for } \Omega\subset\mathbb{R}^2. \tag{3.61}$$

For $\Omega\subset\mathbb{R}^n$ this reads

$$\Lambda_2/\Lambda_1 < 1+4/n. \tag{3.62}$$

Subsequently Hile and Yeh [54] reconsidered this problem obtaining the improved bound

$$\Lambda_2/\Lambda_1 \leq 2.5 \quad \text{for } \Omega \subset \mathbb{R}^2 \tag{3.63}$$

and, in general,

$$\frac{\Lambda_2}{\Lambda_1} \leq \frac{n^2+8n+20}{(n+2)^2} \quad \text{for } \Omega \subset \mathbb{R}^n. \tag{3.64}$$

Both of these inequalities can actually be shown to hold as strict inequalities. It is of note that Λ_2/Λ_1 for a disk in 2 dimensions is 1.796. If the analog of the PPW conjecture held for this problem, then this would be the best possible upper bound for Λ_2/Λ_1 in 2 dimensions.

3.6 More Inequalities for the Low Eigenvalues of the Clamped Plate and Buckling Problems

One can also derive inequalities analogous to (3.41) for the clamped plate and buckling problems. These read

$$\frac{\Lambda_2+\ldots+\Lambda_{n+1}}{\Lambda_1} \leq n+4 \quad \text{for } \Omega \subset \mathbb{R}^n, \tag{3.65}$$

$$\frac{\Gamma_2^{1/2}+\ldots+\Gamma_{n+1}^{1/2}}{\Gamma_1^{1/2}} \leq n+4 \quad \text{for } \Omega \subset \mathbb{R}^n, \tag{3.66}$$

and

$$\frac{\Gamma_2+\ldots+\Gamma_{n+1}}{\Gamma_1} \leq n+24 \quad \text{for } \Omega \subset \mathbb{R}^n. \tag{3.67}$$

Of the last two inequalities, (3.66) is the natural "PPW-analog" and (3.67) is a weaker inequality that derives from it. Note the presence of the PPW factor $\left(1+\frac{4}{n}\right)$ in (3.65) and (3.66) (in the form $n+4 = n\left(1+\frac{4}{n}\right)$). The constant $n+24$ in (3.67) comes from the extreme case of (3.66) where $\Gamma_i^{1/2}/\Gamma_1^{1/2} = 1$ for $2 \leq i \leq n$ and $\Gamma_{n+1}^{1/2}/\Gamma_1^{1/2} = 5$.

Finally, we mention the known results for Γ_2/Γ_1. Obviously (from (3.55)), Payne, Pólya, and Weinberger had

$$\Gamma_2/\Gamma_1 \leq (1+4/n)^2 \quad \text{for } \Omega \subset \mathbb{R}^n \tag{3.68}$$

(and even $\Gamma_{m+1}/\Gamma_1 \leq (1+4/n)^2$ for all m). This was improved upon more recently by fairly elaborate means by Hile and Yeh [54], who, for example, obtained the bounds 7.103 and 4.792 for dimensions 2 and 3 respectively. In general, these upper bounds are determined as the unique root larger than 1 of the cubic

$$(x-1)^3 = \frac{512}{n^2(n+2)}x \tag{3.69}$$

(though Hile and Yeh formulated their result in rather different terms). These values might be compared to those of the ball in 2 and 3 dimensions: 4.3311 and 3.2390, respectively. Again the analog of the PPW conjecture for this problem would project these values as the optimal upper bounds for Γ_2/Γ_1 in 2 and 3 dimensions.

As a final remark, we note that any result mentioned in this section (Section 3) without specific attribution is due to the author and that fuller details will be presented in forthcoming papers.

4 Concluding Remarks and Open Problems

In this concluding section we mention a few open problems and give some further hints to the literature.

Of the original conjectures of Payne, Pólya, and Weinberger [83], only two remain open. These are to show that

$$\frac{\lambda_2+\lambda_3}{\lambda_1} \le \left.\frac{\lambda_2+\lambda_3}{\lambda_1}\right|_{\text{disk}} \approx 5.077 \text{ for } \Omega \subset \mathbb{R}^2, \tag{4.1}$$

and the analogous result for $(\lambda_2 + \ldots + \lambda_{n+1})/\lambda_1$ for $\Omega \subset \mathbb{R}^n$, and that

$$\frac{\lambda_{m+1}}{\lambda_m} \le \left.\frac{\lambda_2}{\lambda_1}\right|_{\text{ball}} \quad \text{for all } \Omega \subset \mathbb{R}^n. \tag{4.2}$$

The latter inequality is known for $m = 1, 2$, and 3 (for $m = 2$ and 3 it follows from the stronger inequality $\lambda_4/\lambda_2 \le (\lambda_2/\lambda_1)|_{\text{ball}}$ proved in [14]), but as yet all higher cases remain open. For further discussion of these problems see [12], [16], and [8] (as well as [14]).

There are also the well-known Pólya conjectures for the Dirichlet and Neumann eigenvalues of the Laplacian on a domain $\Omega \subset \mathbb{R}^n$. For the case of dimension 2, and with notation as in Section 1, these read

$$\lambda_k \ge \frac{4\pi k}{A} \quad \text{for } k = 1, 2, \ldots \tag{4.3}$$

and

$$\mu_k \le \frac{4\pi k}{A} \quad \text{for } k = 0, 1, 2, \ldots, \tag{4.4}$$

where $A = |\Omega|$. There are analogous conjectures for all dimensions $n > 2$. See [16] for their statements and for further discussion.

For the vibrating clamped plate problem there remains the Rayleigh conjecture

$$\Gamma_1(\Omega) \geq \Gamma_1(\Omega^\star) \quad \text{for } \Omega \subset \mathbb{R}^n \tag{4.5}$$

for all $n \geq 4$, and, if one could prove (4.5) (or in any event for $n = 2, 3$), the PPW-type conjecture

$$\frac{\Gamma_2}{\Gamma_1} \leq \left.\frac{\Gamma_2}{\Gamma_1}\right|_{\text{ball}} \quad \text{for all } \Omega \subset \mathbb{R}^n. \tag{4.6}$$

Similarly, for the buckling problem for the clamped plate there remains the Pólya-Szegő conjecture

$$\Lambda_1(\Omega) \geq \Lambda_1(\Omega^\star) \quad \text{for } \Omega \subset \mathbb{R}^n \tag{4.7}$$

for all $n \geq 2$. If this conjecture could be proved then one could also consider the following conjecture for ratios:

$$\frac{\Lambda_2}{\Lambda_1} \leq \left.\frac{\Lambda_2}{\Lambda_1}\right|_{\text{ball}} \quad \text{for all } \Omega \subset \mathbb{R}^n. \tag{4.8}$$

Obviously many other problems could be formulated and investigated. For example, one could consider the ratios Λ_{m+1}/Λ_m, Γ_{m+1}/Γ_m, $(\Lambda_2 + \Lambda_3)/\Lambda_1$, $(\Gamma_2 + \Gamma_3)/\Gamma_1$, and many other combinations analogous to those that have been considered for the eigenvalues of the Dirichlet Laplacian. One could also consider much of what has been discussed in this paper in the more general setting of domains in Riemannian manifolds or for general second-order elliptic operators.

Other more extensive problem lists occur in the review papers of Payne [79], [80], [81], and in Yau's recent problem lists [107], [108] (reprinted in [89]). In addition, one could consult the final section of [16] and also [8].

Acknowledgements

The author is grateful to Brian Davies and Yuri Safarov for the opportunity to participate in the Instructional Conference on Spectral Theory and Geometry in Edinburgh (March 29-April 9, 1998) and to the International Centre for Mathematical Sciences for its generous support. In addition, he gratefully acknowledges his collaborators, Rafael Benguria and Richard Laugesen, with whom a number of the results summarized here were obtained.

References

[1] Abramowitz, M., and I.A. Stegun, editors, **Handbook of Mathematical Functions**, National Bureau of Standards Applied Mathematics Series, vol. **55**, U.S. Government Printing Office, Washington, D.C., 1964.

[2] Adams, R.A., **Sobolev Spaces**, Academic Press, New York, 1975.

[3] Alvino, A., J. I. Diaz, P.-L. Lions, and G. Trombetti, Équations elliptiques et symétrisation de Steiner, Comptes Rendus Acad. Sci. Paris (Ser. I) **314** (1992), 1015-1020.

[4] Alvino, A., P.-L. Lions, and G. Trombetti, A remark on comparison results via symmetrization, Proc. Roy. Soc. Edinburgh **102A** (1986), 37-48.

[5] Alvino, A., P.-L. Lions, and G. Trombetti, Comparison results for elliptic and parabolic equations via Schwarz symmetrization, Ann. Inst. H. Poincaré **7** (1990), 37-65.

[6] Alvino, A., P.-L. Lions, and G. Trombetti, Comparison results for elliptic and parabolic equations via symmetrization: A new approach, Diff. and Integral Eqs. **4** (1991), 25-50.

[7] Alvino, A., G. Trombetti, J. I. Diaz, and P.-L. Lions, Elliptic equations and Steiner symmetrization, Commun. Pure Appl. Math. **49** (1996), 217-236.

[8] Ashbaugh, M.S., Open problems on eigenvalues of the Laplacian, **Analytic and Geometric Inequalities and Applications**, Th. M. Rassias and H.M. Srivastava, editors, Kluwer Academic Publishers, Dordrecht, The Netherlands, to appear.

[9] Ashbaugh, M.S., The universal eigenvalue bounds of Payne-Pólya-Weinberger, Hile-Protter, and H.C. Yang, in preparation.

[10] Ashbaugh, M.S., and R.D. Benguria, Proof of the Payne-Pólya-Weinberger conjecture, Bull. Amer. Math. Soc. **25** (1991), 19-29.

[11] Ashbaugh, M.S., and R.D. Benguria, A sharp bound for the ratio of the first two eigenvalues of Dirichlet Laplacians and extensions, Ann. Math. **135** (1992), 601-628.

[12] Ashbaugh, M.S., and R.D. Benguria, More bounds on eigenvalue ratios for Dirichlet Laplacians in n dimensions, SIAM J. Math. Anal. **24** (1993), 1622-1651.

[13] Ashbaugh, M.S., and R.D. Benguria, A second proof of the Payne-Pólya-Weinberger conjecture, Commun. Math. Phys. **147** (1992), 181-190.

[14] Ashbaugh, M.S., and R.D. Benguria, Isoperimetric bounds for higher eigenvalue ratios for the n-dimensional fixed membrane problem, Proc. Roy. Soc. Edinburgh **123A** (1993), 977-985.

[15] Ashbaugh, M.S., and R.D. Benguria, The range of values of λ_2/λ_1 and λ_3/λ_1 for the fixed membrane problem, Rev. Math. Phys. **6** (1994), 999-1009 (in a special issue dedicated to Elliott H. Lieb). [Also in: **The State of Matter: A Volume Dedicated to E.H. Lieb (Copenhagen, 1992)**, M. Aizenman and H. Araki, editors, Advanced Series in Mathematical Physics, vol. **20**, World Scientific, Singapore, 1994, pp. 167-181.]

[16] Ashbaugh, M.S., and R.D. Benguria, Isoperimetric inequalities for eigenvalue ratios, **Partial Differential Equations of Elliptic Type, Cortona, 1992**, A. Alvino, E. Fabes, and G. Talenti, editors, Symposia Mathematica, vol. **35**, Cambridge University Press, Cambridge, 1994, pp. 1-36.

[17] Ashbaugh, M.S., and R.D. Benguria, Sharp upper bound to the first nonzero Neumann eigenvalue for bounded domains in spaces of constant curvature, J. London Math. Soc. (2) **52** (1995), 402-416.

[18] Ashbaugh, M.S., and R.D. Benguria, On Rayleigh's conjecture for the clamped plate and its generalization to three dimensions, Duke Math. J. **78** (1995), 1-17.

[19] Ashbaugh, M.S., and R.D. Benguria, Bounds for ratios of the first, second, and third membrane eigenvalues, **Nonlinear Problems in Applied Mathematics, in Honor of Ivar Stakgold on his Seventieth Birthday**, T.S. Angell, L. Pamela Cook, R.E. Kleinman, and W.E. Olmstead, editors, Society for Industrial and Applied Mathematics, Philadelphia, Pennsylvania, 1996, pp. 30-42.

[20] Ashbaugh, M.S., and R.D. Benguria, On the Payne-Pólya-Weinberger conjecture on the n-dimensional sphere, **General Inequalities 7 (Oberwolfach, 1995)**, C. Bandle, W.N. Everitt, L. Losonczi, and W. Walter, editors, International Series of Numerical Mathematics, vol. **123**, Birkhäuser, Basel, 1997, pp. 111-128.

[21] Ashbaugh, M.S., and R.D. Benguria, A sharp bound for the ratio of the first two Dirichlet eigenvalues of a domain in a hemisphere of S^n, Trans. Amer. Math. Soc., to appear.

[22] Ashbaugh, M.S., R.D. Benguria, and R.S. Laugesen, Inequalities for the first eigenvalues of the clamped plate and buckling problems, **General Inequalities 7 (Oberwolfach 1995)**, C. Bandle, W.N. Everitt, L. Losonczi, and W. Walter, editors, International Series of Numerical Mathematics, vol. **123**, Birkhäuser, Basel, 1997, pp. 95-110.

[23] Ashbaugh, M.S., and R.S. Laugesen, Fundamental tones and buckling loads of clamped plates, Ann. Scuola Norm. Sup. Pisa Cl. Sci. (4) **23** (1996), 383-402.

[24] Bandle, C., **Isoperimetric Inequalities and Applications**, Pitman Monographs and Studies in Mathematics, vol. **7**, Pitman, Boston, 1980.

[25] Baernstein, A., II, A unified approach to symmetrization, **Partial Differential Equations of Elliptic Type, Cortona, 1992**, A. Alvino, E. Fabes, and G. Talenti, editors, Symposia Mathematics, vol. **35**, Cambridge University Press, Cambridge, 1994, pp. 47-91.

[26] Bérard, P.H., **Spectral Geometry: Direct and Inverse Problems** (with an Appendix by G. Besson), Lect. Notes in Math., vol. **1207**, Springer-Verlag, Berlin, 1986.

[27] Bramble, J.H., and L.E. Payne, Pointwise bounds in the first biharmonic boundary value problem, J. Math. and Phys. **42** (1963), 278-286.

[28] Brands, J.J.A.M., Bounds for the ratios of the first three membrane eigenvalues, Arch. Rational Mech. Anal. **16** (1964), 265-268.

[29] Brezis, H., **Analyse Fonctionnelle: Théorie et applications**, Masson, Paris, 1983.

[30] Chavel, I., **Eigenvalues in Riemannian Geometry**, Academic Press, New York, 1984.

[31] Chavel, I, **Riemannian Geometry: A Modern Introduction**, Cambridge University Press, Cambridge, 1993.

[32] Chen, Z.-C., and C.L. Qian, Estimates for discrete spectrum of Laplacian operator with any order, J. China Univ. Sci. Tech. **20** (1990), 259-266.

[33] Cheng, S.-Y., Eigenfunctions and eigenvalues of Laplacian, Proc. Symp. Pure Math., vol. **27**, part 2, **Differential Geometry**, S.S. Chern and R. Osserman, editors, American Mathematical Society, Providence, Rhode Island, 1975, pp. 185-193.

[34] Chiti, G., Norme di Orlicz delle soluzioni di una classe di equazioni ellitiche, Boll. Un. Mat. Ital. (5) **16-A** (1979), 178-185.

[35] Chiti, G., A reverse Hölder inequality for the eigenfunctions of linear second order elliptic operators, J. Appl. Math. and Phys. (ZAMP) **33** (1982), 143-148.

[36] Chiti, G., An isoperimetric inequality for the eigenfunctions of linear second order elliptic operators, Boll. Un. Mat. Ital. (6) **1-A** (1982), 145-151.

[37] Chiti, G., A bound for the ratio of the first two eigenvalues of a membrane, SIAM J. Math. Anal. **14** (1983), 1163-1167.

[38] Courant, R., and D. Hilbert, **Methods of Mathematical Physics**, vol. **1**, Interscience Publishers, Wiley, New York, 1953.

[39] Davies, E.B., **Spectral Theory and Differential Operators**, Cambridge Studies in Advanced Mathematics, vol. **42**, Cambridge University Press, Cambridge, 1995.

[40] Davies, E. B., L^p spectral theory of higher-order elliptic differential operators, Bull. London Math. Soc. **29** (1997), 513-546.

[41] Duffin, R.J., Nodal lines of a vibrating plate, J. Math. and Phys. **31** (1953), 294-299.

[42] Duffin, R.J., and D.H. Shaffer, On the modes of vibration of a ring-shaped plate, Bull. Amer. Math. Soc. **58** (1952), 652.

[43] Faber, G., Beweis, dass unter allen homogenen Membranen von gleicher Fläche und gleicher Spannung die kriesförmige den tiefsten Grundton gibt, Sitzungberichte der mathematisch-physikalischen Klasse der Bayerischen Akademie der Wissenschaften zu München Jahrgang, 1923, pp. 169-172.

[44] Friedland, S. and W.K. Hayman, Eigenvalue inequalities for the Dirichlet problem on spheres and the growth of subharmonic functions, Comment. Math. Helvetici **51** (1976), 133-161.

[45] Glaser, V., A. Martin, H. Grosse, and W. Thirring, A family of optimal conditions for the absence of bound states in a potential, **Studies in Mathematical Physics: Essays in Honor of Valentine Bargmann**, E. H. Lieb, B. Simon, and A. S. Wightman, editors, Princeton University Press, Princeton, New Jersey, 1976, pp. 169-194.

[46] Gould, S.H., **Variational Methods for Eigenvalue Problems: An Introduction to the Weinstein Method of Intermediate Problems**, second edition, revised and enlarged, University of Toronto Press, Mathematical Expositions, Number **10**, Toronto, 1966.

[47] Gunson, J., Inequalities in mathematical physics, **Inequalities: Fifty Years On from Hardy, Littlewood, and Pólya**, W. N. Everitt, editor, Marcel Dekker, New York, 1991, pp. 53-79 (see especially Section 5, pp. 70-74).

[48] Hardy, G.H., J.E. Littlewood, and G. Pólya, **Inequalities**, second edition, Cambridge University Press, Cambridge, 1952.

[49] Harrell, E.M., II, Some geometric bounds on eigenvalue gaps, Commun. Partial Diff. Eqs. **18** (1993), 179-198.

[50] Harrell, E.M., II, and P.L. Michel, Commutator bounds for eigenvalues, with applications to spectral geometry, Commun. Partial Diff. Eqs. **19** (1994), 2037-2055.

[51] Harrell, E.M., II, and P.L. Michel, Commutator bounds for eigenvalues of some differential operators, **Evolution Equations**, G. Ferreyra, G.R. Goldstein, and F. Neubrander, editors, Lecture Notes in Pure and Applied Mathematics, vol. **168**, Marcel Dekker, New York, 1995, pp. 235-244.

[52] Harrell, E.M., II, and J. Stubbe, On trace identities and universal eigenvalue estimates for some partial differential operators, Trans. Amer. Math. Soc. **349** (1997), 1797-1809.

[53] Hile, G.N., and M.H. Protter, Inequalities for eigenvalues of the Laplacian, Indiana Univ. Math. J. **29** (1980), 523-538.

[54] Hile, G.N., and R.Z. Yeh, Inequalities for eigenvalues of the biharmonic operator, Pac. J. Math. **112** (1984), 115-133.

[55] Hook, S.M., Domain-independent upper bounds for eigenvalues of elliptic operators, Trans. Amer. Math. Soc. **318** (1990), 615-642.

[56] Kawohl, B., **Rearrangements and Convexity of Level Sets in PDE**, Lect. Notes in Math., vol. **1150**, Springer-Verlag, Berlin, 1985.

[57] Kesavan, S., Some remarks on a result of Talenti, Ann. Scuola Norm. Sup. Pisa (4) **15** (1988), 453-465.

[58] Kesavan, S., **Topics in Functional Analysis and Applications**, Wiley, New York, 1989.

[59] Kesavan, S., On a comparison theorem via Schwarz symmetrization, Proc. Roy. Soc. Edinburgh **119A** (1991), 159-167.

[60] Kesavan, S., Comparison theorems via Schwarz symmetrization—a survey, **Partial Differential Equations of Elliptic Type, Cortona, 1992**, A. Alvino, E. Fabes, and G. Talenti, editors, Symposia Mathematica, vol. **35**, Cambridge University Press, Cambridge, 1994, pp. 185-196.

[61] Kesavan, S., Comparison theorems via symmetrization: revisited, Boll. Un. Mat. Ital. (7) **11-A** (1997), 163-172.

[62] Kornhauser, E.T., and I. Stakgold, A variational theorem for $\nabla^2 u + \lambda u = 0$ and its application, J. Math. and Phys. **31** (1952), 45-54.

[63] Krahn, E., Über eine von Rayleigh formulierte Minimaleigenschaft des Kreises, Math. Ann. **94** (1925), 97-100.

[64] Krahn, E., Über Minimaleigenschaften der Kugel in drei und mehr Dimensionen, Acta Comm. Univ. Tartu (Dorpat) **A9** (1926), 1-44. [English translation: Minimal properties of the sphere in three and more dimensions, **Edgar Krahn 1894-1961: A Centenary Volume**, Ü. Lumiste and J. Peetre, editors, IOS Press, Amsterdam, 1994, Chapter 11, pp. 139-174.]

[65] Kozlov, V. A., V. A. Kondrat'ev, and V. G. Maz'ya, On sign variation and the absence of "strong" zeros of solutions of elliptic equations, Izv. Akad. Nauk SSSR, Ser. Mat. **53** (1989), 328-344 (in Russian) [English translation in Math. USSR-Izv. **34** (1990), 337-353].

[66] Lee, J.M., The gaps in the spectrum of the Laplace-Beltrami operator, Houston J. Math. **17** (1991), 1-24.

[67] Leis, R., **Initial Boundary Value Problems in Mathematical Physics**, B. G. Teubner, Stuttgart, and John Wiley and Sons, Chichester, 1986.

[68] Leissa, A.W., **Vibration of Plates**, National Aeronautics and Space Administration (NASA SP-160), U.S. Government Printing Office, Washington, D.C., 1969.

[69] Leung, P.-F., On the consecutive eigenvalues of the Laplacian of a compact minimal submanifold in a sphere, J. Austral. Math. Soc. (Series A) **50** (1991), 409-416.

[70] Li, P., Eigenvalue estimates on homogeneous manifolds, Comment. Math. Helvetici **55** (1980), 347-363.

[71] Lieb, E.H., Existence and uniqueness of the minimizing solution of Choquard's nonlinear equation, Stud. Appl. Math. **57** (1977), 93-105.

[72] Lieb, E.H., Sharp constants in the Hardy-Littlewood-Sobolev and related inequalities, Ann. Math. **118** (1993), 349-374.

[73] Lieb, E.H., and M. Loss, **Analysis**, Graduate Studies in Mathematics, vol. **14**, American Mathematical Society, Providence, Rhode Island, 1997.

[74] Nadirashvili, N.S., An isoperimetric inequality for the principal frequency of a clamped plate, Dokl. Akad. Nauk **332** (1993), 436-439 (in Russian) [English translation in Phys. Dokl. **38** (1993), 419-421].

[75] Nadirashvili, N.S., New isoperimetric inequalities in mathematical physics, **Partial Differential Equations of Elliptic Type, Cortona, 1992**, A. Alvino, E. Fabes, and G. Talenti, editors, Symposia Mathematica, vol. **35**, Cambridge University Press, Cambridge, 1994, pp. 197-203.

[76] Nadirashvili, N.S., Rayleigh's conjecture on the principal frequency of the clamped plate, Arch. Rational Mech. Anal. **129** (1995), 1-10.

[77] Owen, M. P., Asymptotic first eigenvalue estimates for the biharmonic operator on a rectangle, J. Diff. Eqs. **136** (1997), 166-190.

[78] Payne, L.E., Inequalities for eigenvalues of membranes and plates, J. Rational Mech. Anal. **4** (1955), 517-529.

[79] Payne, L.E., Isoperimetric inequalities for eigenvalues and their applications, Autovalori e autosoluzioni, Centro Internazionale Matematico Estivo (C.I.M.E.) 2° Ciclo, Chieti, 1962, pp. 1-58.

[80] Payne, L.E., Isoperimetric inequalities and their applications, SIAM Review **9** (1967), 453-488.

[81] Payne, L.E., Some comments on the past fifty years of isoperimetric inequalities, **Inequalities: Fifty Years On from Hardy, Littlewood, and Pólya**, W.N. Everitt, editor, Marcel Dekker, New York, 1991, pp. 143-161.

[82] Payne, L.E., G. Pólya, and H.F. Weinberger, Sur le quotient de deux fréquences propres consécutives, Comptes Rendus Acad. Sci. Paris **241** (1955), 917-919.

[83] Payne, L.E., G. Pólya, and H.F. Weinberger, On the ratio of consecutive eigenvalues, J. Math. and Phys. **35** (1956), 289-298.

[84] Pólya, G., and G. Szegő, **Isoperimetric Inequalities in Mathematical Physics**, Annals of Mathematics Studies, Number **27**, Princeton University Press, Princeton, New Jersey, 1951.

[85] Protter, M.H., Can one hear the shape of a drum? revisited, SIAM Review **29** (1987), 185-197.

[86] Rayleigh, J.W.S., **The Theory of Sound**, second edition revised and enlarged (in 2 volumes), Dover Publications, New York, 1945 (republication of the 1894/96 edition).

[87] Reed, M., and B. Simon, **Methods of Modern Mathematical Physics, vol. IV: Analysis of Operators**, Academic Press, New York, 1978.

[88] Safarov, Yu., and D. Vassiliev, **The Asymptotic Distribution of Eigenvalues of Partial Differential Equations**, Translations of Mathematical Monographs, vol. **155**, American Mathematical Society, Providence, Rhode Island, 1997.

[89] Schoen, R., and S.-T. Yau, **Lectures on Differential Geometry**, Conference Proceedings and Lecture Notes in Geometry and Topology, vol. **1**, International Press, Boston, 1994.

[90] Simon, B., **Functional Integration and Quantum Physics**, Academic Press, New York, 1979 (see especially pp. 142-143).

[91] Sperner, E., Zur Symmetrisierung von Funktionen auf Sphären, Math. Z. **134** (1973), 317-327.

[92] Szegő, G., On membranes and plates, Proc. Nat. Acad. Sci. (USA) **36** (1950), 210-216.

[93] Szegő, G., Inequalities for certain eigenvalues of a membrane of given area, J. Rational Mech. Anal. **3** (1954), 343-356.

[94] Szegő, G., Note to my paper "On membranes and plates", Proc. Nat. Acad. Sci. (USA) **44** (1958), 314-316.

[95] Talenti, G., Best constant in Sobolev inequality, Ann. Mat. Pura Appl. (Ser. 4) **110** (1976), 353-372.

[96] Talenti, G., Elliptic equations and rearrangements, Ann. Scuola Norm. Sup. Pisa (4) **3** (1976), 697-718.

[97] Talenti, G., Linear elliptic P.D.E's: Level sets, rearrangements and a priori estimates of solutions, Boll. Un. Mat. Ital. (6) **4-B** (1985), 917-949.

[98] Talenti, G., On the first eigenvalue of the clamped plate, Ann. Mat. Pura Appl. (Ser. 4) **129** (1981), 265-280.

[99] Talenti, G., Rearrangements and PDE, **Inequalities: Fifty Years On from Hardy, Littlewood, and Pólya**, W.N. Everitt, editor, Marcel Dekker, New York, 1991, pp. 211-230.

[100] Talenti, G., On isoperimetric theorems of mathematical physics, Chapter 4.4 of **Handbook of Convex Geometry**, vol. **B**, P.M. Gruber and J.M. Wills, editors, North-Holland, Amsterdam, The Netherlands, 1993, pp. 1131-1147.

[101] Thompson, C.J., On the ratio of consecutive eigenvalues in n-dimensions, Stud. Appl. Math. **48** (1969), 281-283.

[102] Weinberger, H.F., An isoperimetric inequality for the n-dimensional free membrane problem, J. Rational Mech. Anal. **5** (1956), 633-636.

[103] Weinberger, H.F., Symmetrization in uniformly elliptic problems, Chapter 58 of **Studies in Mathematical Analysis and Related Topics: Essays in Honor of George Pólya**, G. Szegő, C. Loewner, S. Bergman, M.M. Schiffer, J. Neyman, D. Gilbarg, and H. Solomon, editors, Stanford University Press, Stanford, California, 1962, pp. 424-428.

[104] Weinstein, A., and W. Stenger, **Methods of Intermediate Problems for Eigenvalues: Theory and Ramifications**, Academic Press, New York, 1972.

[105] Yang, H.C., Estimates of the difference between consecutive eigenvalues, 1995 preprint (revision of International Centre for Theoretical Physics preprint IC/91/60, Trieste, Italy, April, 1991).

[106] Yang, P.C., and S.-T. Yau, Eigenvalues of the Laplacian of compact Riemann surfaces and minimal submanifolds, Ann. Scuola Norm. Sup. Pisa Cl. Sci. (4) **7** (1980), 55-63.

[107] Yau, S.-T., Problem section, **Seminar on Differential Geometry**, S.-T. Yau, editor, Annals of Mathematics Studies, Number **102**, Princeton University Press, Princeton, New Jersey, 1982, pp. 669-706 [reprinted as pp. 277-314 of [89]].

[108] Yau, S.-T., Open problems in geometry, **Differential Geometry: Partial Differential Equations on Manifolds**, Proc. Symp. Pure Math., vol. **54**, part 1, R. Greene and S.-T. Yau, editors, American Mathematical Society, Providence, Rhode Island, 1993, pp. 1-28 [reprinted as pp. 365-409 of [89]].

Estimates of heat kernels on Riemannian manifolds

Alexander Grigor'yan
Imperial College
London SW7 2BZ
England
email: a.grigoryan@ic.ac.uk
URL: www.ma.ic.ac.uk/~grigor

May 1999

Contents

1 Introduction

The purpose of these notes is to give introduction to the subject of heat kernels on non-compact Riemannian manifolds. By definition, the heat kernel for the Euclidean space $\mathbb{R}^n$ is the (unique) positive solution of the following Cauchy problem in $(0, +\infty) \times \mathbb{R}^n$

$$\begin{cases} \frac{\partial u}{\partial t} = \Delta u\,, \\ u(0,x) = \delta(x-y), \end{cases}$$

where $u = u(t,x)$ and $y \in \mathbb{R}^n$. It is denoted by $p(t,x,y)$ and is given by the classical formula

$$p(t,x,y) = \frac{1}{(4\pi t)^{n/2}} \exp\left(-\frac{|x-y|^2}{4t}\right). \tag{1.1}$$

In other words, $p(t,x,y)$ is a positive fundamental solution to the heat equation $\frac{\partial u}{\partial t} = \Delta u$. This means, in particular, that the Cauchy problem

$$\begin{cases} \frac{\partial u}{\partial t} = \Delta u, \\ u(0,t) = f(x) \end{cases}$$

is solved by

$$u(t,x) = \int_{\mathbb{R}^n} p(t,x,y) f(y) dy,$$

provided f is a bounded continuous function.

Another definition of the heat kernel (which justifies the letter p) is as follows: it is the transition density of the Brownian motion in $\mathbb{R}^n$ (up to the change of time $t \to t/2$). Given that much, it is not surprising that the heat kernel plays a central role in potential theory in $\mathbb{R}^n$.

Consider now an arbitrary smooth connected Riemannian manifold M. There is a natural generalization of the Laplace operator linked to the Riemannian structure of M. It is called the Riemannian Laplace operator or the Laplace-Beltrami operator and is also denoted by Δ. It turns out that the notion of the heat kernel can be defined on any manifold. Let us denote it also by $p(t, x, y)$, where $t > 0$ and $x, y \in M$. However, explicit formulas for $p(t, x, y)$ exist only for a few classes of manifolds possessing enough symmetries. The simplest explicit heat kernel formula after (1.1) is one for the three-dimensional hyperbolic space $\mathbb{H}^3$ which reads as follows

$$p(t, x, y) = \frac{1}{(4\pi t)^{3/2}} \exp\left(-\frac{d^2}{4t} - t\right) \frac{d}{\sinh d}, \tag{1.2}$$

where $d = d(x, y)$ is the geodesic distance between $x, y \in \mathbb{H}^3$. Clearly, there is certain similarity between (1.1) and (1.2) (note that $|x - y|$ is the geodesic distance in $\mathbb{R}^n$) but there are also two distinctions: the terms $\exp(-t)$ and $\frac{d}{\sinh d}$ in (1.2). They reflect the difference between the geometries of the Euclidean and hyperbolic spaces.

It turns out that the heat kernel is rather sensitive to the geometry of manifolds, which makes the study of the heat kernel interesting and rich from the geometric point of view. On the other hand, there are the properties of the heat kernel which little depend on the geometry and reflect rather structure of the heat equation. For example, the presence of the Gaussian exponential term $\exp\left(-\frac{d^2}{4t}\right)$ in the heat kernel estimates is one of such features.

Most part of these notes is devoted to the heat kernel upper estimates on arbitrary manifolds. We discuss both general techniques of obtaining the heat kernel bounds, presented in Sections 3, 5, 6, and their applications for particular classes of manifolds, in Section 7. In Section 4, we apply the heat kernel bounds to estimate eigenvalues of the Laplace operator. Many results are supplied with proofs whose purpose is to demonstrate the underlying ideas rather than to achieve the full generality.

We do not aim at a detailed account of lower estimates of the heat kernel and touch only one aspect of those in Section 5.3, not the least because this subject is so far not well understood. Other results on the lower bounds of heat kernel on manifolds can be found in [15], [28], [67], [78], [98], [110], [120], [123].

We have to skip some other interesting questions related to the heat kernels such as the Harnack inequality [67], [68], [98], [120], [121], comparison theorems [28], [52], short time asymptotics [62], [85], [105], [112], [119], [131], estimates of time derivatives of the heat kernel [44], [49], [71], gradient estimates [82], [86], [98], [100], [125], [117], [138], [141], discretization techniques [24], [25], [34], [88], homogenization techniques [2], [11], [90], [110], etc. Finally, we do not treat heat kernels on underlying spaces other than Riemannian manifolds and for operators other than the Laplace operator. See the following references for heat kernels

- on symmetric spaces [3];
- on groups and Lie groups [1], [13], [14], [108], [118], [135];
- for random walks on graphs [30], [53], [79], [83], [89], [114], [136];
- for second order elliptic operators with lower order terms [40], [59], [99], [107], [111], [123], [137];
- for higher order elliptic operators [6], [47], [48];
- for subelliptic operators [11], [12], [91], [92];
- for non-linear p-harmonic Laplacian [54];
- for Laplacian on exterior differential forms [56], [57], [58];
- for abstract local Dirichlet forms [127], [128];
- for Brownian motion on fractals [7], [8], [9], [10].

Needless to say that this list of references is very far from being complete.

NOTATION. The letters C, c and their modifications C', c', C_1, c_1 etc. are used for positive constants which may be different in different context.

ACKNOWLEDGMENT. The author thanks with great pleasure E.B.Davies and Yu.Safarov for inviting him to give a series of lectures at the Instructional Conference on Spectral Theory and Geometry held in Edinburgh, April 1998.

2 Construction of the heat kernel on manifolds

2.1 Laplace operator

The Laplace operator in $\mathbb{R}^n$ is defined by

$$\Delta = \sum_{i=1}^{n} \frac{\partial^2}{\partial^2 X^i}$$

where X^1, X^2,..., X^n are the Cartesian coordinates. In order to write down the Laplace operator in an arbitrary curvilinear coordinate system x^1, x^2,..., x^n let us first note that the length element

$$ds^2 = \left(dX^1\right)^2 + \left(dX^2\right)^2 + ... + \left(dX^n\right)^2$$

takes in the coordinates x^1, x^2,..., x^n the following form

$$ds^2 = \sum_{i,j=1}^{n} g_{ij}(x)dx^i dx^j \tag{2.1}$$

where (g_{ij}) is a symmetric positive definite matrix. The change of variables in Δ gives then

$$\Delta = \frac{1}{\sqrt{g}} \sum_{i,j=1}^{n} \frac{\partial}{\partial x^i} \left(\sqrt{g} g^{ij} \frac{\partial}{\partial x^j} \right) \tag{2.2}$$

where $g := \det(g_{ij})$ and $(g^{ij}) = (g_{ij})^{-1}$.

Let M be an arbitrary smooth connected n-dimensional Riemannian manifold M. In general, there is no selected coordinate system on M but one can still define the Laplace operator in any chart $x^1, x^2, ..., x^n$ by using (2.2), where g_{ij} is now the Riemannian metric tensor on M (which determines the length by (2.1)). The definition (2.2) is covariant, that is, in any other chart this operator will have the same form. Hence, Δ is defined on all of M. The Laplace operator can also be represented as $\Delta = \text{div}\nabla$, where the gradient ∇ and the divergence div are defined by

$$(\nabla u)^i = \sum_{j=1}^{n} g^{ij} \frac{\partial u}{\partial x^j}$$

and

$$\text{div} F = \frac{1}{\sqrt{g}} \sum_{i,j=1}^{n} \frac{\partial}{\partial x^i} \left(\sqrt{g} F^i \right).$$

The Riemannian structure allows to introduce on M volumes of all dimensions. Particularly important for us will be the Riemannian n-volume μ defined by

$$d\mu = \sqrt{g} dx^1 dx^2 ... dx^3.$$

The Stokes's theorem implies the following integration-by-parts formula

$$\int_\Omega v \Delta u \, d\mu = -\int_\Omega (\nabla u, \nabla v) \, d\mu, \tag{2.3}$$

where Ω is a pre-compact open subset of M, u and v are C^2 functions in Ω such that one of them vanishes in a neighborhood of the boundary $\partial\Omega$, and

$(\cdot,\cdot)$ means the inner product of the vector fields induced by the Riemannian tensor. More generally, if $u, v \in C^1\left(\overline{\Omega}\right) \cap C^2(\Omega)$ and $\partial\Omega \in C^1$ then

$$\int_\Omega v\Delta u \, d\mu = \int_{\partial\Omega} v\frac{\partial u}{\partial \nu} d\sigma - \int_\Omega (\nabla u, \nabla v)\, d\mu, \tag{2.4}$$

where σ is the surface area, that is, the $(n-1)$-dimensional Riemannian measure on M, and ν is the outward normal vector field on $\partial\Omega$.

See [22] and [119] for a detailed account of the notions of Riemannian geometry related to the Laplace operator.

2.2 Eigenvalues and eigenfunctions of the Laplace operator

Given a precompact open set $\Omega \subset M$, consider the Dirichlet eigenvalue problem in Ω

$$\begin{cases} \Delta u + \lambda u = 0, \\ u|_{\partial\Omega} = 0. \end{cases} \tag{2.5}$$

To be exact, we should define a weak solution to (2.5). Consider the spaces

$$L^2(\Omega) := \left\{ f : \int_\Omega f^2 d\mu < \infty \right\},$$

$$W_1^2(\Omega) := \left\{ f \in L^2(\Omega) : |\nabla f| \in L^2(\Omega) \right\},$$

where ∇f is understood in the sense of distributions, and define $\overset{o}{H}_1(\Omega)$ to be the closure of $C_0^\infty(\Omega)$ in $W_1^2(\Omega)$.

We define a weak solution to (2.5) as a function $u \in \overset{o}{H}_1(\Omega)$ that satisfies the equation $\Delta u + \lambda u = 0$ in the sense of distributions. The latter can be shown to be equivalent to the integral identity

$$\int_\Omega (\nabla u, \nabla v)\, d\mu = \lambda \int uv d\mu, \quad \forall v \in \overset{o}{H}_1(\Omega).$$

The standard technique of the spectral theory of elliptic operators implies that there exists an orthonormal basis $\{\phi_k\}_{k=1}^\infty$ in $L^2(\Omega)$ such that each ϕ_k is a weak eigenvalue of Δ in Ω with an eigenvalue $\lambda_k = \lambda_k(\Omega)$, and

$$0 \le \lambda_1 \le \lambda_2 \le \ldots \le \lambda_k \underset{k\to\infty}{\longrightarrow} \infty.$$

If Ω is connected then $\lambda_1(\Omega)$ is a single eigenvalue, that is, $\lambda_2(\Omega) > \lambda_1(\Omega)$, and $\phi_1(x) \neq 0$ in Ω. It is also possible to prove that if $M\setminus\overline{\Omega}$ is non-empty then $\lambda_1(\Omega) > 0$. On the other hand, if M is compact then we may take $\Omega = M$

in which case $\lambda_1(M) = 0$, with the eigenfunction $\phi_1(x) \equiv \mu(M)^{-1/2} = \text{const}$, but $\lambda_2(M) > 0$.

The operator Δ can be considered as a unbounded operator in $L^2(\Omega)$, with the domain $C_0^\infty(\Omega)$. As such, it turns out to be essentially self-adjoint. Its closure is called the Dirichlet Laplace operator and will be denoted by Δ_Ω. It has the domain

$$Dom(\Delta_\Omega) = \left(f \in \overset{o}{H}_1(\Omega) : \Delta f \in L^2(\Omega) \right)$$

and the spectrum $spec(-\Delta_\Omega) = \{\lambda_k\}_{k=1}^\infty$ (it is sometimes convenient to refer to $-\Delta_\Omega$ rather than to Δ_Ω because the former is positive definite). Clearly, in the basis $\{\phi_k\}$ the operator $-\Delta_\Omega$ is represented by the (infinite) diagonal matrix

$$-\Delta_\Omega = diag\,(\lambda_1, \lambda_2, ..., \lambda_k, ...)\,.$$

By the spectral theory, one can define $f(-\Delta_\Omega)$ where f is a function on $spec\,(-\Delta_\Omega)$. Particularly important is the operator $\exp(t\Delta_\Omega)$ where t is a real parameter. In the basis $\{\phi_k\}$, it has the matrix

$$\exp(t\Delta_\Omega) = diag\left(e^{-t\lambda_1}, e^{-t\lambda_2}, ..., e^{-t\lambda_k}, ...\right). \tag{2.6}$$

Hence, if $t \geq 0$ then $\exp(t\Delta_\Omega)$ is a bounded self-adjoint operator in $L^2(\Omega)$.

2.3 Heat kernel in precompact regions

Consider the following initial-boundary problem in $(0, \infty) \times \Omega$

$$\begin{cases} \frac{\partial u}{\partial t} = \Delta u, \\ u(0,x) = f(x), \\ u(t,x)|_{x\in\partial\Omega} = 0. \end{cases} \tag{2.7}$$

We understand it in a weak sense, as an evolution equation in $L^2(\Omega)$. Namely, we interpret $u(t,x)$ as a function from $[0,\infty)$ to $L^2(\Omega)$ such that

1. u is Fréchet differentiable in $t > 0$ and its Fréchet derivative $\dot{u}$ is equal to Δu (which, in particular, means that $\Delta u \in L^2(\Omega)$);

2. u is L^2-continuous at $t = 0$ and $u(0,\cdot) = f$;

3. for each $t > 0$, $u(t,\cdot) \in \overset{o}{H}_1(\Omega)$.

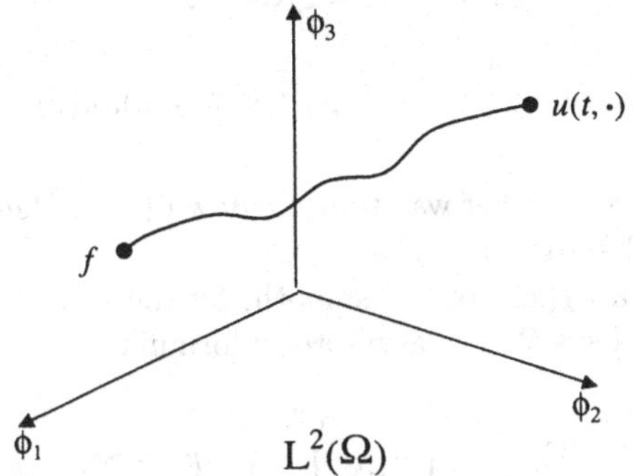

Figure 1 Function $u(t,x)$ as a path in $L^2(\Omega)$

It is easy to verify that the evolution equation $\dot{u} = \Delta u$ has solution

$$u = e^{t\Delta_\Omega} f. \tag{2.8}$$

Let us write this down in the basis $\{\phi_k\}$. The function $f \in L^2(\Omega)$ has the following expansion in this basis

$$f = \sum_{k=1}^{\infty} a_k \phi_k$$

where

$$a_k = \int_\Omega f(y)\phi_k(y)d\mu(y).$$

Then, by (2.6),

$$\begin{aligned} e^{t\Delta_\Omega} f(x) &= \sum_{k=1}^{\infty} a_k e^{-t\lambda_k(\Omega)} \phi_k(x) \\ &= \sum_{k=1}^{\infty} e^{-t\lambda_k(\Omega)} \phi_k(x) \int_\Omega f(y)\phi_k(y)d\mu(y) \\ &= \int_\Omega \left\{ \sum_{k=1}^{\infty} e^{-t\lambda_k(\Omega)} \phi_k(x)\phi_k(y) \right\} f(y)d\mu(y). \end{aligned}$$

The kernel in the curly brackets is called the heat kernel of Ω and will be denoted by $p_\Omega(t,x,y)$. Hence, we have

$$p_\Omega(t,x,y) := \sum_{k=1}^{\infty} e^{-t\lambda_k(\Omega)} \phi_k(x)\phi_k(y) \tag{2.9}$$

and the weak solution $u(t,x)$ to (2.7) is given by

$$u(t,x) = e^{t\Delta_\Omega} f = \int_\Omega p_\Omega(t,x,y) f(y) d\mu(y). \tag{2.10}$$

Note that (2.10) is just another way to write down (2.8). Hence, the operator $e^{t\Delta_\Omega}$ has the integral kernel $p_\Omega(t,x,y)$.

The eigenfunction $\phi_k(x)$ are C^∞-smooth, by the local elliptic regularity. The sequence $\{\lambda_k\}$ obeys Weyl's asymptotic formula

$$\lambda_k(\Omega) \sim c_n \left(\frac{k}{\mu(\Omega)}\right)^{2/n}, \quad k \to \infty,$$

and, hence, is growing fast enough to ensure convergence of (2.9) locally in any $C^m(\Omega)$. Hence, $p_\Omega \in C^\infty((0,\infty) \times \Omega \times \Omega)$. The solution $u(t,x)$ defined by (2.10) is then C^∞-smooth and satisfies the heat equation in the classical sense. If f is continuous then it is possible to show that $u(t,x)$ is continuous in the classical sense in $[0,\infty) \times \Omega$ and $u(t,x) = f(x)$.

Since $\phi_k \in \overset{o}{H}_1(\Omega)$, we obtain that p_Ω is also in $\overset{o}{H}_1(\Omega)$ as a function of x (or y). If the boundary $\partial\Omega$ is smooth, then this implies that $p_\Omega(t,x,y)$ extends continuously to $\overline{\Omega}$ and that p_Ω vanishes on $\partial\Omega$. In particular, the function u defined by (2.10) is also continuous on $(0,\infty) \times \partial\Omega$ and $u(t,x) = 0$ for all $t > 0$ and $x \in \partial\Omega$.

Other simple properties of p_Ω are as follows:

(a) As a function of t and x, the function $p_\Omega(t,x,y)$ satisfies the heat equation

$$\frac{\partial p_\Omega}{\partial t} = \Delta p_\Omega$$

and the initial value

$$p_\Omega(t,\cdot,y) \to \delta_y \quad \text{as} \quad t \to 0+ .$$

(b) The semigroup property: for all $t, s > 0$ and $x, y \in \Omega$,

$$p_\Omega(t+s,x,y) = \int_\Omega p_\Omega(t,x,z) p_\Omega(s,x,y) d\mu(z), \tag{2.11}$$

which is another way to write down the identity $e^{(t+s)\Delta_\Omega} = e^{t\Delta_\Omega} e^{s\Delta_\Omega}$.

(c) The symmetry

$$p_\Omega(t,x,y) = p_\Omega(t,y,x). \tag{2.12}$$

The latter is obvious from (2.9) but would not be so transparent if we were to define p_Ω as a kernel which solves the initial-boundary problem (2.7) by (2.10).

If Ω is connected and $M \setminus \overline{\Omega}$ is non-empty then $\lambda_1(\Omega) > 0$ is a single eigenvalue, $\phi_1(x) \neq 0$ in Ω, and (2.9) implies

$$p_\Omega(t,x,y) \sim e^{-t\lambda_1(\Omega)}\phi_1(x)\phi_1(y), \quad t \to \infty. \tag{2.13}$$

If M is compact then we may take $\Omega = M$ in which case (2.9) yields

$$p_M(t,x,y) = \frac{1}{\mu(M)} + \sum_{k=2}^{\infty} e^{-t\lambda_k(M)}\phi_k(x)\phi_k(y). \tag{2.14}$$

In particular, we have

$$p_M(t,x,y) \to \mu(M)^{-1}, \quad t \to \infty. \tag{2.15}$$

2.4 Maximum principle and positivity of the heat kernel

The properties of the heat kernel p_Ω discussed above follows from the self-adjointness of the Laplace operator in $L^2(\Omega)$. However, there is another aspect of the Laplace operator which cannot be derived only from the spectral properties. For example, it is known that $p_\Omega(t,x,y) > 0$ for all $t > 0$ and $x, y \in \Omega$. However, the positivity of the heat kernel is not at all obvious from the eigenfunction expansion (2.9), because the eigenfunctions ϕ_k are signed (except for ϕ_1).

Here we consider another property of the heat equation which is called the maximum (minimum) principle and which is responsible for the positivity of the heat kernel. Denote $\Omega_T = (0,T) \times \Omega$, which is a cylinder in $[0,\infty) \times M$, and define its *parabolic boundary* $\partial_p\Omega_T$ by

$$\partial_p\Omega_T := \partial\Omega_T \setminus \{(t,x) : t = T\}.$$

In other words, $\partial_p\Omega_T$ is the part of the boundary $\partial\Omega$ without the top of the cylinder.

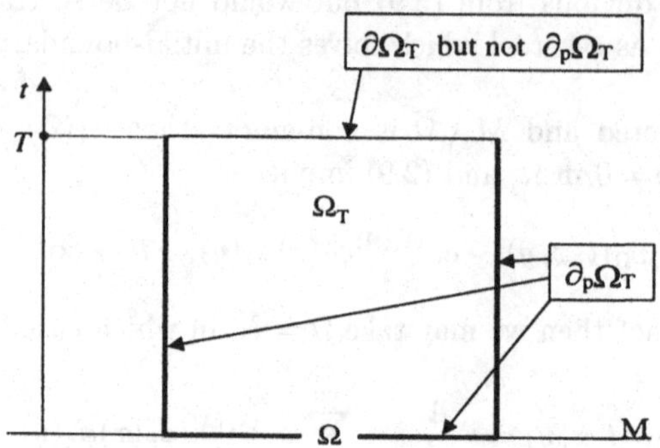

Figure 2 The parabolic boundary $\partial_p\Omega_T$

Proposition 2.1 *(The maximum/minimum principle) Let $u(t,x)\in C^2(\Omega_T)\cap C(\overline{\Omega})$ solve the heat equation in Ω_T. Then*

$$\sup_{\Omega_t} u = \sup_{\partial_p\Omega_T} u \tag{2.16}$$

and

$$\inf_{\Omega_T} u = \inf_{\partial_p\Omega_T} u\,. \tag{2.17}$$

If the initial function f in the initial-boundary problem (2.7) is non-negative then the minimum principle (2.17) implies that a (classical) solution $u(t,x)$ of (2.7) should be non-negative, too. If $\partial\Omega$ is smooth then the function $u = e^{t\Delta_\Omega} f$ is a classical solution to (2.7) whence

$$e^{t\Delta_\Omega} f \geq 0, \quad \forall f\in C_0^\infty(\Omega),\ f\geq 0.$$

Therefore, the kernel p_Ω of the operator $e^{t\Delta_\Omega}$ must be non-negative. By applying the strong version of the minimum principle, one can show that, in fact, p_Ω must be strictly positive in $(0,\infty)\times\Omega$.

A non-smooth boundary $\partial\Omega$ can be handled, too. However, we will always assume that $\partial\Omega$ is smooth if this simplifies the argument.

Another consequence of the maximum principle is the inequality

$$e^{t\Delta_\Omega} 1 \leq 1. \tag{2.18}$$

Indeed, $u = e^{t\Delta_\Omega}1$ solves the problem (2.7) with $f\equiv 1$ and, obviously, $\sup_{\partial_p\Omega_T} u\leq 1$. Hence, by (2.16), we have $\sup_{\Omega_T} u\leq 1$ which means $u\leq 1$ everywhere. Clearly, (2.18) and (2.10) imply

$$\int_\Omega p_\Omega(t,x,y)\,d\mu(y)\leq 1. \tag{2.19}$$

The third consequence of the maximum principle is the monotonicity of p_Ω with respect to Ω: if $\Omega \subset \Omega'$, where Ω' is also a precompact open subset of M then

$$p_\Omega(t,x,y) \le p_{\Omega'}(t,x,y). \tag{2.20}$$

Of course, one should specify the range of t,x,y in (2.20). If we extend $p_\Omega(t,x,y)$ by 0 for $x,y \notin \Omega$ then (2.20) holds for all $t>0$ and $x,y \in M$.

Let us sketch the proof of (2.20). For any function $f \in C_0^\infty(\Omega)$, $f \ge 0$, we compare the functions $u = e^{t\Delta_\Omega} f$ and $u' = e^{t\Delta_{\Omega'}} f$ in Ω_T. Both have the same initial datum, but on the boundary $\partial\Omega$, we have $u(t,x) = 0 \le u'(t,x)$. Hence,

$$\inf_{\partial_p \Omega_T} (u' - u) \ge 0,$$

and the minimum principle (2.17) implies $u' - u \ge 0$ in Ω_T and $u' \ge u$. Clearly, $e^{t\Delta_{\Omega'}} f \ge e^{t\Delta_\Omega} f$ implies $p_{\Omega'} \ge p_\Omega$, which was to be proved.

2.5 Heat semigroup on a manifold

The monotonicity of the heat kernel p_Ω with respect in Ω allows to construct the heat kernel p on the entire manifold M by taking the limit as "$\Omega \to M$". The latter means that we consider an exhaustion sequence $\{\Omega_k\}$ that is a sequence of precompact open sets $\Omega_k \subset M$ such that $\partial\Omega_k$ is smooth, $\overline{\Omega_k} \subset \Omega_{k+1}$ and

$$\bigcup_{k=1}^{\infty} \Omega_k = M.$$

Such sequence can be constructed on any manifold. Then we define

$$p(t,x,y) := \lim_{k\to\infty} p_{\Omega_k}(t,x,y). \tag{2.21}$$

Since $p_{\Omega_{k+1}} \ge p_{\Omega_k}$, the limit exists (finite or infinite) and does not depend on the choice of $\{\Omega_k\}$. As follows from (2.19),

$$\int_M p(t,x,y)d\mu(y) \le 1,$$

so that p is finite almost everywhere. By the convergence properties of solutions to the parabolic equations, $p(t,x,y)$ is finite everywhere and C^∞-smooth.

Clearly, $p(t,x,y)$ inherits all previously discussed properties of $p_\Omega(t,x,y)$ except for the eigenfunction expansion (2.9). Moreover, it is possible to define the Dirichlet extension of the Laplace operator Δ on M (denote it Δ_M) and to show that $p(t,x,y)$ is the kernel of the semigroup $e^{t\Delta_M}$ acting in $L^2(M)$ (see [55]). However, the spectrum of Δ_M is not necessarily discrete as for a precompact region Ω. This is why it is not possible in general to define the heat kernel by the eigenfunction expansion (2.9).

After the heat kernel has been constructed by (2.21), we can give a shorter definition.

Definition 2.2 *The heat kernel $p(t,x,y)$ on M is the smallest positive fundamental solution to the heat equation on $(0,\infty)\times M$.*

A "fundamental solution" means that

$$\begin{cases} \frac{\partial p}{\partial t} = \Delta_x p, \\ p(t,\cdot,y) \underset{t\to 0}{\longrightarrow} \delta_y. \end{cases} \tag{2.22}$$

If $q(t,x,y)$ is another positive fundamental solution then the minimum principle implies $q \geq p_\Omega$ for any precompact region Ω. By (2.21), we obtain $q \geq p$ and p is the smallest one.

The purpose of all constructions in this section was to provide the (sketch of) proof of the existence of the smallest positive fundamental solution and to obtain its most important properties. The full justification of the above constructions can be found in [55], [22].

As an example of application, let us consider a direct Riemannian product $M = M' \times M''$ where M' and M'' are Riemannian manifolds. It is easy to see that $\Delta = \Delta' + \Delta''$ and $\mu = \mu' \times \mu''$, where the dashes refer to the manifolds M' and M'', respectively. The heat kernel on M is also a direct product of the heat kernels in M' and M'', that is,

$$p(t,x,y) = p'(t,x',y')p''(t,x'',y''), \tag{2.23}$$

where $x = (x',x'') \in M$ and $y = (y',y'') \in M$. Indeed, one first proves the obvious modification of (2.23) for a precompact region $\Omega = \Omega' \times \Omega'' \subset M$ directly by (2.9), and then passes to the limit as in (2.21).

In particular, the heat kernel (1.1) in $\mathbb{R}^n$ can be obtained from the heat kernel in $\mathbb{R}^1$ by iterating (2.23). If $M' = \mathbb{R}^m$ and $M'' = K$ where K is a compact manifold then, by (2.23), (1.1) and (2.15), the heat kernel on $M = \mathbb{R}^m \times K$ has the following asymptotic

$$p(t,x,x) \sim \mu''(K)^{-1}(4\pi t)^{-m/2}, \quad t \to \infty. \tag{2.24}$$

3 Integral estimates of the heat kernel

In this section, we introduce an integral version of the maximum principle and apply it to estimate some integrals of the heat kernel.

3.1 Integral maximum principle

Lemma 3.1 *(Aronson [4]) Let $\Omega \subset M$ be a precompact region. Suppose that $u(t,x) \in C^2\left(\overline{\Omega_T}\right)$ solves the heat equation in Ω_T and satisfies the boundary condition $u|_{\partial\Omega} = 0$. Let $\xi(t,x)$ be a locally Lipschitz function on $(0,\infty) \times M$ such that*

$$\xi_t + \frac{1}{2}|\nabla\xi|^2 \leq 0. \tag{3.1}$$

Then the function

$$J(t) := \int_\Omega u^2(t,x) e^{\xi(t,x)} d\mu(x) \tag{3.2}$$

is non-increasing in t.

Why is this called a maximum principle? Indeed, assume $u \geq 0$ and consider another function

$$S(t) = \sup_{x\in\Omega} u(t,x).$$

By applying (2.16) in $\Omega_{s,t} := (s,t) \times \Omega$ where $t > s > 0$, we obtain

$$S(s) = \sup_{\partial_p \Omega_{s,t}} u = \sup_{\Omega_{s,t}} u \geq S(t),$$

that is, $S(t)$ is non-increasing.

It is possible to prove that the following function

$$S_\alpha(t) = \|u(\cdot,t)\|_{L^\alpha(\Omega)}$$

is non-increasing for all $\alpha \in [1,\infty]$. If $\alpha = 2$ then this amounts to Lemma 3.1 with $\xi \equiv 0$. Hence, Lemma (3.1) is a weighted version of the fact that $S_2(t)$ is non-increasing, whereas the classical maximum principle implies that $S_\infty(t)$ is non-increasing.

Non-trivial examples of function ξ satisfying (3.1) are as follows:

$$\xi(t,x) = \frac{d^2(x)}{2t}$$

and

$$\xi(t,x) = ad(x) - \frac{a^2}{2}t, \quad a \in \mathbb{R},$$

provided $d(x)$ is a Lipschitz function such that

$$|\nabla d| \leq 1.$$

Proof of Lemma 3.1. Let us differentiate $J(t)$ and show that $J' \leq 0$. Indeed, we have, by using $\xi_t \leq -\frac{1}{2}|\nabla \xi|^2$, $u_t = \Delta u$ and by (2.3),

$$\begin{aligned} J'(t) &= \int_\Omega u^2 \xi_t e^\xi + 2 \int_\Omega u u_t e^\xi \\ &\leq -\frac{1}{2} \int_\Omega u^2 |\nabla \xi|^2 e^\xi + 2 \int_\Omega u \Delta u \, e^\xi \\ &= -\frac{1}{2} \int_\Omega u^2 |\nabla \xi|^2 e^\xi - 2 \int_\Omega u (\nabla u, \nabla \xi) e^\xi - 2 \int_\Omega |\nabla u|^2 e^\xi \\ &= -\frac{1}{2} \int_\Omega (u \nabla \xi + 2 \nabla u)^2 e^\xi, \end{aligned} \tag{3.3}$$

which is non-positive. ■

One can get from (3.3) a sharper estimate for the decay of $J(t)$. Indeed, let us observe that

$$\frac{1}{2}(u\nabla\xi + 2\nabla u)^2 e^\xi = 2|\nabla(ue^{\xi/2})|^2.$$

By the variational property of $\lambda_1(\Omega)$,

$$\int_\Omega |\nabla(ue^{\xi/2})|^2 d\mu \geq \lambda_1(\Omega) \int_\Omega |ue^{\xi/2}|^2 = \lambda_1(\Omega) J(t).$$

Hence, (3.3) yields $J' \leq -2\lambda_1(\Omega) J$ whence

$$J(t) \leq J(t_0) \exp(-2\lambda_1(\Omega)(t - t_0)), \quad \forall t \geq t_0 > 0. \tag{3.4}$$

If $\xi(t,x) \equiv 0$ and $u(t,x) = p_\Omega(t,x,x_0)$ then, by (2.12) and (2.11),

$$J(t) = \int_\Omega p_\Omega^2(t,x,x_0) d\mu(x) = p_\Omega(2t, x_0, x_0).$$

Therefore, as a consequence of Lemma 3.1, $p_\Omega(t,x_0,x_0)$ is non-increasing in t. By letting $\Omega \nearrow M$ we see that $p(t,x_0,x_0)$ is non-increasing in t either.

3.2 The Davies inequality

The following theorem shows why the Gaussian exponential term is relevant to the heat kernel upper bounds on arbitrary manifolds.

Theorem 3.2 *(Davies [45]) Let M be an arbitrary Riemannian manifold and let A and B be two μ-measurable sets on M. Then*

$$\int_A \int_B p(t,x,y) d\mu(x) d\mu(y) \leq \sqrt{\mu(A)\mu(B)} \exp\left(-\frac{d^2(A,B)}{4t}\right), \tag{3.5}$$

where $d(A,B)$ is the geodesic distance between A and B (if A and B intersect then $d(A,B) = 0$).

The first proof. By the approximation argument, it suffices to prove (3.5) for compact A and B. Furthermore, if A and B are compact then it suffices to prove (3.5) for the heat kernel p_Ω of any precompact open set Ω containing A and B.

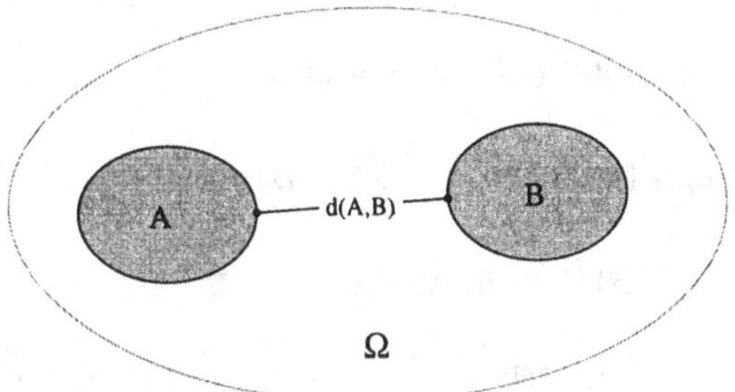

Figure 3 Sets A and B

Consider the function $u(t,x) = e^{t\Delta_\Omega}\mathbf{1}_A$. We can write

$$\begin{aligned}\int_B\int_A p_\Omega(t,x,y)d\mu(y)d\mu(x) &= \int_B\left(\int_\Omega p_\Omega(t,x,y)\mathbf{1}_A d\mu(y)\right)d\mu(x)\\ &= \int_B u(t,x)d\mu(x)\\ &\le \mu(B)^{1/2}\left(\int_B u^2(t,x)d\mu(x)\right)^{1/2}. \qquad (3.6)\end{aligned}$$

Let us set, for some $\alpha > 0$,

$$\xi(t,x) := \alpha d(x,A) - \frac{\alpha^2}{2}t$$

and consider the function

$$J(t) := \int_\Omega u^2(t,x)e^{\xi(t,x)}d\mu(x),$$

which, by Lemma (3.1) is non-increasing in $t > 0$. If $x \in B$ then

$$\xi(t,x) \ge \alpha d(B,A) - \frac{\alpha^2}{2}t,$$

whence

$$\begin{aligned}J(t) &\ge \int_B u^2(t,x)e^{\xi(t,x)}d\mu(x)\\ &\ge \exp\left(\alpha d(A,B) - \frac{\alpha^2}{2}t\right)\int_B u^2(t,x)d\mu(x). \qquad (3.7)\end{aligned}$$

On the other hand, if $x \in A$ then $\xi(0,x) = 0$. By the continuity of $J(t)$ at $t = 0+$, we have

$$J(t) \le J(0) = \int_\Omega e^{\xi(0,x)} \mathbf{1}_A d\mu(x) = \mu(A). \tag{3.8}$$

Combining (3.6), (3.7) and (3.8), we obtain

$$\int_A \int_B p_\Omega(t,x,y) d\mu(x) d\mu(y) \le \sqrt{\mu(A)\mu(B)} \exp\left(-\frac{\alpha}{2} d(A,B) + \frac{\alpha^2}{4} t\right).$$

Setting here $\alpha = d(A,B)/t$ we finish the proof. ■

Remark 3.3 Using (3.4) instead of the monotonicity of J gives the better inequality

$$\int_A \int_B p(t,x,y) d\mu(x) d\mu(y) \le \sqrt{\mu(A)\mu(B)} \exp\left(-\lambda_1(M)t - \frac{d^2(A,B)}{4t}\right), \tag{3.9}$$

where

$$\lambda_1(M) := \inf_{\Omega \subset\subset M} \lambda_1(\Omega). \tag{3.10}$$

It is possible to show that $\lambda_1(M)$ is the bottom of the spectrum of the operator $-\Delta_M$ in $L^2(M,\mu)$. Sometimes $\lambda_1(M)$ is called *the spectral radius* of the manifold M.

The second proof. Assume again that A and B are two compact subsets of a precompact region Ω. Fix some Lipschitz function $\psi(x)$ on Ω and consider the integral

$$\tilde{J}(t) := \int_\Omega u^2(t,x) e^{\psi(x)} d\mu(x),$$

where $u(t,x)$ solves the heat equation in $\mathbb{R}_+ \times \Omega$ and vanishes on $\partial\Omega$. Easy computation shows that

$$\tilde{J}'(t) = 2\int_\Omega u\Delta u\, e^\psi = -2\int_\Omega |\nabla u|^2 e^\psi - 2\int_\Omega u e^\psi (\nabla u, \nabla\psi).$$

Applying the inequality

$$-2u(\nabla u, \nabla\psi) \le 2|\nabla u|^2 + \frac{1}{2}u^2 |\nabla\psi|^2,$$

we obtain

$$\tilde{J}'(t) \le \frac{1}{2}\int_\Omega u^2 e^\psi |\nabla\psi|^2 d\mu. \tag{3.11}$$

Let us set $\psi(x) = \alpha d(x, A)$, for some $\alpha > 0$. Then $|\nabla\psi| \le \alpha$ and (3.11) implies $\tilde{J}' \le \frac{1}{2}\alpha^2 \tilde{J}$ whence

$$\tilde{J}(t) \le \tilde{J}(0)\exp(\frac{1}{2}\alpha^2 t). \tag{3.12}$$

Let us apply the above to the function $u = e^{t\Delta_\Omega}\mathbf{1}_A$. Since

$$\tilde{J}(0) = \int_\Omega \mathbf{1}_A \exp\left(\alpha d(x, A)\right) d\mu(x) = \mu(A)\,,$$

(3.12) implies

$$\int_\Omega u^2(t,x)\exp\left(\alpha d(x,A)\right) d\mu(x) \le \mu(A)\exp(\frac{1}{2}\alpha^2 t)$$

and

$$\int_B u^2(t,x) d\mu(x) \le \mu(A)\exp\left(-\alpha d(A,B) + \frac{1}{2}\alpha^2 t\right).$$

Choosing $\alpha = d(A,B)/t$ and applying (3.6), we finish the proof. ∎

The third proof. This proof is less elementary than the previous two, but it yields the better estimate:

$$\int_A\int_B p(t,x,y)d\mu(x)d\mu(y) \le \sqrt{\mu(A)\mu(B)}\int_\delta^\infty \frac{1}{\sqrt{\pi t}}\exp\left(-\frac{s^2}{4t}\right)ds, \tag{3.13}$$

where $\delta = d(A,B)$. Indeed, it is possible to prove that

$$\int_\delta^\infty \frac{1}{\sqrt{\pi t}}\exp\left(-\frac{s^2}{4t}\right)ds \le \min(1, \frac{2}{\sqrt{\pi}}\frac{\sqrt{t}}{\delta})\exp\left(-\frac{\delta^2}{4t}\right). \tag{3.14}$$

Therefore, (3.13) is better than (3.5) when $\delta >> \sqrt{t}$.

The inequality (3.13) is a particular case of a more general inequality of Cheeger, Gromov Taylor [27, Proposition 1.1], which says the following. Let $\phi \in L^1(\mathbb{R}_+)$ and Φ be its cos-Fourier transform, that is,

$$\Phi(\lambda) = \int_0^\infty \phi(s)\cos(s\lambda)ds. \tag{3.15}$$

The function Φ is bounded and continuous so that we can consider the bounded operator $\Phi(\sqrt{-\Delta_M})$ in $L^2(M,\mu)$ in the sense of the spectral theory. Then, for any function $f \in L^2(M,\mu)$ and any $\delta > 0$, we have

$$\left\|\Phi(\sqrt{-\Delta_M})\,f\right\|_{L^2(M\setminus \mathrm{supp}_\delta f)} \le \|f\| \int_\delta^\infty |\phi(s)|\,ds \tag{3.16}$$

where $\operatorname{supp}_\delta f$ means the δ-neighborhood of $\operatorname{supp} f$ and $\|\cdot\| = \|\cdot\|_{L^2(M)}$.

Given (3.16), let us take another function $g \in L^2(M,\mu)$ and suppose that the distance between the supports of f and g is at least δ. Then (3.16) yields

$$\int_M g\Phi\left(\sqrt{-\Delta_M}\right) f\,d\mu \le \|f\|\,\|g\| \int_\delta^\infty |\phi(s)|\,ds \tag{3.17}$$

(see also [32, Proposition 3.1]). Fix some $t > 0$ and take

$$\phi(s) = \frac{1}{\sqrt{\pi t}} \exp\left(-\frac{s^2}{4t}\right).$$

Then $\Phi(\lambda) = \exp(-t\lambda^2)$ and $\Phi(\sqrt{-\Delta_M}) = \exp(t\Delta_M)$ which is the heat semigroup. Hence, (3.17) implies

$$\int_M\int_M g(x)p(t,x,y)f(y)d\mu(y)d\mu(x) \le \|f\|\,\|g\| \int_\delta^\infty \frac{1}{\sqrt{\pi t}} \exp\left(-\frac{s^2}{4t}\right) ds,$$

and (3.13) follows by taking $f = \mathbf{1}_A$ and $g = \mathbf{1}_B$.

The proof of (3.16) is based on the fact that the function

$$u(t,\cdot) = \cos(t\sqrt{-\Delta_M})f$$

solves the Cauchy problem for *the wave equation*

$$\begin{cases} u_{tt} = \Delta u \\ u|_{t=0} = f \quad \text{and} \quad u_t|_{t=0} = 0. \end{cases}$$

The wave equation possesses *the finite propagation speed* equal to 1, which means that the support of the solution at time t lies in the t-neighborhood of the support of the initial data. Hence,

$$\operatorname{supp} u(t,\cdot) \subset \operatorname{supp}_t f\,. \tag{3.18}$$

Denote $w = \Phi(\sqrt{-\Delta_M})\,f$. Then, by (3.15),

$$w(x) = \int_0^\infty \phi(s)\cos(s\sqrt{-\Delta_M})f(x)ds = \int_0^\infty \phi(s)u(s,x)ds.$$

If $x \notin \operatorname{supp}_t f$ then, by (3.18), $u(s,x) = 0$ for all $s \le t$. Hence, for those x, we have

$$w(x) = \int_t^\infty \phi(s)u(s,x)ds = \int_t^\infty \phi(s)\cos(s\sqrt{-\Delta_M})f(x)ds$$

and, using $|\cos| \le 1$,

$$\begin{aligned}\|w\|_{L^2(M\setminus \operatorname{supp}_t v)} &\le \int_t^\infty \phi(s) \left\|\cos(s\sqrt{-\Delta_M})\right\|_{L^2\to L^2} \|f\|\, ds \\ &= \|f\| \int_t^\infty \phi(s) ds,\end{aligned}$$

which was to be proved. ∎

Lemma 3.1 and Theorem 3.2 can be used for obtaining heat kernel upper and lower bounds, estimating the eigenvalues of the Laplace operator, obtaining conditions for stochastic completeness etc. Some of the applications are show in the next sections.

3.3 Stochastic completeness

A Riemannian manifold M is called *stochastically complete* if, for all $x \in M$ and $t > 0$,

$$\int_M p(t,x,y)d\mu(y) = 1. \tag{3.19}$$

In term of the Brownian motion X_t, (3.19) means that the total probability of X_t to be found on M is equal to 1. The opposite can happen, for example, if M is an open bounded region on $\mathbb{R}^n$ and X_t is the Brownian motion on M with the killing boundary conditions on ∂M. Indeed, the process X_t riches the boundary in finite time with positive probability and then X_t stops existing as a point in M, which makes the integral in (3.19) smaller than 1. However, Azencott [5] showed that even a geodesically complete manifold may be stochastically incomplete. On such a manifold, the Brownian particle moves away extremely fast so that it covers an infinite distance in a finite time. This happens for a geometric reason - the manifold like that has a lot of space in a neighborhood of infinity which "draws" there a Brownian particle.

The following theorem provides a test for stochastic completeness in terms of the volume growth.

Theorem 3.4 *([65]) Let M be a geodesically complete manifold. Assume that, for some point $x \in M$,*

$$\int^\infty \frac{rdr}{\log V(x,R)} = \infty, \tag{3.20}$$

where $V(x,r) = \mu(B(x,r))$. Then M is stochastically complete.

For example, (3.20) holds if $V(x,R) \le C\exp(CR^2)$. In particular, a geodesically complete manifold with bounded below Ricci curvature is stochastically complete. This was first proved by Yau [140].

The proof of Theorem 3.4 can be found in [65] and [74]. It uses the same approach as in the proof of Lemma 3.1 but in a more sophisticated way. A reader interested in further consideration of stochastic completeness and related questions is referred to the survey [74].

4 Eigenvalues estimates

In this section we show an application of Theorem 3.2 for eigenvalue estimates. Let M be a compact connected Riemannian manifold. Denote by

$$0 = \lambda_1 < \lambda_2 \le \lambda_3 \le \ldots \le \lambda_k \le \ldots$$

the eigenvalues of $-\Delta_M$ and by $\phi_k(x)$ their corresponding eigenfunctions forming an orthonormal basis in $L^2(M)$.

Theorem 4.1 *(Chung - Grigor'yan - Yau [31]) Let M be a compact Riemannian manifold. Let A_1, A_2,...,A_k be k disjoint closed set on M. Denote*

$$\delta := \min_{i \ne j} d(A_i, A_j).$$

Then

$$\lambda_k \le \frac{4}{\delta^2} \max_{i \ne j} \left(\log \frac{2\mu(M)}{\sqrt{\mu(A_i)\mu(A_j)}} \right)^2. \tag{4.1}$$

In particular, if we have two sets $A_1 = A$ and $A_2 = B$ then (4.1) becomes

$$\lambda_2(M) \le \frac{4}{\delta^2} \left(\log \frac{2\mu(M)}{\sqrt{\mu(A)\mu(B)}} \right)^2, \tag{4.2}$$

where $\delta = d(A,B)$.

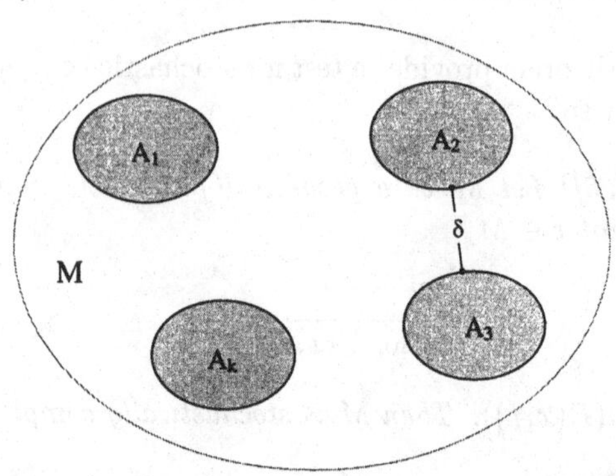

Figure 4 Sets A_i on a compact manifold M

Proof. We first prove (4.2). By the eigenfunction expansion (2.9), we can write, for any $t > 0$,

$$\int_A \int_B p(t,x,y)d\mu(x)d\mu(y) = \sum_{i=1}^{\infty} e^{-t\lambda_i} \int_A \phi_i(x)d\mu(x) \int_B \phi_i(y)d\mu(y).$$

Denote

$$a_i := \int_A \phi_i(x)d\mu(x) = (\mathbf{1}_A, \phi_i)_{L^2(M)}, \quad b_i := (\mathbf{1}_B, \phi_i)_{L^2(M)}$$

and observe that a_i and b_i are the Fourier coefficients of the functions $\mathbf{1}_A$ and $\mathbf{1}_B$ in the basis $\{\phi_i\}$, whence

$$\sum_{i=1}^{\infty} a_i^2 = \|\mathbf{1}_A\|_{L^2(M)}^2 = \mu(A) \quad \text{and} \quad \sum_{i=1}^{\infty} b_i^2 = \mu(B).$$

Since $\phi_1 \equiv 1/\sqrt{\mu(M)}$ (cf. (2.14)), we obtain

$$a_1 = (\mathbf{1}_A, \frac{1}{\sqrt{\mu(M)}})_{L^2(M)} = \frac{\mu(A)}{\sqrt{\mu(M)}} \quad \text{and} \quad b_1 = \frac{\mu(B)}{\sqrt{\mu(M)}}.$$

Thus, we have

$$\begin{aligned}
\int_A \int_B p(t,x,y)d\mu(X)d\mu(y) &= a_1 b_1 + \sum_{i=2}^{\infty} e^{-t\lambda_i} a_i b_i \\
&\geq a_1 b_1 - e^{-t\lambda_2} \left(\sum_{i=2}^{\infty} a_i^2\right)^{1/2} \left(\sum_{i=2}^{\infty} b_i^2\right)^{1/2} \\
&\geq \frac{\mu(A)\mu(B)}{\mu(M)} - e^{-t\lambda_2}\sqrt{\mu(A)\mu(B)}.
\end{aligned}$$

Comparing with the Davies inequality (3.5), we obtain

$$\sqrt{\mu(A)\mu(B)}e^{-\frac{\delta^2}{4t}} \geq \frac{\mu(A)\mu(B)}{\mu(M)} - e^{-t\lambda_2}\sqrt{\mu(A)\mu(B)}$$

and

$$e^{-t\lambda_2} \geq \frac{\sqrt{\mu(A)\mu(B)}}{\mu(M)} - e^{-\frac{\delta^2}{4t}}$$

Choosing t so that

$$e^{-\frac{\delta^2}{4t}} = \frac{1}{2}\frac{\sqrt{\mu(A)\mu(B)}}{\mu(M)},$$

we conclude

$$\lambda_2 \le \frac{1}{t}\log\frac{2\mu(M)}{\sqrt{\mu(A)\mu(B)}} = \frac{4}{\delta^2}\left(\log\frac{2\mu(M)}{\sqrt{\mu(A)\mu(B)}}\right)^2,$$

which is (4.2).

Let us now turn to the case $k > 2$. Consider the following integrals

$$J_{lm} := \int_{A_l}\int_{A_m} p(t,x,y)d\mu(x)d\mu(y)$$

and denote

$$a_i^{(l)} := (\mathbf{1}_{A_l}, \phi_i).$$

Then exactly as above, we have

$$\begin{aligned} J_{lm} &= \sum_{i=1}^{\infty} a_i^{(l)} a_j^{(m)} \\ &= \frac{\mu(A_l)\mu(A_m)}{\mu(M)} + \sum_{i=2}^{k-1} e^{-\lambda_i t} a_i^{(l)} a_i^{(m)} + \sum_{i=k}^{\infty} e^{-\lambda_i t} a_i^{(l)} a_i^{(m)} \\ &\ge \frac{\mu(A_l)\mu(A_m)}{\mu(M)} + \sum_{i=2}^{k-1} e^{-\lambda_i t} a_i^{(l)} a_i^{(m)} - e^{-\lambda_k t}\sqrt{\mu(A_l)\mu(A_m)}. \end{aligned} \tag{4.3}$$

On the other hand, by Theorem 3.2,

$$J_{lm} \le \sqrt{\mu(A_l)\mu(A_m)}e^{-\frac{\delta^2}{4t}}. \tag{4.4}$$

Therefore, we can further argue as in the case $k = 2$ provided the middle term in (4.3) can be discarded, that is,

$$\sum_{i=2}^{k-1} e^{-\lambda_i t} a_i^{(l)} a_i^{(m)} \ge 0. \tag{4.5}$$

Let us show that (4.5) can be achieved by *choosing* l, m. To that end, let us interpret the sequence $a^{(j)} := \left(a_2^{(j)}, a_3^{(j)}, ..., a_{k-1}^{(j)}\right)$ as a $(k-2)$-dimensional vector in $\mathbb{R}^{k-2}$. Here j ranges from 1 to k so that we have k vectors $a^{(j)}$ in $\mathbb{R}^{k-2}$. Let us introduce the inner product of vectors $u = (u_2, ..., u_{k-1})$ and $v = (v_2, ..., v_{k-1})$ in $\mathbb{R}^{k-2}$ by

$$(u,v)_t := \sum_{i=2}^{k-1} e^{-\lambda_i t} u_i v_i \tag{4.6}$$

and apply the following elementary fact:

Lemma 4.2 *From any $n+2$ vectors in n-dimensional Euclidean space, it is possible to choose two vectors with non-negative inner product.*

Note that $n+2$ is the smallest number for which the statement of Lemma 4.2 is true. Indeed, if $e_1, e_2, ..., e_n$ denote an orthonormal basis in the given space, let us set $v := -e_1 - e_2 - ... - e_n$. Then any two of the following $n+1$ vectors

$$e_1 + \varepsilon v,\ e_2 + \varepsilon v,\ ....,\ e_n + \varepsilon v,\ v$$

have a negative inner product, provided $\varepsilon > 0$ is small enough.

Lemma 4.2 is easily proved by induction in n. The inductive step is shown on Fig. 5. Indeed, assume that the $n+2$ vectors $v_1, v_2, ..., v_{n+2}$ form pairwise obtuse angles. Denote by E the hyperplane orthogonal to v_{n+2} and by v_i' the projection of v_i onto E. Each vector v_i with $i \leq n+1$ can be represented as

$$v_i = v_i' - \varepsilon_i v_{n+2}$$

with $\varepsilon_i := -(v_i, v_{n+2}) > 0$. Therefore,

$$(v_i, v_j) = (v_i', v_j') + \varepsilon_i \varepsilon_j |v_{n+2}|^2,$$

and we see that $(v_i, v_j) \geq 0$ provided $(v_i', v_j') \geq 0$. The latter is true by the inductive hypothesis, for some i, j, whence the former holds, too.

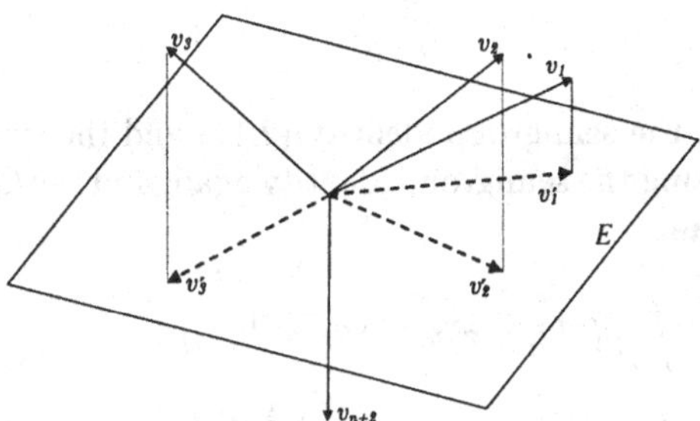

Figure 5 The vectors v_i' are projections of v_i's onto E

Let us finish the proof of Theorem 4.1. Fix some $t > 0$. By Lemma 4.2, we can find l, m so that $(a^{(l)}, a^{(m)})_t \geq 0$ and (4.5) holds. Then (4.3) and (4.4) yield

$$e^{-t\lambda_k} \geq \frac{\sqrt{\mu(A_l)\mu(A_m)}}{\mu(M)} - e^{-\frac{\delta^2}{4t}},$$

and we are left to choose t. However, t should not depend on l, m because we use t to define the inner product (4.6) *before* choosing l, m. So, we first write

$$e^{-t\lambda_k} \geq \min_{i,j} \frac{\sqrt{\mu(A_i)\mu(A_j)}}{\mu(M)} - e^{-\frac{\delta^2}{4t}}$$

and then define t by

$$e^{-\frac{\delta^2}{4t}} = \frac{1}{2} \min_{i,j} \frac{\sqrt{\mu(A_i)\mu(A_j)}}{\mu(M)},$$

whence (4.1) follows. ∎

Somewhat sharper estimates of the eigenvalues can be obtained by using (3.13) instead of (3.5) - see [32].

5 Pointwise estimates of the heat kernel

We discuss here two methods of obtaining the Gaussian upper bounds of the heat kernel $p(t,x,y)$, that is, the estimates containing the factor $\exp\left(-\frac{d^2}{Ct}\right)$ where $d = d(x,y)$. The first approach is based on properties of weighted integrals of the heat kernel in the spirit of Lemma 3.1. The second method is based on Theorem 3.2 and on certain mean-value inequality.

5.1 Gaussian upper bounds for the heat kernel

Let M be so far an arbitrary Riemannian manifold. We start we an observation that

$$p(t,x,x) = \int_M p^2(t/2,x,z)d\mu(z) \tag{5.1}$$

which follows from the semigroup identity (2.11) and the symmetry (2.12) of the heat kernel. Using the semigroup identity again and the Cauchy–Schwartz inequality, we obtain

$$\begin{aligned} p(t,x,y) &= \int_M p(t/2,x,z)p(t/2,y,z)d\mu(z) \\ &\leq \left(\int_M p^2(t/2,x,z)d\mu(z)\right)^{\frac{1}{2}} \left(\int_M p^2(t/2,y,z)d\mu(y)\right)^{\frac{1}{2}}, \end{aligned}$$

whence, by (5.1),

$$p(t,x,y) \leq \sqrt{p(t,x,x)p(t,y,y)}. \tag{5.2}$$

For example, if we knew an *on-diagonal* estimate like

$$p(t,x,x) \leq f(t), \quad \forall x \in M,$$

it would imply the *off-diagonal* estimate

$$p(t,x,y) \leq f(t), \quad \forall x,y \in M.$$

However, the latter does not take into account the distance between x and y. To fix that, we will modify the above argument to introduce the Gaussian factor.

Let us consider the following weighted integral of the heat kernel:

$$E_D(t,x) := \int_M p^2(t,x,z) \exp\left(\frac{d^2(x,z)}{Dt}\right) d\mu(z), \tag{5.3}$$

where $D > 0$ will be specified later. In the limit case $D = \infty$, we obtain by (5.1)

$$E_\infty(t,x) = p(2t,x,x), \tag{5.4}$$

and (5.2) can be rewritten as

$$p(t,x,y) \le \sqrt{E_\infty(t/2,x)E_\infty(t/2,y)}.$$

It turns out that a similar estimate holds for a finite D.

Lemma 5.1 *([69, Proposition 5.1]) We have, for any $D > 0$ and all $x, y \in M$, $t > 0$,*

$$p(t,x,y) \le \sqrt{E_D(t/2,x)E_D(t/2,y)} \exp\left(-\frac{d^2(x,y)}{2Dt}\right). \tag{5.5}$$

Proof. For any points $x, y, z \in M$, let us denote $\alpha = d(y,z)$, $\beta = d(x,z)$ and $\gamma = d(x,y)$. By the triangle inequality, $\alpha^2 + \beta^2 \ge \frac{1}{2}\gamma^2$.

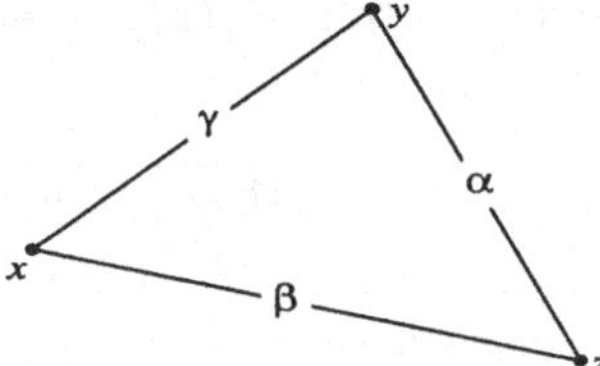

Figure 6 Distances α, β, γ

We have then

$$\begin{aligned} p(t,x,y) &= \int_M p(t/2,x,z)p(t/2,y,z)d\mu(z) \\ &\le \int_M p(t/2,x,z)e^{\frac{\beta^2}{Dt}}p(t/2,y,z)e^{\frac{\alpha^2}{Dt}}e^{-\frac{\gamma^2}{2Dt}}d\mu(z) \\ &\le \left(\int_M p^2(t/2,x,z)e^{\frac{2\beta^2}{Dt}}d\mu(z)\right)^{\frac{1}{2}} \left(\int_M p^2(t/2,y,z)e^{\frac{2\alpha^2}{Dt}}d\mu(y)\right)^{\frac{1}{2}} e^{-\frac{\gamma^2}{2Dt}} \\ &= \sqrt{E_D(t/2,x)E_D(t/2,y)} \exp\left(-\frac{d^2(x,y)}{2Dt}\right), \end{aligned}$$

which was to be proved. ∎

It is not a priori clear that $E_D(t,x)$ is *finite*. Indeed, it is easy to see that in $\mathbb{R}^n$, $E_D = \infty$ for all $D \leq 2$. Nevertheless, the following is true.

Theorem 5.2 *([66], [69]) For any manifold M, $E_D(t,x)$ is finite for all $D > 2$, $t > 0$, $x \in M$. Moreover, $E_D(t,x)$ is non-increasing in t.*

The most non-trivial part of this theorem is the finiteness of E_D. The non-increasing of E_D is an immediate consequence of Lemma 3.1.

Furthermore, the function

$$E_D(t,x)\exp\left(2\lambda_1(M)t\right)$$

is also non-increasing in t, which follows from (3.4) (recall that the spectral radius $\lambda_1(M)$ is defined by (3.10)). Inequality (5.5) implies, for all $t_0 > 0$ and $t > 0$,

$$p(t,x,y) \leq \sqrt{E_D(\tau/2,x)E_D(\tau/2,y)e^{\lambda_1(M)t_0}}\exp\left(-\lambda_1(M)t - \frac{d^2(x,y)}{2Dt}\right), \tag{5.6}$$

where $\tau = \min(t,t_0)$. Indeed, if $t \geq t_0$ then (5.6) follows from (5.5) and

$$E_D\left(t/2,x\right)\exp\left(\lambda_1(M)t\right) \leq E_D(t_0/2)\exp\left(\lambda_1(M)t_0\right). \tag{5.7}$$

If $t < t_0$ then (5.6) follows from (5.5) directly.

If $\lambda_1(M) > 0$ then (5.6) provides already a good upper bound of the heat kernel which can be rewritten as follows, for $t > t_0$,

$$p(t,x,y) \leq \Phi(x,y)\exp\left(-\lambda_1(M)t - \frac{d^2(x,y)}{2Dt}\right), \tag{5.8}$$

where

$$\Phi(x,y) := \sqrt{E_D(t_0/2,x)E_D(t_0/2,y)e^{\lambda_1(M)t_0}}.$$

However, if $\lambda_1(M) = 0$ then (5.6) is of no use and, by Lemma 5.1, the question of obtaining the long time behaviour of $p(t,x,y)$ amounts to the same question for $E_D(t,x)$. The latter is reduced by the following theorem to the *on-diagonal* rate of decay of $p(t,x,x)$ in t.

Theorem 5.3 *(Ushakov [129], Grigor'yan [72]) Assume that, for some $x \in M$ and for all $t > 0$,*

$$p(t,x,x) \leq \frac{C}{f(t)}, \tag{5.9}$$

*where $f(t)$ is an increasing positive function on $(0,+\infty)$ satisfying certain regularity condition (*see below*). Then, for all $D > 2$ and $t > 0$,*

$$E_D(t,x) \leq \frac{C'}{f(\varepsilon t)}, \tag{5.10}$$

for some $\varepsilon > 0$ and C'.

Remark 5.4 If (5.9) holds for $t \le t_0$ then (5.10) also holds for $t \le t_0$. Indeed, extend the function $f(t)$ by the constant $f(t_0)$ for $t > t_0$. Then (5.9) is true for all t because $p(t, x, x)$ is non-increasing in t as was remarked at the end of Section 3.1. Hence, by Theorem 5.3, (5.10) is true, too.

The regularity condition is the following: there are numbers $A \ge 1$ and $a > 1$ such that

$$\frac{f(as)}{f(s)} \le A\frac{f(at)}{f(t)}, \quad \text{for all } 0 < s < t. \tag{5.11}$$

The constants ε and C' in the statement of Theorem 5.3 depend on A and a. There are two simple situations when (5.11) holds:

1. $f(t)$ satisfies *the doubling condition*, that is, for some $A > 1$,

   $$f(2t) \le Af(t), \quad \forall t > 0. \tag{5.12}$$

   Then (5.11) holds with $a = 2$ because

   $$\frac{f(2s)}{f(s)} \le A \le A\frac{f(2t)}{f(t)}.$$

2. $f(t)$ has *at least polynomial growth* in the sense that, for some $a > 1$, the function $f(at)/f(t)$ is increasing in t. Then (5.11) holds for $A = 1$.

If f is differentiable then (5.11) is implied by either of the following properties of the function $l(\xi) := \log f(e^\xi)$ defined in $(-\infty, +\infty)$:

1. l' *is uniformly bounded* (for example, this is the case when $f(t) = t^N$ or $f(t) = \log^N(1+t)$ where $N > 0$);

2. l' *is monotone increasing* (for example, $f(t) = \exp(t^N)$).

On the other hand, (5.11) fails if $l' = \exp(-\xi)$ (it is unbounded as $\xi \to -\infty$) which corresponds to $f(t) = \exp(-t^{-1})$. Also, (5.11) may fail if l' is oscillating.

By putting together Theorem 5.3 and Lemma 5.1, we obtain

Corollary 5.5 *Assume that, for some points $x, y \in M$ and for all $t > 0$,*

$$p(t, x, x) \le \frac{C}{f(t)} \quad \text{and} \quad p(t, y, y) \le \frac{C}{g(t)}, \tag{5.13}$$

where f and g are increasing positive function on $(0, +\infty)$ satisfying the regularity condition (5.11) as above. Then, for all $t > 0$, $D > 2$ and for some $\varepsilon > 0$

$$p(t, x, y) \le \frac{C'}{\sqrt{f(\varepsilon t)g(\varepsilon t)}} \exp\left(-\frac{d^2(x, y)}{2Dt}\right). \tag{5.14}$$

Remark 5.6 By using (5.6) instead of (5.5) we obtain, for all $t_0 > 0$,

$$p(t,x,y) \le \frac{C' e^{\lambda_1(M)t_0}}{\sqrt{f(\varepsilon\tau)g(\varepsilon\tau)}} \exp\left(-\lambda_1(M)t - \frac{d^2(x,y)}{2Dt}\right), \tag{5.15}$$

where $\tau = \min(t, t_0)$. Note that (5.13) may be assumed only for $t \le t_0$. One can always extend $f(t)$ and $g(t)$ for $t > t_0$ by the constants $f(t_0)$ and $g(t_0)$, respectively, and (5.13) will continue to be true by the non-increasing of $p(t,x,x)$ in t.

Hence, the question of obtaining the Gaussian upper bounds of the heat kernel is reduced to obtaining the on-diagonal estimates (5.13), which will be considered in Section 6.

For the proof of Theorem 5.3, the reader is referred to [72]. The proof uses the integral maximum principle of Lemma 3.1. Note that if f and g satisfy the doubling property (5.12) then ε in (5.14) and (5.15) can be absorbed into the constant C'.

The finiteness of $E_D(t,x)$ in Theorem 5.2 can be deduced from Theorem 5.3. All that one needs is the initial upper bound $p(t,x,x) \le C_x t^{-n/2}$, for small t, which can be obtained by Theorem 5.8 from the next Section. See [66] or [69] for details.

Historically, the first method of obtaining the Gaussian upper bounds for the heat kernel of a uniformly elliptic operators in $\mathbb{R}^n$ with variable coefficients was introduced by Aronson [4]. He used the integral maximum principle but in a different way. The estimates of Aronson use the Euclidean distance rather than the Riemannian distance associated with the coefficients. Varadhan [130], [131] first realized that the Riemannian distance should be used instead. His result implies that, on any manifold,

$$\lim_{t\to 0+} t \log p(t,x,y) = -\frac{1}{4} d^2(x,y).$$

The first uniform Gaussian estimates for the heat kernel on manifolds was obtained by Cheng, Li and Yau [29], for manifolds of bounded geometry (see Section 7 below). They were later improved by Cheeger, Gromov and Taylor [27] by using (3.16). The sharp heat kernel estimates for the manifolds of non-negative Ricci curvature was obtained by Li and Yau [98]. Further progress in Gaussian upper bounds (under non-curvature assumptions) is due to Davies [41], [42], [43]. See also [135]. The approach to the Gaussian bounds we have adopted here is due to the author [69], [72].

5.2 Mean-value property

Here we present an alternative method of obtaining Gaussian upper bounds like (5.14), which avoids using $E_D(t,x)$ and, instead, is based on Theorem

3.2 and on the mean-value property. This method was introduced by Davies [45]. The treatment of this section is close to that in [97] and [37].

Fix some distinct points $x, y \in M$ and consider the balls $B(x,r)$, $B(y,r)$. By Theorem 3.2, we have

$$\int_{B(x,r)} \int_{B(y,r)} p(t,\xi,\eta)d\mu(\eta)d\mu(\xi) \leq \sqrt{V(x,r)V(y,r)} \exp\left(-\frac{(d-2r)_+^2}{4t}\right), \tag{5.16}$$

where $V(x,r) := \mu(B(x,r))$ and $d = d(x,y)$. If we knew that the value of the heat kernel at (t,x,y) can be estimated via the integral in (5.16) then we could obtain from (5.16) an upper bound for $p(t,x,y)$. This can be done by using the following *mean-value property.*

Definition 5.7 *We say that the manifold M admits the mean-value property (MV) if, for all $t > \tau > 0$, $\xi \in M$ and for any positive solution $u(s,\eta)$ of the heat equation in the cylinder $(t-\tau, t] \times B(\xi, \sqrt{\tau})$, we have*

$$u(t,\xi) \leq \frac{C}{\tau V(\xi,\sqrt{\tau})} \int_{t-\tau}^{t} \int_{B(\xi,\sqrt{\tau})} u(s,\eta)d\mu(\eta)ds. \tag{5.17}$$

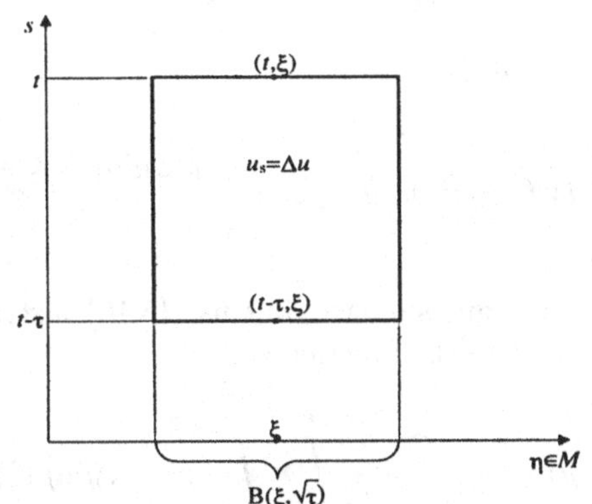

Figure 7 Cylinder $(t-\tau,t) \times B(\xi,\sqrt{\tau})$

The geometric assumptions which imply (MV), will be discussed in Section 6.4. Here we only mention that (5.17) holds, for example, if M is a geodesically complete manifold of nonnegative Ricci curvature.

Theorem 5.8 *(Li - Wang [97], Coulhon - Grigor'yan [37]) Assume that the mean-value property (MV) holds on the manifold M. Then, for all $x \in M$ and $t > 0$,*

$$p(t,x,x) \leq \frac{C}{V(x,\sqrt{t})}. \tag{5.18}$$

Moreover, for all $x, y \in M$, $t > 0$, $D > 2$,

$$p(t,x,y) \leq \frac{C'}{\sqrt{V(x,\sqrt{t/2})V(y,\sqrt{t/2})}} \exp\left(-\frac{d^2(x,y)}{2Dt}\right). \tag{5.19}$$

Hence, (5.19) holds on complete manifolds of non-negative Ricci curvature. For those manifolds, this estimate was first proved by different method by Li and Yau [98]. Moreover, they proved also a matching lower bound for the heat kernel which shows that (5.19) is sharp up to the values of the constants (see Sections 5.3 and 7.8 for the lower bounds of the heat kernel). In $\mathbb{R}^n$, we have $V(x,\sqrt{t}) \asymp t^{n/2}$ so that (5.18) and (5.19) give the correct rate for the long time decay of the heat kernel.

Proof of Theorem 5.8. Let us start with the consequence of (2.19)

$$\int_M p(s,x,z)d\mu(z) \leq 1 \tag{5.20}$$

and integrate it in time s:

$$\int_0^t \int_M p(s,x,z)d\mu(z) \leq t.$$

Applying (5.17) for $u = p(\cdot, x, \cdot)$, we obtain

$$p(t,x,x) \leq \frac{C}{tV(x,\sqrt{t})} \int_0^t \int_{B(x,\sqrt{t})} p(s,x,z)d\mu(z) \leq \frac{C}{V(x,\sqrt{t})},$$

which is exactly (5.18).

To show (5.19), we argue similarly but use (5.16) instead of (5.20). We start with (5.17) applied to the function $u = p(\cdot,\cdot,y)$,

$$p(t,x,y) \leq \frac{C}{\tau V(x,\sqrt{\tau})} \int_{t-\tau}^{t} \int_{B(x,\sqrt{\tau})} p(s,\xi,y)d\mu(\xi)ds, \tag{5.21}$$

for some $\tau \in (0,t)$. On the other hand, also by (5.17) applied to the function $u = p(\cdot,\xi,\cdot)$,

$$p(s,\xi,y) \leq \frac{C}{\tau V(y,\sqrt{\tau})} \int_{s-\tau}^{s} \int_{B(y,\sqrt{\tau})} p(\theta,\xi,\eta)d\mu(\eta)d\theta. \tag{5.22}$$

Combining (5.21) and (5.22), we see that $p(t,x,y)$ is bounded above by

$$\begin{aligned}
& \frac{C^2}{\tau^2 V(x,\sqrt{\tau})V(y,\sqrt{\tau})} \int_{t-\tau}^{t}\int_{s-\tau}^{s}\int_{B(x,\sqrt{\tau})}\int_{B(y,\sqrt{\tau})} p(\theta,\xi,\eta)d\mu(\eta)d\mu(\xi)d\theta ds \\
\leq\ & \frac{C^2}{\tau V(x,\sqrt{\tau})V(y,\sqrt{\tau})} \int_{t-2\tau}^{t}\int_{B(x,\sqrt{\tau})}\int_{B(y,\sqrt{\tau})} p(\theta,\xi,\eta)d\mu(\eta)d\mu(\xi)d\theta\,,
\end{aligned}$$

where we have assumed $\tau \leq t/2$. Using (5.16), we obtain

$$\begin{aligned}
p(t,x,y) &\leq \frac{C^2}{\tau\sqrt{V(x,\sqrt{\tau})V(y,\sqrt{\tau})}} \int_{t-2\tau}^{t} \exp\left(-\frac{(d-2\sqrt{\tau})_+^2}{4\theta}\right)d\theta \\
&\leq \frac{2C^2}{\sqrt{V(x,\sqrt{\tau})V(y,\sqrt{\tau})}} \exp\left(-\frac{(d-2\sqrt{\tau})_+^2}{4t}\right). \qquad (5.23)
\end{aligned}$$

Choose $\tau = t/2$ (which is the maximal τ we can take). If $d \geq C\sqrt{t}$ where C is large enough then

$$\frac{(d-2\sqrt{\tau})_+^2}{4t} \geq \left(1-o(C^{-1})\right)\frac{d^2}{4t},$$

and (5.23) implies (5.19). If $d \leq C\sqrt{t}$ then the Gaussian term in (5.19) is of the order 1, and (5.19) follows again from (5.23) by discarding the Gaussian term in (5.23). ■

Theorem 5.8 admits a localized version. We say that the manifold M admits a *restricted* mean-value property (MV$xy\tau_0$), for some $x, y \in M$ and $\tau_0 \in \mathbb{R}_+$, if the inequality (5.17) holds for all $\tau \in (0,\tau_0]$ and for $\xi = x$ and $\xi = y$. If M admits (MV$xy\tau_0$) then a slight modification of the above proof yields the estimate

$$p(t,x,y) \leq \frac{C'}{\sqrt{V(x,\sqrt{\tau})V(y,\sqrt{\tau})}} \exp\left(-\lambda_1(M)t - \frac{d^2(x,y)}{2Dt}\right) \qquad (5.24)$$

where $\tau = \min(t/2, \tau_0)$. The term $\lambda_1(M)t$ appears if one applies (3.9) instead of (3.5) and uses the boundedness of τ.

Observe that the property (MV$xy\tau_0$) holds on *any* manifold. Namely, for any given $x, y \in M$, there exists τ_0 such that (MV$xy\tau_0$) is true (which provides another proof of (5.8)). However, the constant C in the mean-value inequality (5.17) depends on the certain geometric properties of the balls $B(x,\sqrt{\tau_0})$ and $B(y,\sqrt{\tau_0})$.

If the volume function $V(x,\cdot)$ satisfies the doubling condition (5.12) then (5.19) follows also from (5.18), by Theorem 5.3. In this case, $\sqrt{t/2}$ in (5.19) can be replaced by $\sqrt{t}$. It is not known whether there exists a manifold with (MV) for which $V(x,\cdot)$ is not doubling. Assuming the volume doubling property, one can improve the estimate (5.19) of Theorem 5.8 as follows.

Theorem 5.9 *Assume that the mean-value property (MV) holds on the manifold M and, for all $r' \geq r$ and $x \in M$,*

$$\frac{V(x,r')}{V(x,r)} \leq C\left(\frac{r'}{r}\right)^N, \tag{5.25}$$

with some $N > 0$. Then, for all $x, y \in M$ and $t > 0$,

$$p(t,x,y) \leq \frac{C'}{\sqrt{V(x,\sqrt{t})V(y,\sqrt{t})}}\left(1+\frac{d}{\sqrt{t}}\right)^{N-1}\exp\left(-\frac{d^2}{4t}\right) \tag{5.26}$$

where $d = d(x,y)$.

Remark 5.10 Although (5.25) looks stronger than the doubling property for $V(x,\cdot)$, these two properties are, in fact, equivalent. However, we have preferred (5.25) because the exponent N enters the estimate (5.26) in the sharp way. Indeed, as was shown by Molchanov [105], if M is the sphere $\mathbb{S}^n$ and x and y are conjugate points on $\mathbb{S}^n$ then

$$p(t,x,y) \sim \frac{c}{t^{n/2}}\left(\frac{d}{\sqrt{t}}\right)^{n-1}\exp\left(-\frac{d^2}{4t}\right), \quad t \to 0.$$

Hence, the exponent $N-1$ in the polynomial correction term in (5.26) is sharp. See [51] for further results containing the polynomial correction term.

Proof. If $d^2 < 4t$ then (5.26) follows from (5.19). Assume now $d^2 \geq 4t$ and follow the argument of the previous proof. However, let us use (3.13) and (3.14) instead of (3.5). Then, instead of (5.23), we obtain

$$p(t,x,y) \leq \frac{4C^2}{\sqrt{V(x,\sqrt{\tau})V(y,\sqrt{\tau})}}\frac{\sqrt{t}}{(d-2\sqrt{\tau})_+}\exp\left(-\frac{(d-2\sqrt{\tau})_+^2}{4t}\right), \tag{5.27}$$

for any $\tau \leq t/2$. Let us choose $\tau = \frac{t^2}{d^2}$ which smaller than $t/2$, by $d^2 > 2t$. Then we have

$$\frac{(d-2\sqrt{\tau})^2}{4t} \geq \frac{d^2}{4t} - \frac{d\sqrt{\tau}}{t} = \frac{d^2}{4t} - 1.$$

Also, $d - 2\sqrt{\tau} = d - 2\frac{t}{d} \geq \frac{1}{2}d$, whence

$$\frac{\sqrt{t}}{d - 2\sqrt{\tau}} \leq \frac{2\sqrt{t}}{d}.$$

Finally, by (5.25),

$$V\left(x, \sqrt{\tau}\right) = V(x, \sqrt{t}\frac{\sqrt{t}}{d}) \geq C^{-1}V(x, \sqrt{t})\left(\frac{\sqrt{t}}{d}\right)^N.$$

After substituting all these inequalities into (5.27), we obtain (5.26). ∎

For other application of the mean-value property see [95], [94], [96], [97].

5.3 On-diagonal lower bounds and the volume growth

Here we show how to apply Theorem 5.2 to obtain on-diagonal lower bounds of the heat kernel.

Theorem 5.11 *(Coulhon - Grigor'yan [36]) Let M be a geodesically complete Riemannian manifold. Assume that, for some point $x \in M$ and all $r > r_0$,*

$$V(x, r) \leq Cr^N, \tag{5.28}$$

with some positive constants C and N. Then, for all $t > t_0$,

$$p(t, x, x) \geq \frac{1/4}{V(x, K\sqrt{t\log t})}, \tag{5.29}$$

where $K > 0$ depend on x, r_0, C, N and $t_0 := \max(r_0^2, e)$.

Proof. Take some $\rho > 0$ and denote $\Omega = B(x, \rho)$. By the semigroup identity, we have

$$\begin{aligned} p(2t, x, x) &= \int_M p^2(t, x, y)d\mu(y) \\ &\geq \int_\Omega p^2(t, x, y)d\mu(y) \\ &\geq \frac{1}{\mu(\Omega)}\left(\int_\Omega p(t, x, y)d\mu(y)\right)^2 \\ &= \frac{1}{\mu(\Omega)}\left(1 - \int_{M\setminus\Omega} p(t, x, y)d\mu(y)\right)^2. \end{aligned} \tag{5.30}$$

In the last line, we have used the stochastic completeness of M, that is,

$$\int_M p(t, x, y)d\mu(y) = 1.$$

By Theorem 3.4, this follows from the geodesic completeness of M and from the volume growth hypothesis (5.28).

Next we will choose $\rho = \rho(t)$ so that

$$\int_{M\setminus B(x,\rho)} p(t,x,y)d\mu(y) \le \frac{1}{2}. \tag{5.31}$$

Assume for the moment that (5.31) holds. Then (5.30) yields

$$p(2t,x,x) \ge \frac{1/4}{V(x,\rho(t))}.$$

To match (5.29), we need to estimate $\rho(t)$ as follows

$$\rho(t) \le \mathrm{const}\sqrt{t\log t}. \tag{5.32}$$

Let us prove (5.31) with $\rho(t)$ satisfying (5.32). We apply again the Cauchy–Schwartz inequality as follows, denoting $d = d(x,y)$ and taking some $D > 2$,

$$\begin{aligned}\left[\int_{M\setminus B(x,\rho)} p(t,x,y)d\mu(y)\right]^2 &\le \int_M p^2(t,x,y)\exp\left(\frac{d^2}{Dt}\right)\int_{M\setminus B(x,\rho)}\exp\left(-\frac{d^2}{Dt}\right)\\ &= E_D(t,x)\int_{M\setminus B(x,\rho)}\exp\left(-\frac{d^2}{Dt}\right)d\mu(y), \qquad (5.33)\end{aligned}$$

where $E_D(t,x)$ is defined by (5.3). By Theorem 5.2, we have, for all $t > t_0$,

$$E_D(t,x) \le E_D(t_0,x) < \infty. \tag{5.34}$$

Since x is fixed, we can consider $E_D(t_0,x)$ as a constant. Let us now estimate the integral in (5.33) assuming that $\rho = \rho(t) > r_0$. By splitting the integral over the complement of $B(x,\rho)$ into the sum of the integrals over the annuli $B(x,2^{k+1}\rho)\setminus B(x,2^k\rho)$, $k = 0,1,2,\ldots$, and using the hypothesis (5.28), we obtain

$$\begin{aligned}\int_{M\setminus B(x,\rho)}\exp\left(-\frac{d(x,y)^2}{Dt}\right)d\mu(y) &\le \sum_{k=0}^{\infty}\exp\left(-\frac{4^k\rho^2}{Dt}\right)V(x,2^{k+1}\rho) \qquad (5.35)\\ &\le C2^N\rho^N\sum_{k=0}^{\infty}2^{Nk}\exp\left(-\frac{4^k\rho^2}{Dt}\right) \qquad (5.36)\end{aligned}$$

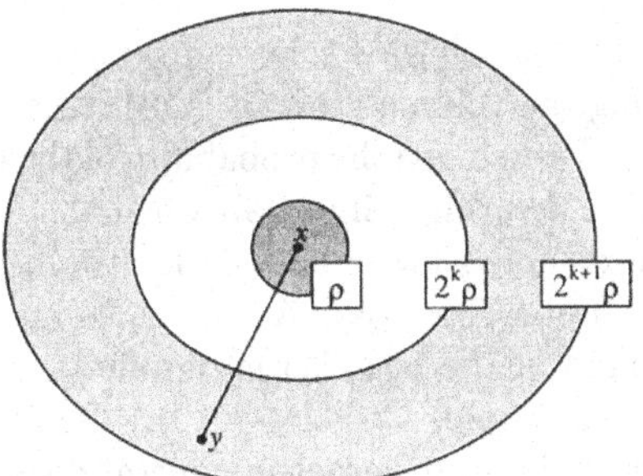

Figure 8 Annulus $B(x, 2^{k+1}\rho) \setminus B(x, 2^k\rho)$

Assuming $\rho^2/t \geq 1$, the sum in the line above is majorized by a geometric series whence

$$\int\limits_{M \setminus B(x,\rho)} \exp\left(-\frac{d(x,y)^2}{Dt}\right) d\mu(y) \leq C' \rho^N \exp\left(-\frac{\rho^2}{Dt}\right). \tag{5.37}$$

By setting $\rho(t) = K\sqrt{t \log t}$ with K large enough, we make the integral above arbitrarily small, whence (5.31) follows by (5.33) and (5.34). To finish the proof, we have to make sure that $\rho(t) > r_0$. Indeed, this follows from $t > t_0 = \max(r_0^2, e)$ and $K > 1$.

■

One may wonder what is geometric background of the quantity $E_D(t_0, x)$ which we have interpreted as a constant. In fact, an upper bound of it can be proved in terms of an intrinsic geometric property of the ball $B(x, \varepsilon)$, for arbitrarily small ε - see [66] (this can be extracted also from Theorems 5.3 and 6.7). The geometric property in question is a Sobolev inequality in $B(x, \varepsilon)$ which holds because the geometry of $B(x, \varepsilon)$ is nearly Euclidean. In particular, the constant K does not depend on x if the manifold M has bounded geometry (see Section 7.5).

Note that no off-diagonal lower bound of the heat kernel can be proved under such a mild assumption as (5.28). Indeed, the manifold M may consist of two large parts connected by a thin tube. Suppose that x belongs to one part and y - to another.

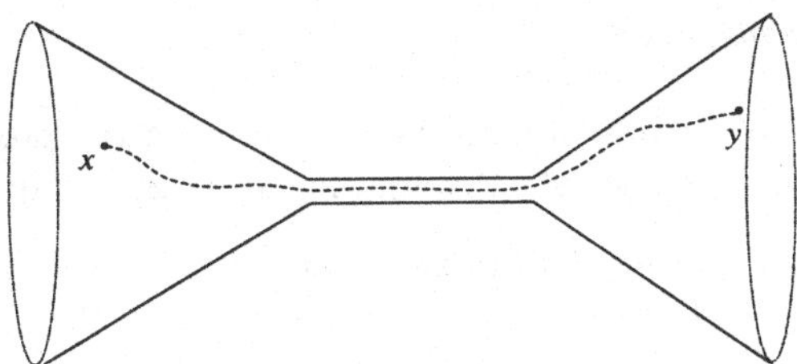

Figure 9 Manifold with a bottleneck

Then by making the tube thinner, one can get $p(t,x,y)$ to be arbitrarily small, without violating the volume growth (5.28). It is especially clear from the probabilistic point of view since the probability of the Brownian motion X_t getting from x to y can be arbitrarily small when the tube shrinks. Hence, the situation with off-diagonal lower bounds for the heat kernel is entirely different than that of upper bounds. As we have seen in Section 5.1, an on-diagonal upper bound of the heat kernel implies a Gaussian off-diagonal upper bound (see, for example, Corollary 5.5). On the contrary, the on-diagonal lower bound of the heat kernel in general does not imply anything about the off-diagonal values of the heat kernel.

Comparing the lower bound (5.29) with the upper bounds (5.18) and (5.19) (which hold, for example, on non-negatively curved manifolds) we see that both are governed by the volume of balls but with different radii[1]. Indeed, the former radius is of the order $\sqrt{t\log t}$ whereas the later is of the order $\sqrt{t}$. The radius $\sqrt{t}$ matches the heat kernel behaviour in $\mathbb{R}^n$ where we have

$$p(t,x,x) = \frac{1}{(4\pi t)^{n/2}} = \frac{\text{const}}{V(x,\sqrt{t})}.$$

There is an example [36] showing that in the lower bound (5.29), one cannot in general get rid of $\log t$ assuming only the hypotheses of Theorem 5.11. However, under certain additional hypotheses, it is possible as is shown by the following statement (cf. Theorem 5.8).

Theorem 5.12 *(Coulhon - Grigor'yan [36]) Let M be a geodesically complete Riemannian manifold. Assume that, for some point $x \in M$ and all $r > 0$*

$$V(x,2r) \le CV(x,r) \tag{5.38}$$

and, for all $t > 0$,

$$p(t,x,x) \le \frac{C}{V(x,\sqrt{t})}. \tag{5.39}$$

Then, for all $t > 0$,

$$p(t,x,x) \ge \frac{c}{V(x,\sqrt{t})}, \tag{5.40}$$

where $c > 0$ depends on C.

Proof. The proof follows almost the same line as the proof of Theorem 5.11. The difference comes when estimating $E_D(t,x)$. Instead of using the

[1] The function $\rho(t)$ satisfying (5.31) is closely related to the escape rate of the Brownian motion - see [75], [73] and [76].

monotonicity of $E_D(t,x)$, we apply Theorem 5.3. Indeed, by Theorem 5.3, the hypotheses (5.39) and (5.38) yield

$$E_D(t,x) \le \frac{C'}{V(x,\sqrt{t})}.$$

By substituting this into (5.33) and applying the doubling property (5.38) to estimate the sum in (5.35), we obtain instead of (5.37)

$$\int_{M\setminus B(x,\rho)} \exp\left(-\frac{d(x,y)^2}{Dt}\right) d\mu(y) \le C'' \exp\left(-\frac{\rho^2}{Dt}\right). \tag{5.41}$$

Hence, the integral in (5.41) can be made arbitrarily small by choosing $\rho = K\sqrt{t}$ with K large enough.

Finally, one uses again the doubling property to write

$$p(2t,x,x) \ge \frac{1/4}{V(x,\rho(t))} \ge \frac{c}{V(x,\sqrt{2t})},$$

finishing the proof. ∎

Theorem 5.11 can be extended to a more general volume growth assumption as follows.

Theorem 5.13 *([36, Theorem 6.1]) Let M be a geodesically complete Riemannian manifold. Assume that, for some point $x \in M$ and all $r > r_0$,*

$$V(x,r) \le \mathcal{V}(r),$$

where $\mathcal{V}(r) > 2$ is a continuous increasing function on (r_0,∞) such that $\frac{r^2}{\log \mathcal{V}(r)}$ is strictly decreasing in r. Define the function $\rho(t)$ by

$$t = \frac{\rho^2(t)}{\log \mathcal{V}(\rho(t))},$$

for $t > t_0 = t_0(r_0)$. Then, for all $t > t_0$,

$$p(t,x,x) \ge \frac{1/4}{V(x,\rho(Kt))},$$

where $K > 1$ depends on x and r_0.

For example, of $\mathcal{V}(r) = \exp(r^\alpha)$, $0 < \alpha < 2$, then we obtain $\rho(t) \asymp t^{\frac{1}{2-\alpha}}$ and

$$p(t,x,x) \ge c\exp\left(-Ct^{\frac{\alpha}{2-\alpha}}\right). \tag{5.42}$$

As we will see in Section 7.7, if $\alpha \leq 1$ then the exponent $\frac{\alpha}{2-\alpha}$ in (5.42) is sharp - cf. (7.56).

However, (5.42) is not sharp if $\alpha > 1$, that is, if $V(x,r)$ grows superexponentially in r. Indeed, $p(t,x,x)$ cannot decay faster than exponentially in t as is said by the following statement.

Proposition 5.14 *For any manifold M, for any $x \in M$ and $\varepsilon > 0$, there exists $c = c_x > 0$ such that*

$$p(t,x,x) \geq c_x \exp\left(-\left(\lambda_1(M) + \varepsilon\right) t\right), \quad \forall t > 0, \tag{5.43}$$

where $\lambda_1(M)$ is the spectral radius defined by (3.10).

Proof. Take a precompact region Ω containing x and such that

$$\lambda_1(\Omega) \leq \lambda_1(M) + \varepsilon.$$

We have $p(t,x,x) \geq p_\Omega(t,x,x)$ and, by the eigenfunction expansion (2.9),

$$p_\Omega(t,x,x) = \sum_{k=1}^{\infty} e^{-\lambda_k(\Omega)t}\phi_k^2(x) \geq e^{-\lambda_1(\Omega)t}\phi_1^2(x),$$

whence (5.43) follows. ∎

Combining Proposition 5.14 with the upper bound (5.8), we obtain

Corollary 5.15 *(Li [93]) For any manifold M and for all $x \in M$,*

$$\lim_{t\to\infty} \frac{\log p(t,x,x)}{t} = -\lambda_1(M).$$

6 On-diagonal upper bounds and Faber-Krahn inequalities

In this section, we discuss mainly on-diagonal upper bounds of the heat kernel of the type

$$p(t,x,x) \leq \frac{C}{f(t)}. \tag{6.1}$$

As we know from Section 5.1, the on-diagonal upper bound implies the off-diagonal Gaussian upper bound (5.14). The main emphasis will be made on geometric background of the estimate (6.1). We will consider two situations when the estimate (6.1) is well understood:

1. a uniform estimate when (6.1) is meant to hold for all $t > 0$ and $x \in M$ with *the same* function f;
2. a "relative" estimate when the function $f(t)$ depends on x as follows: $f(t) = V(x, \sqrt{t})$ (cf. (5.18)).

6.1 Polynomial decay of the heat kernel

Here we describe in the historical order the results related to the heat kernel upper bound

$$p(t,x,x) \le \frac{C}{t^{n/2}}, \quad \forall x \in M,\ t > 0, \tag{6.2}$$

which is obviously motivated by the heat kernel in $\mathbb{R}^n$. One may ask under what geometric assumptions (6.2) holds? Historically the first result was obtained by Nash [109] who discovered a simple method of deducing (6.2) from the Sobolev inequality. The latter is the following assertion

for any $f \in C_0^\infty(M)$, $f \ge 0$,

$$\left(\int_M f^{\frac{2n}{n-2}} d\mu\right)^{\frac{n-2}{n}} \le C \int_M |\nabla f|^2 \, d\mu \tag{6.3}$$

(of course, we assume $n > 2$ here).

It is well known that the Sobolev inequality (6.3) holds in $\mathbb{R}^n$. However, for a general manifold, it may not be true. In fact, the Sobolev inequality is quite sensitive to the geometry of the manifold and can be regarded itself as a geometric condition. It can be deduced from the following isoperimetric inequality: for all precompact regions Ω with smooth boundary

$$\sigma(\partial\Omega) \ge c\mu(\Omega)^{\frac{n-1}{n}} \tag{6.4}$$

(see [61] and [101]). The inequality (6.4) is well-known in $\mathbb{R}^n$. It is an obvious consequence of the isoperimetric property of a ball in $\mathbb{R}^n$: any region of the same volume as the ball has larger boundary area unless it is the ball. We will discuss the isoperimetric type inequalities in more details in Section 7.

Following Nash's argument, let us prove the following Theorem.

Theorem 6.1 *If the Sobolev inequality (6.3) is true on M then (6.2) holds, too.*

Proof. By the exhaustion argument, it suffices to prove (6.2) for p_Ω where Ω is a precompact open subset of M with smooth boundary. Fix $y \in \Omega$ and denote $u(t,x) = p(t,x,y)$ and

$$J(t) := \int_\Omega u^2(t,x) d\mu(x).$$

Arguing as in Section 3.1, we obtain

$$J'(t) = 2\int u\, u_t \, d\mu = 2\int u\Delta u \, d\mu = -2\int |\nabla u|^2 \, d\mu. \tag{6.5}$$

As in Section 3.1, we can conclude from (6.5) that $J(t)$ is non-increasing. However, we can go further by estimating the right hand side of (6.5) by using (6.3). Indeed, it is easy to see that the Sobolev inequality extends to functions like $u(\cdot,t)$ vanishing on $\partial\Omega$ whence

$$\int |\nabla u|^2 \, d\mu \geq c \left(\int u^{\frac{2n}{n-2}} d\mu \right)^{\frac{n-2}{n}}. \tag{6.6}$$

We would like to have on the right hand side of (6.6) $\int u^2$ in order to be able to create a differential inequality for $J(t)$. To that ends, we use the Hölder interpolation inequality

$$\left(\int u^{\alpha} d\mu \right)^{\frac{1}{\alpha-1}} \left(\int u d\mu \right)^{\frac{\alpha-2}{\alpha-1}} \geq \int u^2 d\mu, \tag{6.7}$$

which is true for all $\alpha > 2$. Naturally, we take here $\alpha = \frac{2n}{n-2}$ and obtain from (6.6) *the Nash inequality*

$$\int |\nabla u|^2 \, d\mu \geq c \left(\int u^2 d\mu \right)^{\frac{n+2}{n}} \left(\int u d\mu \right)^{-\frac{4}{n}}. \tag{6.8}$$

Observing that

$$\int u(t,x) d\mu(x) = \int p_\Omega(t,x,y) d\mu(x) \leq 1,$$

we deduce from (6.5) and (6.8) the differential inequality

$$J' \leq -cJ^{\frac{n+2}{n}}.$$

By integrating it from 0 to t, we find $J(t) \leq Ct^{-n/2}$. We are left to observe that by the semigroup property $J(t) = p_\Omega(2t,y,y)$, whence (6.2) follows. ■

In 1967, Aronson [4] proved his famous two-sided Gaussian estimates for the heat kernel associated with a uniformly elliptic operator in $\mathbb{R}^n$ (see also [116], [60], [124]). In our notation, the Aronson upper bound can be written in the form

$$p(t,x,y) \leq \frac{C}{t^{n/2}} \exp\left(-\frac{d^2(x,y)}{Ct} \right), \tag{6.9}$$

assuming that manifold M is $\mathbb{R}^n$ equipped with a Riemannian metric that is quasi-isometric to the Euclidean one. Now we know that (6.9) follows from (6.2) by Theorem 5.3. The proof of Aronson was different and used the integral maximum principle (see Lemma 3.1). Some versions of his proof can be found in [116], [66], [69].

In 1985, Varopoulos [133] proved that the Sobolev inequality is not only sufficient but also necessary for the on-diagonal upper bound (6.2). Another proof of that will follow from the results of Section 6.2 (cf. Theorem 6.5).

Two years later, Carlen, Kusuoka and Stroock [19] proved that (6.2) is also equivalent to the Nash inequality (6.8). They were also able to localize the heat kernel estimates for small and large time t so that the exponent n could be different for $t \to 0$ and for $t \to \infty$. Another method of doing so will be considered in Sections 6.3 and 6.2.

In 1987-89, Davies [41], [42], [43] proved that the on-diagonal upper bound (6.2) is equivalent to the log-*Sobolev inequality*: for any $f \in C_0^\infty(M)$, $f \geq 0$, and for any $\varepsilon > 0$,

$$\int f^2 \log \frac{f}{\|f\|_2} d\mu \leq \varepsilon \int |\nabla f|^2 \, d\mu + \beta(\varepsilon) \int f^2 d\mu \tag{6.10}$$

where $\|f\|_2 = \left(\int f^2 d\mu\right)^{1/2}$ and $\beta(\varepsilon) = C - \frac{n}{4} \log \varepsilon$. Davies also created a powerful method of proving the off-diagonal upper bounds like (6.9) using (6.10), which is called the semigroup perturbation method. A detailed account of it can be found in [43]. In the present paper, we have focused on two more recent methods of obtaining the Gaussian bounds - one based on the Davies inequality (3.5) and on the mean value property (5.17), and the other based on Ushakov's argument, which was stated in Corollary 5.5.

In 1994, Carron [20] and the author [69] proved that the on-diagonal upper bound (6.2) is equivalent to the *Faber-Krahn inequality:* for all precompact open sets $\Omega \subset M$,

$$\lambda_1 (\Omega) \geq c \mu (\Omega)^{-2/n}, \tag{6.11}$$

where $\lambda_1 (\Omega)$ is the lowest eigenvalue of the Dirichlet Laplace operator in Ω. The classical theorem of Faber and Krahn says that (6.11) holds in $\mathbb{R}^n$ with the constant c such that the equality in (6.11) is attained when Ω is a ball. In general, we do not need a sharp constant in (6.11) to obtain the heat kernel estimates.

By the variational principle, we have

$$\lambda_1(\Omega) = \inf_{\substack{f \in C_0^\infty(\Omega) \\ f \not\equiv 0}} \frac{\int_\Omega |\nabla f|^2 \, d\mu}{\int_\Omega f^2 d\mu}. \tag{6.12}$$

Hence, (6.11) can be rewritten as

$$\int_\Omega |\nabla f|^2 \, d\mu \geq c \mu (\Omega)^{-2/n} \int_\Omega f^2 d\mu, \quad \forall f \in C_0^\infty (\Omega). \tag{6.13}$$

It is not difficult to deduce the Nash inequality (6.8) directly from (6.13) - see Lemma 6.3 below.

Hence, we have the following equivalences:

log-Sobolev inequality (6.10)	$\Leftrightarrow$	Sobolev inequality (6.3)
$\Updownarrow$		$\Updownarrow$
Off-diagonal Gaussian bound (6.9)	$\Leftrightarrow$	On-diagonal bound (6.2)
$\Updownarrow$		$\Updownarrow$
Faber-Krahn inequality (6.11)	$\Leftrightarrow$	Nash inequality (6.8)

In the next sections, we will discuss similar relationships between more general heat kernel upper bounds and modifications of the Faber-Krahn inequality.

6.2 Arbitrary decay of the heat kernel

It is natural to ask what geometric or functional-analytic properties of the manifold M are responsible for the heat kernel bound as follows:

$$p(t,x,x) \leq \frac{C}{f(t)}, \quad \forall x \in M,\ t>0, \tag{6.14}$$

where $f(t)$ is a prescribed[2] increasing function on $(0,\infty)$. The case $f(t) = t^{n/2}$ was considered above. However, there are plenty of simple examples of manifolds where such a function is not enough to describe the heat kernel behaviour. To start with, let us consider the manifolds $M = \mathbb{R}^m \times \mathbb{S}^k$ of the dimension $n = m+k$. Since the local structure of M is similar to that of $\mathbb{R}^n$, one may expect that, for short time t, we have $p(t,x,x) \asymp t^{-n/2}$ like in $\mathbb{R}^n$ (cf. (5.24)). However, in the large scale, M resembles $\mathbb{R}^m$ and, by (2.24), the long time asymptotic of $p(t,x,x)$ also looks like in $\mathbb{R}^m$. This motivates considering the following function

$$f(t) = \begin{cases} t^{n/2}, & t \leq 1, \\ t^{m/2}, & t>1. \end{cases} \tag{6.15}$$

On the hyperbolic space, the heat kernel decays exponentially in time as is seen from (1.2). One may presume that there are manifolds with super-polynomial but subexponential decay of $p(t,x,x)$ as $t \to \infty$, and this is true. This motivates us to consider the function

$$f(t) = \begin{cases} t^{n/2}, & t \leq 1, \\ \exp\left(t^{\alpha}\right), & t > 1. \end{cases} \tag{6.16}$$

[2]As Proposition 5.14 says, $p(t,x,x)$ decays at most exponentially as $t \to \infty$. Therefore, $f(t)$ should grow at most exponentially, too.

It is natural to try and extend the results of the preceding section to a wider class of functions f . The extension of the log-Sobolev inequality matching rather general $f(t)$ was obtained by Davies and can be found in his book [43]. A generalized Faber-Krahn inequality equivalent in some sense to (6.14), was obtained by the author [69] and will be discussed below. Finally, a generalized Nash inequality, also equivalent to (6.14), is due to Coulhon [35]. It seems that a proper generalization of the Sobolev inequality is not know yet (see [21], though).

Suppose that M is connected, non-compact and geodesically complete, and let Ω be a precompact region in M with smooth boundary. By (2.13), the long time asymptotic of $p_\Omega(t,x,x)$ reads as follows:

$$p_\Omega(t,x,x) \sim \exp\left(-\lambda_1(\Omega)\, t\right) \phi_1^2(x), \quad t \to \infty. \tag{6.17}$$

One may want to pass to the limit in (6.17) as $\Omega \to M$. Since $\lambda_1(\Omega)$ is decreasing on enlargement of Ω, the limit $\lim_{\Omega\to M} \lambda_1(\Omega)$ exists and coincides with the spectral radius $\lambda_1(M)$ (see (3.10)). If $\lambda_1(M) > 0$ then one may expect that $p(t,x,x)$ behaves like $\exp\left(-\lambda_1(M)t\right)$ as $t \to \infty$. Indeed, (5.7) and (5.4) imply

$$p(t,x,x) \leq \exp\left(-\lambda_1(M)(t-t_0)\right) p(t_0,x,x)\,. \tag{6.18}$$

This estimate is good when $\lambda_1(M) > 0$ (cf. (5.43)) but becomes trivial if $\lambda_1(M) = 0$. As the matter of fact, $\lambda_1(M) = 0$ for all geodesically complete manifolds with subexponential volume growth (which follows from the theorem of Brooks [17]). The latter means that, for some $x \in M$,

$$V(x,r) = \exp\left(o(r)\right), \quad r \to \infty.$$

Hence, the case $\lambda_1(M) = 0$ is most interesting from our point of view. One may wonder, if the rate of convergence of $\lambda_1(\Omega)$ to 0 as $\Omega \to M$ affects the rate of convergence of $p(t,x,x)$ to 0 as $t \to \infty$. In fact, it does if one understands the former as *a Faber-Krahn type inequality*

$$\lambda_1(\Omega) \geq \Lambda(\mu(\Omega)), \tag{6.19}$$

where Λ is a positive decreasing function on $(0,\infty)$. As we have mentioned in the previous section, (6.19) is true on $\mathbb{R}^n$ with the function $\Lambda(v) = cv^{-2/n}$. It turns out that inequality (6.19) can be proved in many interesting cases with various functions Λ. We will call Λ a *Faber-Krahn function* of M, assuming that (6.19) holds for all precompact $\Omega \subset M$.

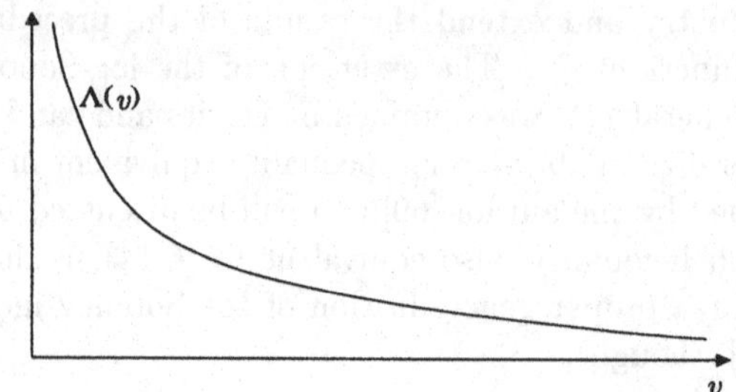

Figure 10 Example of a Faber-Krahn function

Theorem 6.2 *([69]) Assume that manifold M admits a Faber-Krahn function Λ. Let us define the function $f(t)$ by*

$$t = \int_0^{f(t)} \frac{dv}{v\Lambda(v)}, \tag{6.20}$$

assuming the convergence of the integral in (6.20) at 0. Then, for all $t > 0$, $x \in M$, and $\varepsilon > 0$,

$$p(t, x, x) \leq \frac{2\varepsilon^{-1}}{f((1-\varepsilon)t)}. \tag{6.21}$$

Examples: 1. If $\Lambda(v) = cv^{-2/n}$ then (6.20) yields $f(t) = c't^{n/2}$. Hence, (6.21) amounts to (6.2).

2. Let

$$\Lambda(v) \asymp \begin{cases} v^{-2/n}, & v \leq 1, \\ v^{-2/m}, & v > 1. \end{cases} \tag{6.22}$$

For example, the manifold $M = K \times \mathbb{R}^m$, where K is a compact manifold of the dimension $n - m$, admits the Faber-Krahn function (6.22) (see Section 7.5). Then (6.20) gives

$$f(t) \asymp \begin{cases} t^{n/2}, & t \leq 1, \\ t^{m/2}, & t > 1, \end{cases}$$

and

$$\sup_x p(t, x, x) \leq \frac{C}{t^{m/2}}, \quad \forall t > 1.$$

3. Assume

$$\Lambda(v) \asymp \log^{-\alpha} v, \quad v > 2,$$

and $\Lambda(v) \asymp v^{-2/n}$ for $v < 2$ (see Section 7.6 for examples of manifolds with this Λ). Then, for large t,

$$f(t) \asymp c_1 \exp\left(c_2 t^{\frac{1}{1+\alpha}}\right)$$

and

$$\sup_x p(t,x,x) \le C \exp\left(-ct^{\frac{1}{1+\alpha}}\right).$$

4. Let us take

$$\Lambda(v) \equiv \lambda_1(M), \quad v > 1,$$

and $\Lambda(v) \asymp v^{-2/n}$ for $v < 1$ (note that the constant function $\Lambda(v) \equiv \lambda_1(M)$ satisfies (6.19) but the integral (6.20) diverges, so we have to modify it near $v = 0$). Then, for large t,

$$f(t) \asymp \exp(\lambda_1(M)t)$$

and

$$\sup_x p(t,x,x) \le C \exp\left(-\left(\lambda_1(M) - \varepsilon\right)t\right).$$

In fact, in this case ε can be taken 0 (this can be seen from the proof below or from (6.18)).

Proof of Theorem 6.2. Fix a point $y \in M$. We will prove that, for any precompact open set Ω with smooth boundary,

$$p_\Omega(t,y,y) \le \frac{C_\varepsilon}{f((1-\varepsilon)t)},$$

provided $y \in \Omega$. Let us start as in the proof of Theorem 6.1: denote $u(t,x) = p_\Omega(t,x,y)$,

$$J(t) := \int_\Omega u^2(t,x)d\mu(x)$$

and obtain

$$J'(t) = -2\int_\Omega |\nabla u|^2 \, d\mu. \tag{6.23}$$

Next, we have to estimate $\int |\nabla u|^2 d\mu$ from below via $\int u^2 d\mu$. The simplest way to do so is by using the variational property of the first eigenvalue which gives

$$\int_\Omega |\nabla u|^2 d\mu \ge \lambda_1(\Omega)\int_\Omega u^2 d\mu \ge \Lambda(\mu(\Omega))\int_\Omega u^2 d\mu.$$

However, this is not suitable for us because the resulting estimate of u will depend on Ω. The following lemma provides a more sophisticated way of applying (6.19).

Lemma 6.3 *Assuming that (6.19) holds, we have, for any non-negative function $u \in C^2(\Omega) \cap C(\overline{\Omega})$ vanishing on $\partial\Omega$,*

$$\int |\nabla u|^2 \ge (1-\varepsilon)\left(\int u^2\right)\Lambda\left(\frac{2}{\varepsilon}\frac{\left(\int u\right)^2}{\int u^2}\right), \tag{6.24}$$

for any $\varepsilon \in (0,1)$.

Remark 6.4 If $\Lambda(v) = cv^{-2/n}$ then (6.24) becomes the Nash inequality (6.8). Hence, (6.24) can be considered as a generalized Nash inequality.

Proof. The proof follows the argument of Gushchin [80]. Denote for simplicity $A = \int u d\mu$ and $B = \int u^2 d\mu$. For any positive s, we have the obvious inequality

$$u^2 \leq (u-s)_+^2 + 2su,$$

which implies, by integration,

$$B \leq \int_{\{u>s\}} (u-s)^2 d\mu + 2sA. \tag{6.25}$$

Applying the Faber-Krahn inequality (6.19) in the region $\Omega_s := \{u > s\}$ (observe that $u - s$ vanishes on the boundary $\partial\Omega_s$), we get

$$\int_{\Omega_s} (u-s)^2 d\mu \leq \frac{\int_{\Omega_s} |\nabla u|^2 d\mu}{\Lambda(\mu(\Omega_s))}. \tag{6.26}$$

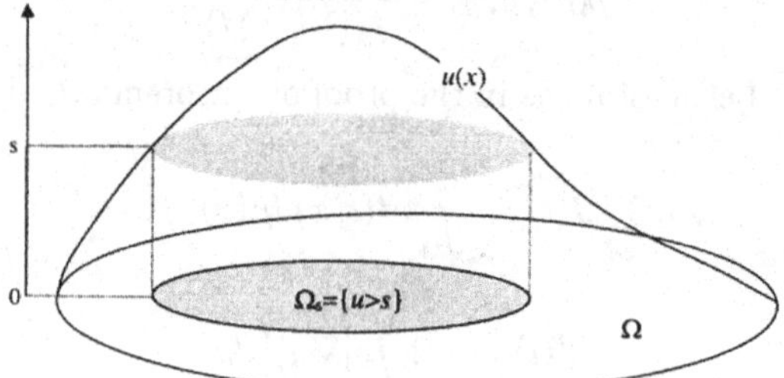

Figure 11 Applying a Faber-Krahn inequality for the region Ω_s

Unlike Ω, the region Ω_s admits estimating of its volume via the function u, as follows

$$\mu(\Omega_s) \leq \frac{1}{s}\int_\Omega u\, d\mu = s^{-1}A.$$

Hence, (6.25) and (6.26) imply

$$B \leq \frac{\int_{\Omega_s} |\nabla u|^2 d\mu}{\Lambda(s^{-1}A)} + 2sA$$

whence

$$\int_\Omega |\nabla u|^2 d\mu \geq (B - 2sA)\Lambda(s^{-1}A).$$

Taking here $s = \frac{\varepsilon B}{2A}$, we finish the proof. ∎

Applying Lemma 6.3 to estimate the right hand side of (6.23) and taking into account that

$$\int u(t,x)d\mu(x) \leq 1,$$

we obtain

$$J' \leq -2(1-\varepsilon) J\Lambda\left(\frac{2}{\varepsilon}\frac{1}{J}\right).$$

Dividing this inequality by the right hand side, integrating it against dt from 0 to t and changing the variables $v = 2\varepsilon^{-1}J^{-1}$, we obtain

$$\int_0^{2\varepsilon^{-1}J^{-1}} \frac{dv}{v\Lambda(v)} \geq 2(1-\varepsilon)t$$

whence, by the definition (6.20) of function f,

$$J(t) \leq \frac{2\varepsilon^{-1}}{f(2(1-\varepsilon)t)}.$$

We are left to notice that $J(t) = p_\Omega(2t, y, y)$, and (6.21) follows. ■

If the function Λ satisfying the hypotheses of Theorem 6.2 is continuous then $f(t) \in C^1(\mathbb{R}_+)$ and

$$f' > 0, \quad f(0) = 0, \quad f(\infty) = \infty \quad \text{and} \quad \frac{f'}{f} \text{ is non-increasing.} \tag{6.27}$$

Conversely, if $f \in C^1(\mathbb{R}_+)$ satisfies (6.27) then Λ from (6.20) can be recovered by

$$\Lambda(f(t)) = \frac{f'}{f}. \tag{6.28}$$

The following Theorem is almost converse to Theorem 6.2.

Theorem 6.5 *([69]) Let the heat kernel on the manifold M admit the following estimate*

$$\sup_x p(t,x,x) \leq \frac{1}{f(t)}, \quad \forall t > 0, \tag{6.29}$$

where $f \in C^1(\mathbb{R}_+)$ satisfies (6.27) and certain regularity condition below. Then M admits the Faber-Krahn function $c\Lambda(v)$ where Λ is defined by (6.28).

Moreover, for any precompact region $\Omega \subset M$ and any integer $k \geq 1$,

$$\lambda_k(\Omega) \geq c\Lambda\left(\frac{\mu(\Omega)}{k}\right). \tag{6.30}$$

We say that a function $g(t)$ has *at most polynomial decay* if, for some $\alpha > 0$ and $a \in [1,2]$,

$$g(at) \geq \alpha g(t), \quad \forall t > 0. \tag{6.31}$$

The the regularity condition in the statement of Theorem 6.5 is as follows: the function $g := (\log f)'$ has at most polynomial decay. For example, the latter holds if, for some large $A > 0$,

$$\frac{f''}{f'} \geq \frac{f'}{f} - \frac{A}{t}.$$

All examples of f considered above, satisfy this condition. On the contrary, $f(t) = 1 - e^{-t}$ does not satisfy it.

Proof. The hypotheses (6.29) implies $p_\Omega(t,x,x) \leq \frac{1}{f(t)}$ whence

$$\int_\Omega p_\Omega(t,x,x)d\mu(x) \leq \frac{\mu(\Omega)}{f(t)}.$$

On the other hand, by the eigenfunction expansion (2.9),

$$\int_\Omega p_\Omega(t,x,x)d\mu(x) = \int_\Omega \sum_{i=1}^\infty e^{-\lambda_i(\Omega)t}\phi_i^2(x)d\mu(x) = \sum_{i=1}^\infty e^{-\lambda_i(\Omega)t}.$$

The right hand side here is bounded from below by $ke^{-\lambda_k(\Omega)t}$, whence

$$\frac{\mu(\Omega)}{f(t)} \geq ke^{-\lambda_k(\Omega)t}$$

and

$$\lambda_k(\Omega) \geq \frac{1}{t}\log\frac{kf(t)}{\mu(\Omega)}. \tag{6.32}$$

This inequality holds for all $t > 0$ so that we can choose t. Let us find t from the equation

$$f(t/2) = \mu(\Omega)/k.$$

For this t, we obtain from (6.32)

$$\lambda_k(\Omega) \geq \frac{1}{t}\left(\log f(t) - \log f(t/2)\right) = \frac{1}{2}g(\theta),$$

where $g := (\log f)'$ and $\theta \in (t/2, t)$. By the regularity condition (6.31), we have $g(\theta) \geq \alpha g(t/2)$. Finally, we apply (6.28):

$$g(t/2) = \frac{f'(t/2)}{f(t/2)} = \Lambda\left(f(t/2)\right) = \Lambda(\frac{\mu(\Omega)}{k})$$

whence

$$\lambda_k(\Omega) \geq \frac{\alpha}{2}\Lambda(\frac{\mu(\Omega)}{k}),$$

which was to be proved. ■

It is follows from Theorems 6.2 and 6.5 that the heat kernel upper bound (6.29) is equivalent to the certain Faber-Krahn type inequality, up to some constant multiples. Clearly, the generalized Nash inequality (6.24) involved in the proof of Theorem 6.2, is also equivalent to each of these hypotheses. The latter was further developed in an abstract semigroup setting by Coulhon [35]. He also gave another proof of the first part of Theorem 6.5 avoiding usage of the eigenfunction expansion.

Putting together Theorems 6.2 and 6.5, we obtain

Corollary 6.6 *Suppose that the function Λ satisfies the hypotheses of Theorems 6.2 and 6.5. If, for any precompact open set $\Omega \subset M$,*

$$\lambda_1(\Omega) \geq \Lambda(\mu(\Omega)) \tag{6.33}$$

then, for all integers $k \geq 1$,

$$\lambda_k(\Omega) \geq c\Lambda(C\mu(\Omega)). \tag{6.34}$$

For example, for the Euclidean function $\Lambda(v) = cv^{-2/n}$, we obtain

$$\lambda_k(\Omega) \geq c' \left(\frac{k}{\mu(\Omega)}\right)^{2/n}.$$

Note that (6.34) does not follow from (6.33) for an *individual* set Ω: it is essential that (6.33) holds for *all* Ω.

By reverting the Faber-Krahn inequality (6.33), it is possible to prove some lower bounds of the heat kernel - see [36, Theorem 3.2].

6.3 Localized upper estimate

We consider here the localized version of Theorem 6.1.

Theorem 6.7 *([66], [69]) Suppose that, for some $x \in M$ and $r > 0$, the following Faber-Krahn inequality holds: for any precompact open set $\Omega \subset B(x,r)$*

$$\lambda_1(\Omega) \geq a\mu(\Omega)^{-2/n}, \tag{6.35}$$

where $a > 0$ and $n > 0$. Then, for any $t > 0$,

$$p(t,x,x) \leq \frac{Ca^{-n/2}}{\min(t,r^2)^{n/2}}, \tag{6.36}$$

where $C = C(n)$.

Remark 6.8 This theorem contains Theorem 6.1 as $r \to \infty$, taking into account that (6.35) is equivalent to (6.3). However, the method of proof of Theorems 6.1 and 6.2 does not work in the setting of Theorem 6.7 because it requires (6.35) for all Ω, not only for those in $B(x,r)$.

The coefficient a in (6.36) can be absorbed into the constant C. However, by varying the ball $B(x,r)$, we may have different a for different balls so that the exact dependence on a in (6.36) may be crucial. In the next section we will consider a setting where a depends explicitly on $B(x,r)$.

Proof. We start with the following mean-value type inequality.

Lemma 6.9 *Suppose that the Faber-Krahn inequality (6.35) holds for any precompact open set $\Omega \subset B(x,r)$. Then, for any $\tau \in (0,r^2]$, $t \geq \tau$ and for any positive solution u of the heat equation in the cylinder $(t-\tau,t] \times B(x,\sqrt{\tau})$, we have*

$$u(t,x) \leq \frac{Ca^{-n/2}}{\tau^{1+n/2}} \int_{t-\tau}^{t} \int_{B(x,\sqrt{\tau})} u(s,y)d\mu(y)ds, \tag{6.37}$$

where $C = C(n)$.

Remark 6.10 The term $\tau^{1+n/2}$ is proportional to the volume of the cylinder of the height τ and of the base being a ball of radius $\sqrt{\tau}$ in $\mathbb{R}^n$. This reflects the fact that the Faber-Krahn inequality (6.35) is optimal in $\mathbb{R}^n$ but may not be optimal in M. A different type of the mean-value property (5.17) related to the volume $V(x,\sqrt{\tau})$ on M was considered in Section 5.2 (see also the next section).

The proof of (6.37) consists of two steps. The first step is the L^2-mean-value inequality, that is, (6.37) for u^2 instead of u, which was proved in [67, Theorem 3.1]. Alternatively, the L^2-mean-value inequality follows from the first part of the Moser iteration argument [106], given the equivalence of (6.35) and certain Sobolev inequality (see [20]). The second step is to derive (6.37) from the L^2-mean-value inequality. This is done by using the argument of Li and Schoen [95] (see also [97] and [37]).

The estimate (6.36) follows from Lemma 6.9 similarly to the first part of the proof of Theorem 5.8. Indeed, integrating in time the inequality

$$\int_M p(s,x,y)d\mu(y) \leq 1$$

we obtain, for all $t \geq \tau > 0$,

$$\int_{t-\tau}^{t} \int_M p(s,x,y)d\mu(y) \leq \tau.$$

Hence (6.37) yields, for $u = p(\cdot, x, \cdot)$ and for $\tau \in (0, r^2]$,

$$p(t, x, x) \leq \frac{Ca^{-n/2}}{\tau^{1+n/2}} \int_{t-\tau}^{t} \int_M p(s, x, y) d\mu(y) \leq \frac{Ca^{-n/2}}{\tau^{n/2}}.$$

Clearly, (6.36) follows if we choose $\tau = \min(t, r^2)$. ∎

Corollary 6.11 *Suppose that, for all $x \in M$ and some $r > 0$, the Faber-Krahn inequality (6.35) holds, for any precompact open set $\Omega \subset B(x, r)$. Then, for any $D > 2$ and all $x, y \in M$, $t > 0$,*

$$p(t, x, y) \leq \frac{Ca^{-n/2}}{\min(t, r^2)^{n/2}} \exp\left(-\frac{d^2(x, y)}{2Dt}\right), \tag{6.38}$$

where $C = C(n, D)$.

The estimate (6.38) follows from (6.36) and from the inequality (5.14) of Corollary 5.5. If we use (5.15) instead of (5.14) then we obtain

$$p(t, x, y) \leq \frac{Ca^{-n/2} e^{\lambda_1(M) t_0}}{\min(t, t_0)^{n/2}} \exp\left(-\lambda_1(M) t - \frac{d^2(x, y)}{2Dt}\right) \tag{6.39}$$

where $t_0 = r^2$. By absorbing a and t_0 into C, we can rewrite (6.39) as follows:

$$p(t, x, y) \leq \frac{C'}{\min(t, 1)^{n/2}} \exp\left(-\lambda_1(M) t - \frac{d^2(x, y)}{2Dt}\right). \tag{6.40}$$

6.4 Relative Faber-Krahn inequality and the decay of the heat kernel as $V(x, \sqrt{t})^{-1}$.

We return here to heat kernel upper bound

$$p(t, x, x) \leq \frac{C}{V(x, \sqrt{t})}, \tag{6.41}$$

which was discussed already in Section 5.17. Similarly to the equivalence

On-diagonal bound (6.2)$\Longleftrightarrow$Faber-Krahn inequality (6.11)

mentioned in Section 6.1, we will show that (6.41) is "almost" equivalent to *the relative Faber-Krahn inequality* defined as follows.

Definition 6.12 *We say that M admits* the relative Faber-Krahn inequality *if, for any ball $B(x, r) \subset M$ and for any precompact open set $\Omega \subset B(x, r)$,*

$$\lambda_1(\Omega) \geq \frac{b}{r^2} \left(\frac{V(x, r)}{\mu(\Omega)}\right)^{\nu}, \tag{6.42}$$

with some positive constants b, ν.

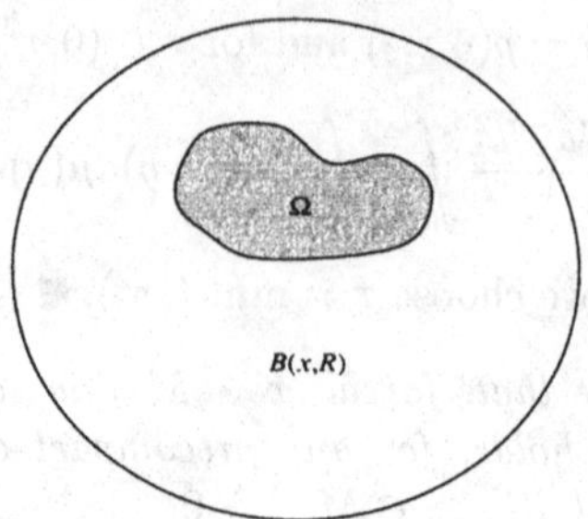

Figure 12 Subset Ω of the ball $B(x,R)$

It is easy to see that (6.42) holds in $\mathbb{R}^n$ with $\nu = 2/n$. It is possible to prove that the relative Faber-Krahn inequality holds on any geodesically complete manifold of non-negative Ricci curvature - see [67]. In general, a non-negatively curved manifold admits no *uniform* Faber-Krahn function in the spirit of Section 6.2. The inequality (6.42) was designed to overcome this difficulty. It provides a lower bound for $\lambda_1(\Omega)$, which takes into account not only volume $\mu(\Omega)$ but also *location* of the set Ω, via the ball $B(x,r)$.

We say that the volume function $V(x,r)$ satisfies *the doubling property* if, for some constant C,

$$V(x,2r) \le CV(x,r), \quad \forall x \in M,\ r > 0. \tag{6.43}$$

Now we can state the main theorem of this section.

Theorem 6.13 *([69, Proposition 5.2]) Let M be a geodesically complete manifold.*

If M admits the relative Faber-Krahn inequality then the heat kernel satisfies the upper bound (6.41), for all $x \in M$ and $t > 0$, and the volume function $V(x,r)$ satisfies the doubling property (6.43).

Conversely, the heat kernel upper bound (6.41) and the doubling volume property (6.43) imply (6.42).

Proof. The implication (6.42)$\Longrightarrow$(6.41) follows from Theorem 6.7. Indeed, given a ball $B(x,r)$, we have, by (6.35), for any precompact open set $\Omega \subset B(x,r)$,

$$\lambda_1(\Omega) \ge a\mu(\Omega)^{-2/n}, \tag{6.44}$$

where $n = 2/\nu$ and

$$a = \frac{b}{r^2} V(x,r)^{2/n}. \tag{6.45}$$

Hence, Theorem 6.7 implies, for any $r > 0$,

$$p(t,x,x) \le \frac{Ca^{-n/2}}{\min(t,r^2)^{n/2}}.$$

Taking $r = \sqrt{t}$ and substituting a from (6.45) we obtain (6.41).

Another proof of (6.42)$\Longrightarrow$(6.41) follows by Theorem 5.8. Indeed, given (6.44), Lemma 6.9 implies that, for any positive solution u of the heat equation in the cylinder $(t - r^2, t] \times B(x, r)$ (assuming $\tau := r^2 \leq t$),

$$u(t,x) \leq \frac{Ca^{-n/2}}{r^{2+n}} \int_{t-r^2}^{t} \int_{B(x,r)} u(s,y)d\mu(y)ds.$$

Substituting a from (6.45), we obtain the mean-value property (MV) (see Definition 5.7). Hence, Theorem 5.8 can be applied and yields (6.41).

The implication (6.42)$\Longrightarrow$(6.43) is proved by the argument of Carron [20] - see [69, p.442]. The second part of Theorem 6.13 – the implication (6.41)+(6.43)$\Longrightarrow$(6.42), is proved similarly to Theorem 6.5 - see [69, p.443]. ∎

The above proof together with Theorems 5.8 and 5.12 gives the following

Corollary 6.14 *The following implications hold*

Relative Faber-Krahn inequality (6.42)

$\Updownarrow$

Mean-value property (MV) *and volume doubling (6.43).*

$\Updownarrow$

Gaussian upper bound (5.19) and volume doubling (6.43)

$\Updownarrow$

On-diagonal upper bound (6.41) and volume doubling (6.43)

$\Downarrow$

On-diagonal lower bound (5.40)

Another (direct) proof of the second equivalence was obtained by Li and Wang [97].

7 Isoperimetric inequalities

7.1 Isoperimetric inequalities and $\lambda_1(\Omega)$

Isoperimetric inequality relates the boundary area of regions to their volume. We say that manifold M admits *the isoperimetric function* I if, for any precompact open set $\Omega \subset M$ with smooth boundary,

$$\sigma(\partial\Omega) \geq I(|\Omega|), \tag{7.1}$$

where

$$|\Omega| := \mu(\Omega).$$

For example, $\mathbb{R}^n$ admits the isoperimetric function $I(v) = c_n v^{\frac{n-1}{n}}$. Indeed, let $\Omega^* \subset \mathbb{R}^n$ be a ball of the same volume as Ω and let r be its radius. Then, by the classical isoperimetric inequality,

$$\sigma(\partial\Omega) \geq \sigma(\partial\Omega^*) = \omega_n r^{n-1} = c_n \left(\frac{\omega_n}{n} r^n\right)^{\frac{n-1}{n}} = c\,|\Omega^*|^{\frac{n-1}{n}} = c_n\,|\Omega|^{\frac{n-1}{n}}.$$

Other examples of isoperimetric functions will be shown below.

It turns out that the isoperimetric inequality (7.1) implies a Faber-Krahn inequality (6.19). The next statement is a version of Cheeger's inequality [26].

Proposition 7.1 *Let $I(v)$ be a non-negative continuous function on $\mathbb{R}_+$ such that $I(v)/v$ is non-increasing. Assume that M admits the isoperimetric function I. Then M admits the Faber-Krahn function*

$$\Lambda(v) := \frac{1}{4}\left(\frac{I(v)}{v}\right)^2. \tag{7.2}$$

For example, if $I(v) = cv^{\frac{n-1}{n}}$ then (7.2) yields $\Lambda(v) = \frac{c^2}{4} v^{-2/n}$.

Proof. Given a non-negative function $f \in C_0^\infty(\Omega)$, we denote

$$\Omega_s = \{x : f(x) > s\}.$$

By Sard's theorem, the boundary $\partial\Omega_s$ is smooth, for almost all s, so that we can apply the isoperimetric inequality (7.1) for Ω_s and obtain

$$\sigma(\partial\Omega_s) \geq I(|\Omega_s|) \tag{7.3}$$

for almost all s. Next, we use the co-area formula

$$\int_M |\nabla f|\, d\mu = \int_0^\infty \sigma(\partial\Omega_s)\, ds, \tag{7.4}$$

which implies with (7.3) and the non-increasing of $I(v)/v$,

$$\begin{aligned}
\int_M |\nabla f|\, d\mu &\geq \int_0^\infty I(|\Omega_s|) ds \\
&= \int_0^\infty \frac{I(|\Omega_s|)}{|\Omega_s|} |\Omega_s|\, ds \\
&\geq \frac{I(|\Omega|)}{|\Omega|} \int_0^\infty |\Omega_s|\, ds \\
&= \frac{I(|\Omega|)}{|\Omega|} \int_M f\, d\mu.
\end{aligned} \tag{7.5}$$

By the Cauchy-Schwarz inequality, we have

$$\int_M |\nabla f^2| \, d\mu = 2 \int_M f \, |\nabla f| \leq 2 \left[\int_M f^2 \, d\mu \int_M |\nabla f|^2 \, d\mu \right]^{1/2}. \tag{7.6}$$

Applying (7.5) to f^2 instead of f and by (7.6), we obtain

$$\frac{I(|\Omega|)}{|\Omega|} \int_M f^2 \, d\mu \leq \int_M |\nabla f^2| \, d\mu \leq 2 \left[\int_M f^2 \, d\mu \int_M |\nabla f|^2 \, d\mu \right]^{1/2}$$

whence

$$\left(\frac{I(|\Omega|)}{|\Omega|} \right)^2 \int_M f^2 \, d\mu \leq 4 \int_M |\nabla f|^2 \, d\mu$$

and

$$\lambda_1(\Omega) \geq \frac{1}{4} \left(\frac{I(|\Omega|)}{|\Omega|} \right)^2,$$

which was to be proved. ∎

Combining Proposition 7.1 with Theorem 6.2 and Corollary 5.5, we obtain

Corollary 7.2 *Assume that manifold M admits a non-negative continuous isoperimetric function $I(v)$ such that $I(v)/v$ is non-increasing. Let us define the function $f(t)$ by*

$$t = 4 \int_0^{f(t)} \frac{v dv}{I^2(v)}, \tag{7.7}$$

assuming the convergence of the integral in (7.7) at 0. Then, for all $x \in M$, $t > 0$ and $\varepsilon > 0$,

$$p(t, x, x) \leq \frac{2\varepsilon^{-1}}{f((1 - \varepsilon) t)}. \tag{7.8}$$

Furthermore, if the function f satisfies in addition the regularity condition (5.11) then, for all $x, y \in M$, $t > 0$, $D > 2$ and some $\varepsilon > 0$,

$$p(t, x, y) \leq \frac{C}{f(\varepsilon t)} \exp \left(-\frac{d^2(x, y)}{2Dt} \right). \tag{7.9}$$

In the next sections, we will show examples of manifolds satisfying certain isoperimetric and Faber-Krahn inequalities where the heat kernel estimates given by Theorems 6.2, 6.7, 6.13 and Corollaries 6.11, 7.2 can be applied.

7.2 Isoperimetric inequalities and the distance function

Here we mention a certain method of proving isoperimetric inequalities, which was introduced by Michael and Simon [104]. Suppose that we have a distance function $r(x,\xi)$ on M. This may be the Riemannian distance or a general distance function satisfying the usual axioms of the metric space, in particular, the triangle inequality. We will denote $r(x,\xi)$ by $r_\xi(x)$ to emphasize that it will be regarded as a function of x with a fixed (but arbitrary) ξ. It turns out that an isoperimetric inequality on M can be proved if one knows certain bounds for $|\nabla r_\xi|$ and Δr_ξ.

Theorem 7.3 *(Michael–Simon [104], Chung–Grigor'yan–Yau [33]) Let M be a geodesically complete Riemannian manifold of dimension $n>1$. Suppose that $r_\xi(x)$ is a distance function on M such that, for all $\xi, x \in M$,*

$$|\nabla r_\xi(x)| \leq 1 \tag{7.10}$$

and

$$\Delta r_\xi^2(x) \geq 2n \tag{7.11}$$

(assuming that $r_\xi^2 \in C^2(M)$). Then M admits the isoperimetric function $I(v) = cv^{\frac{n-1}{n}}$ where $c = c_n > 0$.

Let us explain why (7.10) and (7.11) should be related to isoperimetric inequalities. Indeed, integrating (7.10) over a precompact region $\Omega \subset M$ with smooth boundary and using the Green formula (2.4), we obtain

$$2n\,|\Omega| = \int_\Omega \Delta(r_\xi^2)d\mu = 2\int_{\partial\Omega} r_\xi \frac{\partial r_\xi}{\partial\nu} d\sigma\,, \tag{7.12}$$

where ν is the outward normal vector field on $\partial\Omega$. With the obvious inequality $\frac{\partial r_\xi}{\partial \nu} \leq |\nabla r_\xi| \leq 1$, (7.12) implies

$$n\,|\Omega| \leq \left(\sup_\Omega r_\xi\right)\sigma(\partial\Omega). \tag{7.13}$$

This is already a weak form of isoperimetric inequality. A certain argument allows to extend (7.13) as to show the isoperimetric inequality

$$\sigma(\partial\Omega) \geq c\,|\Omega|^{\frac{n-1}{n}}.$$

See [104] or [33] for further details.

7.3 Minimal submanifolds

Let M be an n-dimensional submanifold of $\mathbb{R}^N$ with the Riemannian metric inherited from $\mathbb{R}^N$. Submanifold M is called *minimal* if its normal mean curvature vector $H(x) = (H_1(x), H_2(x), ..., H_N(x))$ vanishes for all $x \in M$ (see [113]). It turns out that already the minimality of M implies certain isoperimetric property.

Theorem 7.4 *(Bombieri - de Giorgi - Miranda [16]) Any n-dimensional minimal submanifold M admits the isoperimetric function $I(v) = cv^{\frac{n-1}{n}}$, $c > 0$.*

Proof. Consider the coordinate functions X_i in $\mathbb{R}^N$ as functions on M. Then $\Delta X_i = H_i(x)$ where Δ is the Laplacian on M (see [113, Theorem 2.4]). Let us denote by $r_\xi(x)$ the (extrinsic) Euclidean distance in $\mathbb{R}^N$ between the points $x, \xi \in M$. Shifting the coordinates $X_1, X_2, ..., X_N$ in $\mathbb{R}^N$ to have the origin ξ, we have

$$\Delta r_\xi^2 = \sum_{i=1}^{N} \Delta(X_i^2) = 2\sum_{i=1}^{N} X_i H_i + 2\sum_{i=1}^{N} |\nabla X_i|^2 .$$

The term $\sum X_i H_i$ identically vanishes if M is minimal. The sum $\sum_{i=1}^N |\nabla X_i|^2$ is equal to n for *any* n-dimensional submanifold, which can be verified by a direct computation. Hence, we conclude

$$\Delta r_\xi^2 = 2n. \tag{7.14}$$

On the other hand, it is obvious that $|\nabla r_\xi| \leq 1$ because the extrinsic distance is majorized by the intrinsic Riemannian distance on M. Hence, we can apply Theorem 7.3 and conclude the proof. ∎

Theorem 7.4 and Corollary 7.2 imply the following uniform upper bound for the heat kernel on M

$$p(t, x, y) \leq \frac{C}{t^{n/2}} \exp\left(-\frac{d^2(x, y)}{2Dt}\right). \tag{7.15}$$

7.4 Cartan-Hadamard manifolds

A manifold M is called *a Cartan-Hadamard manifold* if M is geodesically complete simply connected non-compact Riemannian manifold with non-positive sectional curvature. For example, $\mathbb{R}^n$ and $\mathbb{H}^n$ are Cartan-Hadamard manifolds.

Theorem 7.5 *(Hoffman - Spruck [84]) Any Cartan-Hadamard manifold M of the dimension n admits the isoperimetric function $I(v) = cv^{\frac{n-1}{n}}$, $c > 0$.*

This theorem can also be derived from Theorem 7.3. Indeed, denote by $r_\xi(x)$ the geodesic distance between the point x and ξ on M. If $M = \mathbb{R}^n$ then $\Delta r_\xi^2 = 2n$. For general Cartan-Hadamard manifold, the comparison theorem for the Laplacian implies $\Delta r_\xi^2 \geq 2n$ (see [122]), whereas $|\nabla r_\xi| \leq 1$ holds on any manifold.

Hence, the heat kernel estimate (7.15) is valid on Cartan-Hadamard manifolds, too. On the other hand, by Proposition 7.1, Cartan-Hadamard manifolds satisfy the hypotheses of Corollary 6.11. Therefore, (6.40) holds, that is,

$$p(t,x,y) \leq \frac{C}{\min\left(1, t^{n/2}\right)} \exp\left(-\lambda_1(M)t - \frac{d^2(x,y)}{2Dt}\right), \tag{7.16}$$

which can be better than (7.15) if $\lambda_1(M) > 0$.

McKean's theorem [103] says that if the sectional curvature of the Cartan-Hadamard manifold is bounded from above by $-K^2$ then

$$\lambda_1(M) \geq \frac{(n-1)^2}{4} K^2. \tag{7.17}$$

Indeed, consider again the Riemannian distance function $r_\xi(x)$. As Yau [139] showed, on such manifolds

$$\Delta r_\xi \geq (n-1)K, \tag{7.18}$$

away from ξ. Given a precompact open set $\Omega \subset M$ with smooth boundary, we choose $\xi \notin \overline{\Omega}$ and integrate (7.18) over Ω. By the Green formula 2.4, we obtain

$$\int_{\partial\Omega} \frac{\partial r_\xi}{\partial \nu} d\sigma \geq (n-1)K\,|\Omega|\,.$$

Since $\frac{\partial r_\xi}{\partial \nu} \leq |\nabla r_\xi| \leq 1$, we arrive to the isoperimetric inequality

$$\sigma\left(\partial\Omega\right) \geq (n-1)K\,|\Omega|\,,$$

whence (7.17) follows by Proposition 7.1.

Hence, (7.16) implies, for all $x, y \in M$, $t > 0$ and $D > 2$,

$$p(t,x,y) \leq \frac{C}{\min(1, t^{n/2})} \exp\left(-\frac{(n-1)^2}{4}K^2 t - \frac{d^2(x,y)}{2Dt}\right). \tag{7.19}$$

Let us compare (7.19) with the sharp uniform estimate of the heat kernel on the hyperbolic space $\mathbb{H}_K^n$ of the constant negative curvature $-K^2$ obtained by Davies and Mandouvalos [50]:

$$p(t,x,y) \asymp \frac{(1+d+t)^{\frac{n-3}{2}}(1+d)}{t^{n/2}} \exp\left(-\frac{(n-1)^2}{4}K^2 t - \frac{d^2}{4t} - \frac{n-1}{2}Kd\right), \tag{7.20}$$

where $d = d(x,y)$ (see [43] and [77] for the exact formula for the heat kernel on hyperbolic spaces; the estimate (7.20) admits a far reaching generalization for symmetric spaces - see [3]).

If $t \to \infty$ then (7.20) yields

$$p(t,x,x) \asymp t^{-3/2} \exp\left(-\frac{(n-1)^2}{4}K^2 t\right),$$

which is better than (7.19) by the factor $t^{-3/2}$. The geometric nature of this factor is still unclear.

Another difference between (7.19) and (7.20) is the constant $D > 2$ in the Gaussian term. It is possible to put $D = 2$ in (7.19) at expense of the polynomial correction term, as in Theorem 5.9 – see [68, p.254] or [69, Theorem 5.2].

Yet another difference between (7.19) and (7.20) is the third term $\frac{n-1}{2}Kd$ in the exponential. It does not play a significant role for the heat kernel in the hyperbolic space because it is dominated by the sum of two other terms in the exponential (7.20). However, it is possible to introduce a similar term for general Cartan-Hadamard manifolds, and it may be leading if the curvature goes to $-\infty$ fast enough as $x \to \infty$.

Fix a point $o \in M$ and denote[3]

$$L(r) := \lambda_1(M \setminus \overline{B(o,r)}). \tag{7.21}$$

Clearly, $L(r)$ is increasing function and $L(0) = \lambda_1(M)$. If the sectional curvature outside the ball $B(o,r)$ is bounded above by $-K^2(r)$ (where $K(r)$ is positive and increasing) then a modification of (7.17) says that

$$L(r) \geq \frac{(n-1)^2}{4}K(r)^2. \tag{7.22}$$

Theorem 7.6 *([69, Theorem 5.3]) Let M be a Cartan-Hadamard manifold of the dimension n and $o \in M$. Then, for all $t > 0$, $x \in M \setminus B(o,\sqrt{t})$, $c \in (0,1)$ and some $\varepsilon = \varepsilon(c) > 0$,*

$$p(t,o,x) \leq \frac{C}{t^{n/2}} \exp\left(-c\lambda_1(M)t - c\frac{d^2}{4t} - \varepsilon d\sqrt{L(cd)}\right), \tag{7.23}$$

where $d = \operatorname{dist}(o,x)$ and $C = C(c,n)$.

Proof of Theorem 7.6 is similar to that of Theorem 6.7 but uses a more general mean-value type inequality than Lemma 6.9 – see [69] for details. It is not clear whether one can put $c = 1$ here.

[3] The notation $\lambda_1(\Omega)$ (where Ω is not necessarily compact) is defined by (6.12). This is the bottom of the spectrum of the operator $-\Delta_\Omega$ in $L^2(\Omega,\mu)$.

As a consequence we see that if the sectional curvature *outside* the ball $B(o,r)$ is bounded *above* by $-K^2(r)$ then (7.23) and (7.22) yield

$$p(t,o,x) \le \frac{C}{t^{n/2}} \exp\left(-c\lambda_1(M)t - c\frac{d^2}{4t} - \varepsilon K(cd)d\right). \qquad (7.24)$$

In particular, if $K(r) \gg r$ then the term $K(cd)d$ is leading as $d \to \infty$.

It is possible to prove the following matching lower bound. Assume that the sectional curvature *inside* the ball $B(o,r)$ is bounded *below* by $-K^2(r)$. Then we claim that, for all $t>0$, $x \in M$ and $\varepsilon > 0$,

$$p(t,o,x) \ge \frac{c}{t^{n/2}} \exp\left(-(\lambda_1(M)+\varepsilon)\,t - C\frac{d^2}{4t} - CK(d+C)d\right), \qquad (7.25)$$

where $c = c(o,\varepsilon) > 0$ (cf. (7.28) below).

Comparison of (7.24) and (7.25) shows that there is a big gap in the values of the constants in the upper and lower bounds. The problem of obtaining optimal heat kernel estimates when the curvature grows to $-\infty$ faster than quadratically in r, is not well understood.

Another interesting consequence of (7.23) is related to *the essential spectrum* of the operator Δ_M. Denote by $\lambda_{ess}(M)$ the bottom of the essential spectrum of $-\Delta_M$ in $L^2(M,\mu)$. It is known and is due to Donnelly [58] that, on any complete manifold,

$$\lambda_{ess}(M) = \lim_{r\to\infty} L(r), \qquad (7.26)$$

where $L(r)$ is defined by (7.21). It is possible to derive from (7.23) and (7.26) that

$$\limsup_{d\to\infty} \frac{1}{d} \sup_{t>0} \log p(t,o,x) \le -C\left(\sqrt{\lambda_1(M)} + \sqrt{\lambda_{ess}(M)}\right), \qquad (7.27)$$

where $d = d(x,o)$ (see [69, Corollary 5.1]) and $C>0$ is an absolute constant. It would be interesting to understand to what extent this inequality is sharp, in particular, what is a sharp value of C. For the hyperbolic space, equality is attained in (7.27) with $C=1$..

7.5 Manifolds of bounded geometry

We say that a manifold M has *bounded geometry* if its Ricci curvature is uniformly bounded below and if its injectivity radius is bounded away from 0. The hypothesis of bounded geometry implies that, for some $r_0 > 0$, all balls $B(x,r_0)$, $x \in M$, are uniformly quasi-isometric to the Euclidean ball $B_E(r_0) \subset \mathbb{R}^n$, where $n = \dim M$. The term "quasi-isometric" means that there is a diffeomorphism between $B(x, r_0)$ and $B_E(r_0)$ which changes the distances at most by a constant factor, and the word "uniform" refers to the fact that this constant factor is the same for all $x \in M$.

In other words, the manifold M looks like being patched from slightly distorted copies of the ball $B_E(r_0)$. Clearly, a manifold of bounded geometry is geodesically complete.

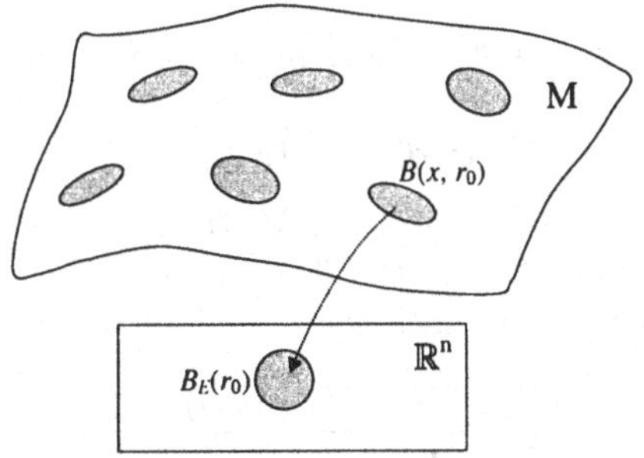

Figure 13 Manifold of bounded geometry is "patched" from slightly distorted Euclidean balls

The hypotheses of Corollary 6.11 are satisfied on the manifold M of bounded geometry. Hence, the upper bound (6.40) (the same as (7.16)) holds on M. Using Proposition 5.14 and the standard chain argument with the local Harnack inequality, it is possible to prove the following lower bound

$$p(t,x,y) \geq c_{x,\varepsilon} \exp\left(-(\lambda_1(M)+\varepsilon)t - C\frac{d^2(x,y)}{t}\right), \tag{7.28}$$

for any $\varepsilon > 0$ and all $t > 0$, $x, y \in M$. Hence, if $\lambda_1(M) > 0$ then (6.40) provides a correct rate of decay of $p(t,x,y)$ as $t \to \infty$ and as $y \to \infty$.

It turns out that even in the case $\lambda_1(M) = 0$, there is a priori rate of decay of the heat kernel as $t \to \infty$. This is based on the following isoperimetric property of manifolds of bounded geometry.

Theorem 7.7 *Any n-dimensional manifold of bounded geometry admits the following isoperimetric function, for some $c > 0$,*

$$I(v) = c \min\left(v^{\frac{n-1}{n}}, 1\right). \tag{7.29}$$

The meaning of (7.29) is the following. For small v, the function $I(v)$ is nearly the Euclidean one which corresponds to the fact that the small scale geometry of M is uniformly quasi-isometric to that of $\mathbb{R}^n$. For large v, (7.29) gives $I(v) = c$, which is attained on the cylinder $M = \mathbb{R}^{n-1} \times \mathbb{S}^1$ (see Fig. 14). Roughly speaking, (7.29) says that a manifold of bounded geometry expands at ∞ at least as fast as a cylinder.

Figure 14 The surface area of $\partial\Omega$ does not increase when the set $\Omega \subset \mathbb{R}^{n-1} \times \mathbb{S}^1$ is stretching.

Proof. We need to show that, for any precompact open set Ω with smooth boundary,

$$\sigma(\partial\Omega) \geq I(|\Omega|).$$

CASE 1. Assume that, for any $x \in M$, the set Ω covers less than a half of the volume of the ball $B(x, r_0/3)$. Then, for any point $x \in \overline{\Omega}$, there exists a positive number $r(x) \leq r_0/3$ such that

$$\frac{1}{4} \leq \frac{\sigma(\partial\Omega \cap B(x, r(x)))}{V(x, r(x))} \leq \frac{1}{2}. \tag{7.30}$$

All balls $B(x, r(x))$, $x \in \overline{\Omega}$, cover $\overline{\Omega}$. Choose a finite number of them also covering $\overline{\Omega}$. Then, by using the Banach argument, choose out of them a finite family of disjoint balls $B(x_i, r_i)$ (where $r_i = r(x_i)$) so that the concentric balls $B(x_i, 3r_i)$ cover Ω.

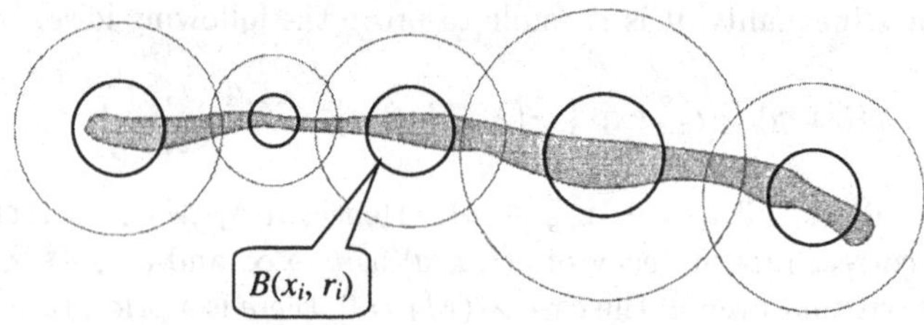

Figure 15 Set Ω (shaded) takes in each ball $B(x_i, r_i)$ nearly one half of its volume, whereas the balls $B(x_i, 3r_i)$ cover all of Ω.

In particular, we have

$$\sum_i V(x_i, 3r_i) \geq |\Omega|. \tag{7.31}$$

On the other hand, we use the following isoperimetric property of partitions of balls.

Proposition 7.8 *For any smooth hypersurface Γ in $B(x, r)$ (where $r \leq r_0$) dividing $B(x, r)$ into two open subsets each having volume at least v,*

$$\sigma(\Gamma) \geq c v^{\frac{n-1}{n}}. \tag{7.32}$$

If $B(x,r)$ is a Euclidean ball then (7.32) is a classical inequality, the best constant c in which was found by Maz'ya - see [102]. If $B(x,r)$ is a ball on manifold of bounded geometry and $r \le r_0$ then (7.32) follows from its Euclidean version and from the fact that the measures of all dimensions in $B(x,r_0)$ differ from their Euclidean counterparts at most by a constant factor.

Applying (7.32) to $\Gamma = \partial\Omega \cap B(x_i,r_i)$ and using (7.30), we obtain

$$\sigma\left(\partial\Omega \cap B(x_i,r_i)\right) \ge cV(x_i,r_i)^{\frac{n-1}{n}}.$$

Adding up these inequalities over all i and applying the elementary inequality

$$\sum_i a_i^\nu \ge \left(\sum_i a_i\right)^\nu, \tag{7.33}$$

which is valid for $a_i \ge 0$ and $0 \le \nu \le 1$, we obtain

$$\sigma(\partial\Omega) \ge c\left(\sum_i V(x_i,r_i)\right)^{\frac{n-1}{n}}. \tag{7.34}$$

Finally, we use $V(x,3r) \le CV(x,r)$, which is true for all $r \le r_0/3$. Therefore, (7.31) and (7.34) imply

$$\sigma(\partial\Omega) \ge c\,|\Omega|^{\frac{n-1}{n}}. \tag{7.35}$$

which was to be proved.

CASE 2. Assume that, for some $x \in M$, the set Ω covers at least a half of the volume of the ball $B(x,r_0/3)$. By moving the point x away, we may assume that

$$|\Omega \cap B(x,r_0/3)| = \frac{1}{2}V(x,r_0/3). \tag{7.36}$$

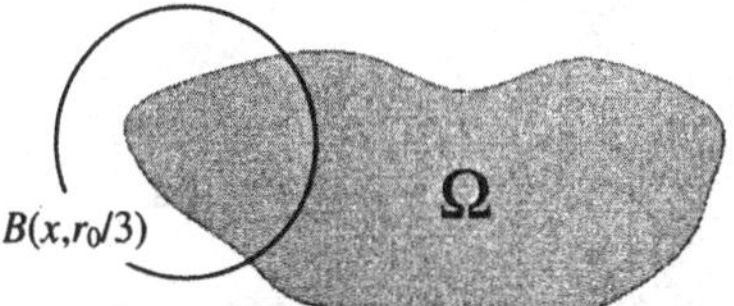

Figure 16 Set Ω covers exactly one half of the volume of the ball $B(x,r_0/3)$.

By Proposition 7.8 and (7.36), we have

$$\sigma\left(\partial\Omega\right) \ge \sigma\left(\partial\Omega \cap B(x,r_0/3)\right) \ge cV(x,r_0/3)^{\frac{n-1}{n}} \ge \text{const}, \tag{7.37}$$

which together with (7.35) finishes the proof. ■

Theorem 7.7 and Corollary 7.2 imply

Theorem 7.9 *On any manifold of bounded geometry,*

$$p(t,x,x) \le C \max\left(t^{-n/2}, t^{-1/2}\right),$$

for all $x \in M$ and $t > 0$. In particular, for all $x \in M$ and $t > 1$,

$$p(t,x,x) \le Ct^{-1/2}. \tag{7.38}$$

The exponent $\frac{1}{2}$ in (7.38) is sharp as is shown by (2.24) for the manifold $M = \mathbb{S}^1 \times \mathbb{R}^{n-1}$. The estimate (7.38) was proved by different methods by Varopoulos [132], Chavel and Feldman [25], Coulhon and Saloff-Coste [39], [34], Grigor'yan [70].

Sharper estimates of the heat kernel can be obtained assuming a *modified* isoperimetric inequality on M. Let us say that M admits a *modified isoperimetric* function $I(v)$ if $\sigma(\partial\Omega) \ge I(|\Omega|)$, for all precompact regions $\Omega \subset M$ with smooth boundary such that Ω contains a ball of the radius r_0. The purpose of this notion introduced by Chavel and Feldman [25] is to separate the large scale isoperimetric properties of the manifold from its local properties. For example, the Riemannian product $M = K \times \mathbb{R}^m$ where K is a compact manifold of the dimension $n - m$, admits the modified isoperimetric function $I(v) = cv^{\frac{m-1}{m}}$ (see [64]). As follows from Theorem 7.7, every manifold of bounded geometry admits the modified isoperimetric function $I(v) = \text{const}$.

Theorem 7.10 *(Chavel - Feldman [25]) If M is a manifold of bounded geometry admitting the modified isoperimetric function $I(v) = cv^{\frac{m-1}{m}}$ then, for all $x \in M$ and $t > 1$,*

$$p(t,x,x) \le Ct^{-m/2}.$$

Proof. It suffices to verify that M admits the following (not modified!) isoperimetric function

$$v \mapsto c' \begin{cases} v^{\frac{n-1}{n}}, & v \le 1, \\ v^{\frac{m-1}{m}}, & v \ge 1. \end{cases}$$

The proof of that follows the same line of reasoning as the proof of Theorem 7.7, with the following modification. No change is required for Case 1. In Case 2, take again the point x for which (7.36) is satisfied and consider the region $\Omega_0 = \Omega \cup B(x, r_0)$. Since Ω_0 contains a ball of radius r_0, the modified isoperimetric inequality gives

$$\sigma(\partial\Omega_0) \ge c\,|\Omega_0|^{\frac{m-1}{m}} \ge c\,|\Omega|^{\frac{m-1}{m}}.$$

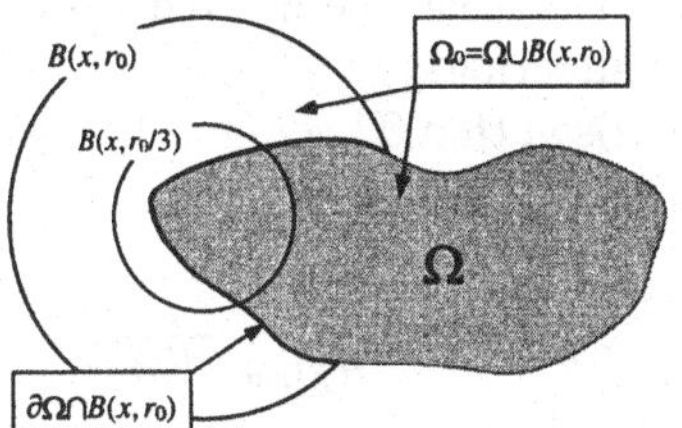

Figure 17 Sets Ω and Ω_0

To finish the proof, it suffices to show

$$\sigma(\partial\Omega_0) \le C\sigma(\partial\Omega). \tag{7.39}$$

The idea is that by adding the ball to Ω, we do not increase considerably the surface area of $\partial\Omega$ because the part of $\partial\Omega$ covered by the ball is comparable to the boundary of the ball. Formally, we write

$$\begin{aligned}\sigma(\partial\Omega_0) &\le \sigma(\partial\Omega) + \sigma(\partial B(x, r_0)) \\ &\le \sigma(\partial\Omega) + Cr_0^{n-1}. \end{aligned} \tag{7.40}$$

On the other hand, as follows from (7.37),

$$\sigma(\partial\Omega \cap B(x, r_0)) \ge cr_0^{n-1}.$$

We see that the first term dominates in (7.40) whence (7.39) follows. ■

See [70] for extension of Theorem 7.10 to a more general setup of manifolds of weak bounded geometry.

The following theorem improves (7.38) assuming the volume growth instead of an isoperimetric inequality.

Theorem 7.11 *(Coulhon - Saloff-Coste [38]) Assume that M has bounded geometry and that*

$$V(x, r) \asymp r^N, \quad \forall x \in M,\ r > 1. \tag{7.41}$$

Then, for all $x \in M$ and $t > 1$,

$$p(t, x, x) \le Ct^{-\frac{N}{N+1}}. \tag{7.42}$$

Note that $\frac{1}{2} \le \frac{N}{N+1} < 1$ so that (7.42) is better than (7.38) whenever $N > 1$. The proof of Theorem 7.11 in [38, Theorem 8] contains implicitly the fact that the manifold in question admits the Faber-Krahn function

$$\Lambda(v) = c\begin{cases} v^{-2/n}, & v \le 1, \\ v^{-\frac{N+1}{N}}, & v \ge 1. \end{cases}$$

Given that much, Theorem 7.11 can be derived from Theorem 6.2. See [35] for further results in this direction.

Let us recall for comparison that Theorem 5.11 yields, under the hypotheses of Theorem 7.11, the following lower bound, for all $x \in M$ and t large enough,

$$p(t,x,x) \geq \frac{c}{(t\log t)^{N/2}}.$$

It seems that the entire range between these two extreme behaviors of the heat kernel given by $t^{-\frac{N}{N+1}}$ and $(t\log t)^{-N/2}$, is actually possible.

Any manifold M of bounded geometry has at most exponential volume growth, that is, for all $x \in M$ and $r > 1$,

$$V(x,r) \leq C\exp(Cr).$$

This follows from the fact that M can be covered by a countable family of balls of radius $r_0/2$, which has a uniformly finite multiplicity (see, for example, [87]). By Theorem 5.13, one obtains the heat kernel lower estimate, for all $x \in M$ and $t \geq t_0$,

$$p(t,x,x) \geq c\exp(-Kt),$$

for some $c > 0$ and $K > 0$. By the chain argument involving the local Harnack inequality, this estimate can be extended to

$$p(t,x,y) \geq c\exp\left(-Kt - C\frac{d^2(x,y)}{t}\right), \tag{7.43}$$

for all $x, y \in M$ and $t \geq t_0$. The difference between (7.28) and (7.43) is that the former is *not* uniform with respect to x. On the other hand, (7.28) provides the sharp rate $e^{-\lambda_1(M)t}$ of the heat kernel decay as $t \to \infty$ whereas the constant K is (7.43) may be much larger than $\lambda_1(M)$ - see [126].

7.6 Covering manifolds

Let Γ be a discrete group of isometries of the manifold M. We say that the manifold M is a *regular cover* of the manifold K with the deck transformation group Γ if K is isometric to the quotient M/Γ. Intuitively, one can imagine M as a manifold glued from many copies of K moving from one to another by using the group action of Γ.

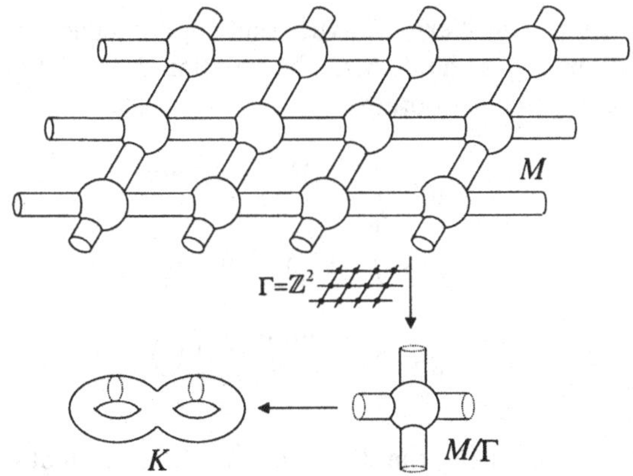

Figure 18 Manifold M covers K by the group $\Gamma = \mathbb{Z}^2$

If M is a regular cover of a compact manifold K then M has bounded geometry so that the results of the previous section apply. However, much more can be said about the heat kernel given the volume growth of M. The following isoperimetric inequality of Coulhon and Saloff-Coste plays the crucial role.

Theorem 7.12 *(Coulhon – Saloff-Coste [38, Theorem 4]) Let a non-compact manifold M be a regular cover of a compact manifold K. Let $\mathcal{V}(r)$ be a positive increasing function on $\mathbb{R}_+$ possessing certain regularity and such that, for some (fixed) $o \in M$ and all $r > 0$,*

$$V(o,r) \geq \mathcal{V}(r). \tag{7.44}$$

Then, for some (large) constant $C > 0$, the manifold M admits the isoperimetric function

$$I(v) := \frac{v}{C\mathcal{V}^{-1}(Cv)}, \tag{7.45}$$

where $\mathcal{V}^{-1}$ is the inverse function.

Next theorem provides the heat kernel upper bound on covering manifolds.

Theorem 7.13 *Referring to Theorem 7.12, we have, for all $x \in M$, $t > 0$ and some $\varepsilon > 0$,*

$$p(t,x,x) \leq \frac{C}{\mathcal{V}(\rho(\varepsilon t))}, \tag{7.46}$$

where ρ is defined by

$$t = \int_0^{\rho(t)} r^2 \frac{d}{dr} \log \mathcal{V}(r)\, dr. \tag{7.47}$$

Proof. Let us apply Corollary 7.2 with the isoperimetric function (7.45). Then the upper bound (7.8) holds with the function f defined by (7.7). Substituting (7.45) into (7.7), we obtain

$$\begin{aligned} t &= 4C^2 \int_0^{f(t)} \frac{\left[\mathcal{V}^{-1}(Cv)\right]^2 dv}{v} \\ &= 4C^2 \int_0^{Cf(t)} \frac{\left[\mathcal{V}^{-1}(v)\right]^2 dv}{v} \\ &= 4C^2 \int_0^{\mathcal{V}^{-1}(Cf(t))} \frac{r^2 d\mathcal{V}(r)}{\mathcal{V}(r)}. \end{aligned}$$

Setting $c = 1/(4C^2)$ and using the definition (7.47) of ρ, we obtain

$$\mathcal{V}^{-1}\left(Cf(t)\right) = \rho\left(ct\right)$$

and

$$f(t) = C^{-1}\mathcal{V}(\rho(ct)).$$

Hence, (7.46) follows from (7.8). ■

For example, if $V(o,r) \geq cr^N$ then take $\mathcal{V}(r) = cr^N$ and, by (7.47), $\rho(t) \asymp \sqrt{t}$. Hence, (7.46) implies $p(t,x,x) \leq Ct^{-N/2}$.

If $V(o,r) \geq \exp(r^\alpha) =: \mathcal{V}(r)$ (for large r) then we obtain $\rho(t) \asymp t^{\frac{1}{2+\alpha}}$ and

$$p(t,x,x) \leq C \exp\left(-ct^{\frac{\alpha}{2+\alpha}}\right), \tag{7.48}$$

for large t. In the particular case $\alpha = 1$, we have

$$p(t,x,x) \leq C \exp\left(-ct^{1/3}\right). \tag{7.49}$$

It turns out that the exponent 1/3 here is sharp. Indeed, Alexopoulos [1] showed that a similar on-diagonal lower bound holds provided the deck transformation group Γ is polycyclic and $V(o,r) \asymp \exp(r)$. See [115] for further results of this type.

7.7 Spherically symmetric manifolds

Let us fix the origin $o \in \mathbb{R}^n$, some positive smooth function ψ on $\mathbb{R}_+$ such that

$$\psi(0) = 0 \quad \text{and} \quad \psi'(0) = 1, \tag{7.50}$$

and define a *spherically symmetric* (or *model*) Riemannian manifold M_ψ as follows

1. as a set of points, M_ψ is $\mathbb{R}^n$;

2. in the polar coordinates (r,θ) at o (where $r \in \mathbb{R}_+$ and $\theta \in \mathbb{S}^{n-1}$) the Riemannian metric on $M_\psi \backslash \{o\}$ is defined as

$$ds^2 = dr^2 + \psi^2(r)d\theta^2\,, \tag{7.51}$$

where $d\theta$ denotes the standard Riemannian metric on $\mathbb{S}^{n-1}$;

3. the Riemannian metric at o is a smooth extension of (7.51) possibility of that is ensured by (7.50).

For instance, if $\psi(r) \equiv r$ then (7.51) coincides with the Euclidean metric of $\mathbb{R}^n$ and M_ψ is isometric to $\mathbb{R}^n$. If $\psi(r) = \sinh r$ then M_ψ is isometric to $\mathbb{H}^n$.

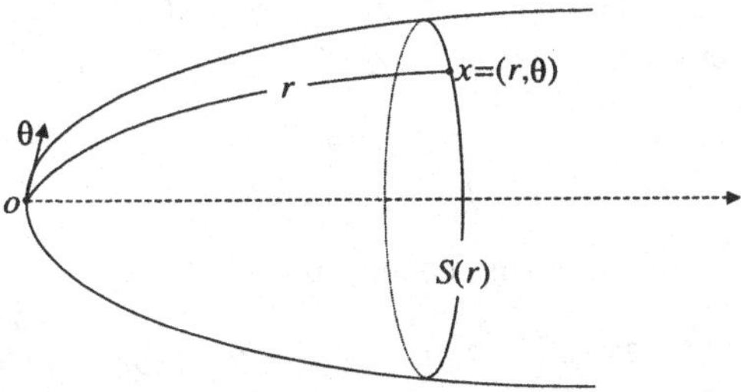

Figure 19 Model manifold as a surface of revolution

Clearly, the surface area $S(r)$ of the geodesic sphere $\partial B(o,r)$ on M_ψ is computed as

$$S(r) = \omega_n \psi^{n-1}(r),$$

and the volume $V(r)$ of the ball $B(o,r)$ is given by

$$V(r) = \int_0^r S(\xi)d\xi = \omega_n \int_0^r \psi^{n-1}(\xi)d\xi.$$

The Laplace operator on M_ψ can be written as follows (see [63], [122, p.97])

$$\Delta = \frac{\partial^2}{\partial^2 r} + \frac{S'}{S}\frac{\partial}{\partial r} + \frac{1}{\psi^2}\Delta_\theta\,, \tag{7.52}$$

where Δ_θ denotes the Laplace operator on the sphere $\mathbb{S}^{n-1}$.

We would like to estimate the heat kernel $p(t,x,x)$ on M_ψ by using Corollary 7.2. Isoperimetric function $I(v)$ seems to be unknown for general ψ. However, if we restrict our talk to estimating $p(t,o,o)$ then there is a simple way out. Careful analysis of the proof of Corollary 7.2 shows that we need to know the isoperimetric inequality

$$\sigma(\partial\Omega) \geq I(|\Omega|) \tag{7.53}$$

only for those sets Ω which are level sets of the function $p(t,o,\cdot)$, that is, for the two-parameter family of regions

$$\Omega_{s,t} = \{x \in M : p(t,o,x) > s\}$$

(see [36, Proposition 8.1]).

Given the rotation symmetry of M_ψ with respect to the point o, it is easy to prove that all $\Omega_{s,t}$ are ball centered at o. Hence, we need to find the function I such that (7.53) holds for all balls $\Omega = B(o,r)$. Since there is only one ball $B(o,r)$ with a prescribed volume, the isoperimetric function I can be defined by

$$I(v) = S(r) \quad \text{if } V(r) = v.$$

In order to apply Corollary 7.2, $I(v)/v$ should be non-increasing which amounts to the non-increasing of $S(r)/V(r)$. The equation (7.7) for $f(t)$ becomes

$$t = 4\int_0^{V^{-1}(f(t))} \frac{V(r)dr}{S(r)}.$$

Hence, we arrive to the following conclusion.

Theorem 7.14 *([36, Theorem 8.3]) Suppose that, for a model manifold M_ψ, the function $S(r)/V(r)$ is non-increasing. Let us define the function $\rho(t)$ by*

$$t = 4\int_0^{\rho(t)} \frac{dr}{\frac{d}{dr}\log V(r)}. \tag{7.54}$$

Then, for all $t > 0$ and $\varepsilon \in (0,1)$,

$$p(t,o,o) \le \frac{C_\varepsilon}{V(\rho(\varepsilon t))}. \tag{7.55}$$

For example, if $V(r) = Cr^N$ then (7.54) gives $\rho(t) \asymp \sqrt{t}$ and (7.55) implies

$$p(t,o,o) \le \frac{C'}{V(\sqrt{t})} = \frac{C''}{t^{N/2}}.$$

Note that, by Theorem 5.12, we have in this case the matching lower bound for $p(t,o,o)$. More generally, if $V(r)$ is doubling (see (5.38)) then one obtains from Theorems 7.14 and 5.12

$$p(t,o,o) \asymp \frac{1}{V(\sqrt{t})}$$

(see [36, Corollary 8.5]).

If $V(r) = \exp(r^\alpha)$, $0 < \alpha \le 1$, then we obtain $\rho(t) \asymp t^{\frac{1}{2-\alpha}}$ and

$$p(t,o,o) \le C \exp\left(-ct^{\frac{\alpha}{2-\alpha}}\right). \tag{7.56}$$

Theorem 5.13 yields for this volume growth the matching lower bound - cf. (5.42).

It is interesting that, for covering manifold M with the same volume growth function, the upper bound (7.48) is *weaker* than (7.56). In some sense, model manifolds possess the smallest heat kernel per volume growth function. This happens because all directions from o to the infinity are equivalent, which maximizes the capability of the Brownian motion to escape to the infinity and, thereby, minimizes the heat kernel $p(t,o,o)$. On covering manifolds, there may exists two non-equivalent ways of escaping (this is the case, for example, if the desk transformation group Γ is polycyclic) one of them being "narrow" in some sense and providing for a higher probability of return.

7.8 Manifolds of non-negative Ricci curvature

The main result of this section is the following isoperimetric inequality on non-negatively curved manifolds.

Theorem 7.15 *Let M be a geodesically complete non-compact Riemannian manifold of non-negative Ricci curvature. Then, for any ball $B(z,R) \subset M$ and any open set $\Omega \subset B(z,R)$ with smooth boundary,*

$$\frac{\sigma(\partial\Omega)}{|\Omega|} \ge \frac{c}{R}\left(\frac{V(z,R)}{|\Omega|}\right)^{1/n} \tag{7.57}$$

where $n = \dim M$ and $c = c(n) > 0$.

By Proposition 7.1, the isoperimetric inequality (7.57) implies the relative Faber-Krahn inequality (6.42) (the latter was proved for manifolds of non-negative Ricci curvature in [67]). Hence, Theorems 6.13 and 5.9 imply the following upper bound of the heat kernel on M

$$p(t,x,y) \le \frac{C}{\sqrt{V(x,\sqrt{t})V(y,\sqrt{t})}} \exp\left(-\frac{d^2(x,y)}{2Dt}\right), \tag{7.58}$$

where $D > 2$. This estimate was first obtained by Li and Yau [98]. They also proved the matching lower bound

$$p(t,x,y) \ge \frac{c}{\sqrt{V(x,\sqrt{t})V(y,\sqrt{t})}} \exp\left(-\frac{d^2(x,y)}{2Dt}\right), \tag{7.59}$$

for $D < 2$. Some improvements of these estimates can be found in [93]. See [134], [46] for heat kernel estimates on manifolds with $Ric_M \geq -K$, and [81] for similar heat kernel estimates in unbounded regions in $\mathbb{R}^n$ with the Neumann boundary condition.

We precede the proof of (7.57) by two properties of non-negatively curved manifolds.

Lemma 7.16 *Let M be a geodesically complete non-compact manifold of non-negative Ricci curvature. Then, for all balls intersecting $B(x,r)$ and $B(y,r')$ with $r \leq r'$,*

$$c\left(\frac{r'}{r}\right)^{\varepsilon} \leq \frac{V(y,r')}{V(x,r)} \leq C\left(\frac{r'}{r}\right)^{n}, \tag{7.60}$$

where ε, c, C are positive and depend on n.

Proof. We use Gromov's volume comparison theorem which says that if $Ric_M \geq 0$ then, for all $x \in M$ and $r' \geq r > 0$,

$$\frac{V(x,r')}{V(x,r)} \leq \left(\frac{r'}{r}\right)^{n} \tag{7.61}$$

(see [27], [23], [22]).

Denote $\delta = d(x,y)$. Then $\delta \leq r' + r \leq 2r'$, and the right hand inequality in (7.60) follows by (7.61)

$$\frac{V(y,r')}{V(x,r)} \leq \frac{V(x,r'+\delta)}{V(x,r)} \leq \left(\frac{r'+\delta}{r}\right)^{n} \leq \left(\frac{3r'}{r}\right)^{n} = 3^n\left(\frac{r'}{r}\right)^{n}.$$

To prove the left hand inequality in (7.60), let us first verify it in the particular case $r' = 3r$ and $y = x$. Find a point ξ such that $d(x,\xi) = 2r$ (here we use the completeness and non-compactness of M). Then $B(\xi,r)$ is contained in $B(x,3r)$ but does not intersect $B(x,r)$. Hence, we obtain

$$V(x,3r) \geq V(x,r) + V(\xi,r) \geq V(x,r)(1 + C^{-1}).$$

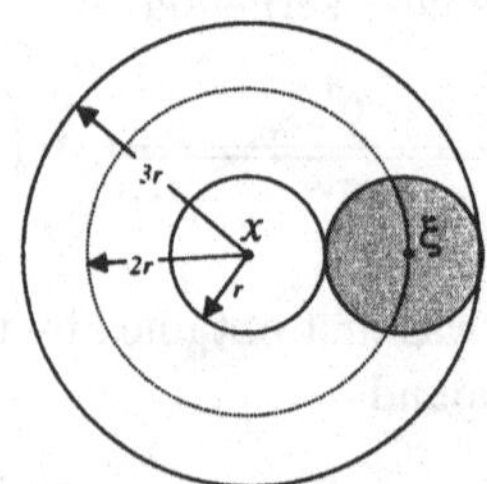

Figure 20 Ball $B(\xi,r)$ is contained in $B(x,3r)$ but does not intersect $B(x,r)$

In general, let us find an integer k such that

$$3^k \le \frac{r'}{r} < 3^{k+1}.$$

Then

$$V(y,r') \ge C^{-1}V(x,r') \ge C^{-1}V(x,3^k r) \ge C^{-1}(1+C^{-1})^k V(x,r)$$

and

$$\frac{V(y,r')}{V(x,r)} \ge C^{-1}\left(1+C^{-1}\right)^k \ge c\left(\frac{r'}{r}\right)^{\log_3\left(1+C^{-1}\right)},$$

which was to be proved. ■

Lemma 7.17 *(Buser [18]) Let M be a geodesically complete manifold of non-negative Ricci curvature. Then, for any ball $B(x,r)$ and for any smooth hypersurface Γ in $B(x,r)$ dividing $B(x,r)$ into two sets both having volume at least v,*

$$\sigma(\Gamma) \ge c\frac{v}{r}, \tag{7.62}$$

where $c = c(n) > 0$.

Observe that inequality (7.62) follows from (7.32) if $M = \mathbb{R}^n$. We refer the reader to [18, Lemma 5.1] or [67, Theorem 2.1] for the proof in general case.

Proof of Theorem 7.15. The proof is similar to Theorem 7.7. For any point $x \in \Omega$, let us find a positive $r(x)$ so that Ω covers exactly one half of the volume of $B(x, r(x))$. To that end, consider the function

$$h(r) := \frac{|\Omega \cap B(x,r)|}{V(x,r)}.$$

For r small enough, we have $h(r) = 1$. If $r > R$ then, by Lemma 7.16,

$$h(r) \le \frac{|\Omega|}{V(x,r)} \le \frac{V(z,R)}{V(x,r)} \le c^{-1}\left(\frac{R}{r}\right)^{\varepsilon}.$$

In particular, if the ratio r/R is large enough then $h(r) < 1/2$. Hence, for some $r \le C'R$, we have $h(r) = 1/2$.

The family of balls $B(x, r(x))$, $x \in \Omega$, covers Ω. By using the Banach ball covering argument, we can select at most countable subset $B(x_i, r_i)$ (where $r_i = r(x_i)$) so that the balls $B(x_i, r_i)$ are disjoint whereas the union of $B(x_i, 5r_i)$ covers Ω. The latter implies

$$\sum_i V(x_i, r_i) \ge 5^{-n}\sum_i V(x_i, 5r_i) \ge 5^{-n}|\Omega|. \tag{7.63}$$

On the other hand, Lemma 7.17 and the fact that

$$|\Omega \cap B(x_i, r_i)| = \frac{1}{2} V(x_i, r_i),$$

imply

$$\sigma\left(\partial\Omega \cap B(x_i, r_i)\right) \geq \frac{c}{2r_i} V(x_i, r_i). \tag{7.64}$$

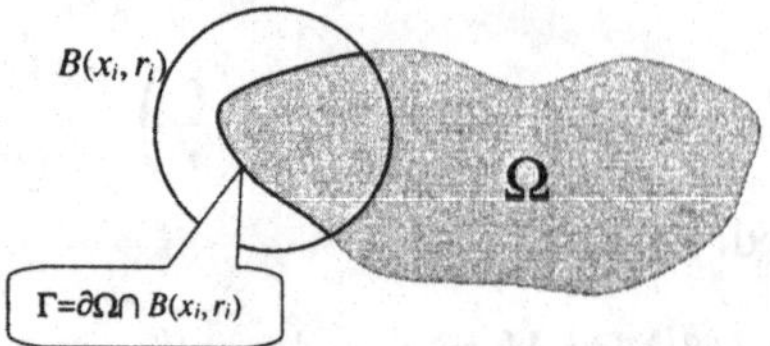

Figure 21 Estimating the area of Γ via the volumes of the sets $\Omega \cap B(x_i, r_i)$ and $B(x_i, r_i) \setminus \Omega$

Since

$$\frac{V(z,R)}{V(x,r_i)} \leq \frac{V(x, R+r_i)}{V(x,r_i)} \leq \frac{V(x, R(1+C'))}{V(x,r_i)} \leq C'' \left(\frac{R}{r_i}\right)^n,$$

we have

$$\frac{1}{r_i} \geq \frac{c'}{R} \left(\frac{V(z,R)}{V(x,r_i)}\right)^{1/n}.$$

By substituting into (7.64), we obtain

$$\sigma\left(\partial\Omega \cap B(x_i, r_i)\right) \geq \frac{c'}{R} V(z,R)^{1/n} V(x_i, r_i)^{1-1/n}.$$

Finally, summing up over all i and applying (7.33) and (7.63), we conclude

$$\sigma\left(\partial\Omega\right) \geq \frac{c''}{R} V(z,R)^{1/n} |\Omega|^{1-1/n},$$

which is equivalent to (7.57). ■

By Corollary 6.14, the upper bound (7.58) is equivalent to the relative Faber-Krahn inequality, under the standing assumption of the doubling volume property. It turns out that the conjunction of both upper and lower bounds (7.58) and (7.59) is equivalent to *the Poincaré inequality*

$$\int_{B(x,2r)} |\nabla f|^2 d\mu \geq \frac{c}{r^2} \inf_{\xi \in \mathbb{R}} \int_{B(x,r)} (f-\xi)^2 d\mu, \tag{7.65}$$

which is meant to hold for all $x \in M$, $r > 0$ and $f \in C^1(B(x, 2r))$ (see [67], [68], [120], [121]). One can regard (7.65) as the L^2-version of the isoperimetric inequality (7.62). Indeed, the later is equivalent to the functional inequality

$$\int_{B(x,2r)} |\nabla f| \, d\mu \geq \frac{c}{r^2} \inf_{\xi \in \mathbb{R}} \int_{B(x,r)} |f - \xi| \, d\mu. \tag{7.66}$$

Theorem 7.15 can be stated as the implication (7.66)$\Rightarrow$(7.57). Similarly, one can prove that (7.65) implies the relative Faber-Krahn inequality (6.42) - see [67, Theorem 1.4] and [120, Theorem 2.1].

References

[1] **Alexopoulos G.K.**, A lower estimate for central probabilities on polycyclic groups, *Can. J. Math.*, **44** (1992) no.5, 897-910.

[2] **Alexopoulos G.K.**, An application of Homogenization theory to Harmonic analysis on solvable Lie groups of polynomial growth, *Pacific J. Math.*, **159** (1993) 19-45.

[3] **Anker J-Ph., Ji L.**, Heat kernel and Green function estimates on noncompact symmetric spaces, preprint

[4] **Aronson D.G.**, Bounds for the fundamental solution of a parabolic equation, *Bull. of AMS*, **73** (1967) 890-896.

[5] **Azencott R.**, Behavior of diffusion semi-groups at infinity, *Bull. Soc. Math. (France)*, **102** (1974) 193-240.

[6] **Barbatis G., Davies E.B.**, Sharp bounds on heat kernels of higher order uniformly elliptic operators, *J. Operator Theory*, **36** (1996) 179–198.

[7] **Barlow M.**, Diffusion on fractals, Lecture notes

[8] **Barlow M., Bass R.**, Transition densities for Brownian motion on the Sierpinski carpet, *Probab. Th. Rel. Fields*, **91** (1992) 307–330.

[9] **Barlow M., Bass R.**, Brownian motion and harmonic analysis on Sierpinski carpets, to appear in *Canad. J. Math.*

[10] **Barlow M., Perkins A.**, Brownian motion on the Sierpinski gasket, *Probab. Th. Rel. Fields*, **79** (1988) 543–623.

[11] **Batty C.J.K., Bratelli O., Jorgensen P.E.T., Robinson D.W.**, Asymptotics of periodic subelliptic operators, preprint

[12] **Ben Arous G., Léandre R.**, Décroissance exponentielle du noyau de la chaleur sur la diagonale 1,2, *Prob. Th. Rel. Fields*, **90** (1991) 175-202 and 377-402.

[13] **Bendikov A.**, "Potential theory on infinite dimensional Ablelian groups" De Gruyter, 1995.

[14] **Bendikov A., Saloff-Coste L.**, Elliptic operators on infinite products, *J. Reine Angew. Math.*, **493** (1997) 171–220.

[15] **Benjamini I., Chavel I., Feldman E.A.**, Heat kernel lower bounds on Riemannian manifolds using the old ideas of Nash, *Proceedings of London Math. Soc.*, **72** (1996) 215-240.

[16] **Bombieri E., de Giorgi E., Miranda M.**, Una maggiorazioni a-priori relative alle ipersuperfici minimali non parametriche, *Arch. Rat. Mech. Anal.*, **32** (1969) 255-367.

[17] **Brooks R.**, A relation between growth and the spectrum of the Laplacian, *Math. Z.*, **178** (1981) 501-508.

[18] **Buser P.**, A note on the isoperimetric constant, *Ann. Scient. Ec. Norm. Sup.*, **15** (1982) 213-230.

[19] **Carlen E.A., Kusuoka S., Stroock D.W.**, Upper bounds for symmetric Markov transition functions, *Ann. Inst. H. Poincaré, Prob. et Stat.*, (1987) suppl. au no.2, 245-287.

[20] **Carron G.**, Inégalités isopérimétriques de Faber-Krahn et conséquences, *in:* "Actes de la table ronde de géométrie différentielle en l'honneur de Marcel Berger", *Collection SMF Séminaires et Congrès*, no.1, 1994.

[21] **Carron G.**, Inégalités de Sobolev-Orlicz non-uniformes, preprint 1996.

[22] **Chavel I.**, "Eigenvalues in Riemannian geometry" Academic Press, New York, 1984.

[23] **Chavel I.**, "Riemannian geometry : a modern introduction" Cambridge Tracts in Mathematics 108, Cambridge University Press, 1993.

[24] **Chavel I.**, "Large time heat diffusion on Riemannian manifolds" Lecture notes 1998.

[25] **Chavel I., Feldman E.A.**, Modified isoperimetric constants, and large time heat diffusion in Riemannian manifolds, *Duke Math. J.*, **64** (1991) no.3, 473-499.

[26] **Cheeger J.**, A lower bound for the smallest eigenvalue of the Laplacian, *in:* "Problems in Analysis: A Symposium in honor of Salomon Bochner", Princeton University Press. Princeton, 1970. 195-199.

[27] **Cheeger J., Gromov M., Taylor M.**, Finite propagation speed, kernel estimates for functions of the Laplace operator, and the geometry of complete Riemannian manifolds, *J. Diff. Geom.*, **17** (1982) 15-53.

[28] **Cheeger J., Yau S.-T.**, A lower bound for the heat kernel, *Comm. Pure Appl. Math.*, **34** (1981) 465-480.

[29] **Cheng S.Y., Li P., Yau S.-T.**, On the upper estimate of the heat kernel of a complete Riemannian manifold, *Amer. J. Math.*, **103** (1981) no.5, 1021-1063.

[30] **Chung F.R.K.**, "Spectral Graph Theory" AMS publications, 1996.

[31] **Chung F.R.K., Grigor'yan A., Yau S.-T.**, Upper bounds for eigenvalues of the discrete and continuous Laplace operators, *Advances in Math.*, **117** (1996) 165-178.

[32] **Chung F.R.K., Grigor'yan A., Yau S.-T.**, Eigenvalues and diameters for manifolds and graphs, *in:* "Tsing Hua Lectures on Geometry and Analysis", ed. S.-T.Yau, International Press, 1997. 79-105.

[33] **Chung F.R.K., Grigor'yan A., Yau S.-T.**, Higher eigenvalues and isoperimetric inequalities on Riemannian manifolds and graphs, to appear in *Comm. Anal. Geom*

[34] **Coulhon T.**, Noyau de la chaleur et discrétisation d'une variété riemannienne, *Israël J. Math.*, **80** (1992) 289-300.

[35] **Coulhon T.**, Ultracontractivity and Nash type inequalities, *J. Funct. Anal.*, **141** (1996) 510-539.

[36] **Coulhon T., Grigor'yan A.**, On-diagonal lower bounds for heat kernels on non-compact manifolds and Markov chains, *Duke Math. J.*, **89** (1997) no.1, 133-199.

[37] **Coulhon T., Grigor'yan A.**, Random walks on graphs with regular volume growth, *Geom. and Funct. Analysis*, **8** (1998) 656-701.

[38] **Coulhon T., Saloff-Coste L.**, Isopérimétrie pour les groupes et les variétés, *Revista Mathemática Iberoamericana*, **9** (1993) no.2, 293-314.

[39] **Coulhon T., Saloff-Coste L.**, Minorations pour les chaînes de Markov unidimensionnelles, *Prob. Theory Relat. Fields*, **97** (1993) 423-431.

[40] **Davies B., Simon B.**, Ultracontractivity and the heat kernel for Schrödinger operators and Dirichlet Laplacians, *J. Funct. Anal.*, **59** (1984) 335-395.

[41] **Davies E.B.**, Explicit constants for Gaussian upper bounds on heat kernels, *Amer. J. Math.*, **109** (1987) 319-334.

[42] **Davies E.B.**, Gaussian upper bounds for the heat kernel of some second-order operators on Riemannian manifolds, *J. Funct. Anal.*, **80** (1988) 16-32.

[43] **Davies E.B.**, "Heat kernels and spectral theory" Cambridge University Press, Cambridge, 1989.

[44] **Davies E.B.**, Pointwise bounds on the space and time derivatives of the heat kernel, *J. Operator Theory*, **21** (1989) 367-378.

[45] **Davies E.B.**, Heat kernel bounds, conservation of probability and the Feller property, *J. d'Analyse Math.*, **58** (1992) 99-119.

[46] **Davies E.B.**, The state of art for heat kernel bounds on negatively curved manifolds, *Bull. London Math. Soc.*, **25** (1993) 289-292.

[47] **Davies E.B.**, Uniformly elliptic operators with measurable coefficients, *J. Func. Anal.*, **132** (1995) 141-169.

[48] **Davies E.B.**, L^p spectral theory of higher-order elliptic differential operators, *Bull. London Math. Soc.*, **29** (1997) 513-546.

[49] **Davies E.B.**, Non-Gaussian aspects of heat kernel behaviour, *J. London Math. Soc.*, **55** (1997) no.1, 105-125.

[50] **Davies E.B., Mandouvalos N.**, Heat kernel bounds on hyperbolic space and Kleinian groups, *Proc. London Math. Soc.(3)*, **52** (1988) no.1, 182-208.

[51] **Davies E.B., Pang M.M.H.**, Sharp heat kernel bounds for some Laplace operators, *Quart. J. Math.*, **40** (1989) 281-290.

[52] **Debiard A., Gaveau B., Mazet E.**, Théorèmes de comparison in gèomètrie riemannienne, *Publ. Kyoto Univ.*, **12** (1976) 391-425.

[53] **Delmotte T.**, Parabolic Harnack inequality and estimates of Markov chains on graphs, preprint

[54] **DiBenedetto E.**, "Degenerate parabolic equations" Universitext, Springer,

[55] **Dodziuk J.**, Maximum principle for parabolic inequalities and the heat flow on open manifolds, *Indiana Univ. Math. J.*, **32** (1983) no.5, 703-716.

[56] **Dodziuk J., Mathai V.**, Approximating L^2 invariants of amenable covering spaces: a heat kernel approach, preprint

[57] **Donnelly H.**, Heat equation asymptotics with torsion, *Indiana Univ. Math. J.*, **34** (1985) no.1, 105–113.

[58] **Donnelly H.**, Essential spectrum and heat kernel, *J. Funct. Anal.*, **75** (1987) no.2, 362–381.

[59] **Erdös L.**, Estimates on stochastic oscillatory integrals and on the heat kernel of the magnetic Schrödinger operator, *Duke Math. J.*, **76** (1994) 541-566.

[60] **Fabes E.B., Stroock D.W.**, A new proof of Moser's parabolic Harnack inequality via the old ideas of Nash, *Arch. Rat. Mech. Anal.*, **96** (1986) 327-338.

[61] **Federer H., Fleming W.H.**, Normal and integral currents, *Ann. Math.*, **72** (1960) 458-520.

[62] **Gilkey P.B.**, "The index theorem and the heat equation" Publish or Perish, Boston MA, 1974.

[63] **Greene R., Wu W.**, "Function theory of manifolds which possess a pole" Lecture Notes in Math. 699, Springer, 1979.

[64] **Grigor'yan A.**, Isoperimetric inequalities for Riemannian products, (in Russian) *Mat. Zametki*, **38** (1985) no.4, 617-626. Engl. transl. *Math. Notes*, **38** (1985) 849-854.

[65] **Grigor'yan A.**, On stochastically complete manifolds, (in Russian) *DAN SSSR*, **290** (1986) no.3, 534-537. Engl. transl. *Soviet Math. Dokl.*, **34** (1987) no.2, 310-313.

[66] **Grigor'yan A.**, On the fundamental solution of the heat equation on an arbitrary Riemannian manifol, (in Russian) *Mat. Zametki*, **41** (1987) no.3, 687-692. Engl. transl. *Math. Notes*, **41** (1987) no.5-6, 386-389.

[67] **Grigor'yan A.**, The heat equation on non-compact Riemannian manifolds, (in Russian) *Matem. Sbornik*, **182** (1991) no.1, 55-87. Engl. transl. *Math. USSR Sb.*, **72** (1992) no.1, 47-77.

[68] **Grigor'yan A.**, Heat kernel on a non-compact Riemannian manifold, *in:* "1993 Summer Research Institute on Stochastic Analysis", ed. M.Pinsky et al., *Proceedings of Symposia in Pure Mathematics,* **57** (1994) 239-263.

[69] **Grigor'yan A.**, Heat kernel upper bounds on a complete non-compact manifold, *Revista Mathemática Iberoamericana,* **10** (1994) no.2, 395-452.

[70] **Grigor'yan A.**, Heat kernel on a manifold with a local Harnack inequality, *Comm. Anal. Geom.*, **2** (1994) no.1, 111-138.

[71] **Grigor'yan A.**, Upper bounds of derivatives of the heat kernel on an arbitrary complete manifold, *J. Funct. Anal.*, **127** (1995) no.2, 363-389.

[72] **Grigor'yan A.**, Gaussian upper bounds for the heat kernel on arbitrary manifolds, *J. Diff. Geom.*, **45** (1997) 33-52.

[73] **Grigor'yan A.**, Escape rate of Brownian motion on weighted manifolds, to appear in *Applicable Analysis*

[74] **Grigor'yan A.**, Analytic and geometric background of recurrence and non-explosion of the Brownian motion on Riemannian manifolds, *Bull. Amer. Math. Soc.*, **36** (1999) 135-249.

[75] **Grigor'yan A., Kelbert M.**, Range of fluctuation of Brownian motion on a complete Riemannian manifold, *Ann. Prob.*, **26** (1998) 78-111.

[76] **Grigor'yan A., Kelbert M.**, On Hardy-Littlewood inequality for Brownian motion on Riemannian manifolds, to appear in *J. London Math. Soc.*

[77] **Grigor'yan A., Noguchi M.**, The heat kernel on hyperbolic space, *Bull. London Math. Soc.*, **30** (1998) 643-650.

[78] **Grigor'yan A., Saloff-Coste L.**, Heat kernel on connected sums of Riemannian manifolds, to appear in *Math. Research Letters*

[79] **Guivarc'h Y.**, Sur la loi des grands nombres et le rayon spectral d'une marche aléatoire, *Journée sur les marches aléatoires, Astérisque,* **74** (1980) 47-98.

[80] **Gushchin A.K.**, On the uniform stabilization of solutions of the second mixed problem for a parabolic equation, (in Russian) *Matem. Sbornik,* **119(161)** (1982) no.4, 451-508. Engl. transl. *Math. USSR Sb.*, **47** (1984) 439-498.

[81] **Gushchin A.K., Michailov V.P., Michailov Ju.A.**, On uniform stabilization of the solution of the second mixed problem for a second order parabolic equation, (in Russian) *Matem. Sbornik*, **128(170)** (1985) no.2, 147-168. Engl. transl. *Math. USSR Sb.*, **56** (1987) 141-162.

[82] **Hamilton R.S.**, A matrix Harnack estimate for the heat equation, *Comm. Anal. Geom.*, **1** (1983) no.1, 113–126.

[83] **Hebisch W., Saloff-Coste, L.**, Gaussian estimates for Markov chains and random walks on groups, *Ann. Prob.*, **21** (1993) 673–709.

[84] **Hoffman D., Spruck J.**, Sobolev and isoperimetric inequalities for Riemannian submanifolds, *Comm. Pure Appl. Math.*, **27** (1974) 715–727. See also "A correction to: Sobolev and isoperimetric inequalities for Riemannian submanifolds", *Comm. Pure Appl. Math.*, **28** (1975) no.6, 765–766.

[85] **Hsu E.P.**, Short-time asymptotics of the heat kernel on a concave boundary, *SIAM J. Math. Anal.*, **20** (1989) no.5, 1109-1127.

[86] **Hsu E.P.**, Estimates of derivatives of the heat kernel on a compact Riemannian manifolds, preprint

[87] **Kanai M.**, Rough isometries, and combinatorial approximations of geometries of noncompact Riemannian manifolds, *J. Math. Soc. Japan*, **37** (1985) 391-413.

[88] **Kanai M.**, Rough isometries and the parabolicity of manifolds, *J. Math. Soc. Japan*, **38** (1986) 227-238.

[89] **Kotani M., Shirai T., Sunada T.**, Asymptotic behavior of the transition probability of a random walk on an infinite graph, *J. Func. Anal.*, **159** (1998) 664-689.

[90] **Kotani M., Sunada T.**, Albanese maps and off diagonal long time asymptotics for the heat kernel, preprint

[91] **Kusuoka S., Stroock D.W.**, Long time estimates for the heat kernel associated with uniformly subelliptic symmetric second order operators, *Ann. Math.*, **127** (1989) 165-189.

[92] **Léandre R.**, Uniform upper bounds for hypoelliptic kernels with drift, *J. Math. Kyoto Univ.*, **34** no.2, (1994) 263-271.

[93] **Li P.**, Large time behavior of the heat equation on complete manifolds with nonnegative Ricci curvature, *Ann. Math.*, **124** (1986) 1–21.

[94] **Li P.**, Harmonic sections of polynomial growth, *Math. Research Letters*, **4** (1997) no.1, 35–44.

[95] **Li P., Schoen R.**, L^p and mean value properties of subharmonic functions on Riemannian manifolds, *Acta Math.*, **153** (1984) 279-301.

[96] **Li P., Wang J.**, Convex hull properties of harmonic maps, *J. Diff. Geom.*, **48** (1998) 497-530.

[97] **Li P., Wang J.**, Mean value inequalities, preprint

[98] **Li P., Yau S.-T.**, On the parabolic kernel of the Schrödinger operator, *Acta Math.*, **156** (1986) no.3-4, 153-201.

[99] **Liskevich V., Semenov Yu.**, Two-sided estimates of the heat kernel of the Schrödinger operator, *Bull. London Math. Soc.*, **30** (1998) 596-602.

[100] **Malliavin P., Stroock D.W.**, Short time behavior of the heat kernel and its logarithmic derivatives, *J. Diff. Geom.*, **44** (1996) 550-570.

[101] **Maz'ya V.G.**, Classes of domains and embedding theorems for functional spaces, (in Russian) *Dokl. Acad. Nauk SSSR*, **133** no.3, (1960) 527-530. Engl. transl. *Soviet Math. Dokl.*, **1** (1961) 882-885.

[102] **Maz'ya V.G.**, "Sobolev spaces" (in Russian) Izdat. Leningrad Gos. Univ. Leningrad, 1985. Engl. transl. Springer, 1985.

[103] **McKean H.P.**, An upper bound to the spectrum of Δ on a manifold of negative curvature, *J. Diff. Geom.*, **4** (1970) 359–366.

[104] **Michael J.H., Simon L.M.**, Sobolev and mean-value inequalities on generalized submanifolds of R^n, *Comm. Pure Appl. Math.*, **26** (1973) 361-379.

[105] **Molchanov S.A.**, Diffusion processes and Riemannian geometry, (in Russian) *Uspekhi Matem. Nauk*, **30** (1975) no.1, 3-59. Engl. transl. *Russian Math. Surveys*, **30** (1975) no.1, 1-63.

[106] **Moser J.**, A Harnack inequality for parabolic differential equations, *Comm. Pure Appl. Math.*, **17** (1964) 101-134.

[107] **Murata M.**, Large time asymptotics for fundamental solutions of diffusion equations, *Tôhoku Math. J.*, **37** (1985) 151-195.

[108] **Mustapha S.**, Gaussian estimates for heat kernels on Lie groups, preprint

[109] **Nash J.**, Continuity of solutions of parabolic and elliptic equations, *Amer. J. Math.*, **80** (1958) 931-954.

[110] **Norris J.R.**, Heat kernel bounds and homogenization of elliptic operators, *Bull. London Math. Soc.*, **26** (1994) 75-87.

[111] **Norris J.R.**, Long time behaviour of heat flow: global estimates and exact asymptotics, preprint

[112] **Norris J.R.**, Heat kernel asymptotics and the distance function for measurable Riemannian metrics, preprint

[113] **Osserman R.**, Minimal varieties, *Bull. Amer. Math. Soc.*, **75** (1969) 1092-1120.

[114] **Pittet Ch., Saloff-Coste L.**, A survey on the relationship between volume growth, isoperimetry, and the behavior of simple random walk on Cayley graphs, with examples, preprint

[115] **Pittet Ch., Saloff-Coste L.**, Amenable groups, isoperimetric profiles and random walks, *in:* "Geometric group theory down under. Proceedings of a special year in geometric group theory, Canberra, Australia, 1996", ed. J. Cossey, C.F. Miller III, W.D. Neumann and M. Shapiro, Walter De Gruyter, 1999.

[116] **Porper F.O., Eidel'man S.D.**, Two-side estimates of fundamental solutions of second-order parabolic equations and some applications, (in Russian) *Uspekhi Matem. Nauk*, **39** (1984) no.3, 101-156. Engl. transl. *Russian Math. Surveys*, **39** (1984) no.3, 119-178.

[117] **Qian Z.**, Gradient estimates and heat kernel estimates, *Proc. Roy. Soc. Edinburgh*, **125A** (1995) 975-990.

[118] **Robinson D.W.**, "Elliptic operators and Lie groups" Oxford Math. Mono., Clarenton Press, Oxford New York Tokyo, 1991.

[119] **Rosenberg S.**, "The Laplacian on a Riemannian manifold" Student Texts 31, London Mathematical Society, 1991.

[120] **Saloff-Coste L.**, A note on Poincaré, Sobolev, and Harnack inequalities, *Duke Math J., I.M.R.N.*, **2** (1992) 27-38.

[121] **Saloff-Coste L.**, Parabolic Harnack inequality for divergence form second order differential operators, *Potential Analysis*, **4** (1995) 429-467.

[122] **Schoen R., Yau S.-T.**, "Lectures on Differential Geometry" Conference Proceedings and Lecture Notes in Geometry and Topology 1, International Press, 1994.

[123] **Semenov Yu.A.**, On perturbation theory for linear elliptic and parabolic operators; the method of Nash, *Contemporary Math.*, **221** (1999) 217-284.

[124] **Stroock D.W.**, Estimates on the heat kernel for the second order divergence form operators, *in:* "Probability theory. Proceedings of the 1989 Singapore Probability Conference held at the National University of Singapore, June 8-16 1989", ed. L.H.Y. Chen, K.P. Choi, K. Hu and J.H. Lou, Walter De Gruyter, 1992. 29-44.

[125] **Stroock D.W., Turetsky J.**, Upper bounds on derivatives of the logarithm of the heat kernel, preprint

[126] **Sturm K-Th.**, Heat kernel bounds on manifolds, *Math. Ann.*, **292** (1992) 149-162.

[127] **Sturm K-Th.**, Analysis on local Dirichlet spaces II. Upper Gaussian estimates for the fundamental-solutions of parabolic equations, *Osaka J. Math*, **32** (1995) no.2, 275-312.

[128] **Sturm K-Th.**, Analysis on local Dirichlet spaces III. The parabolic Harnack inequality, *Journal de Mathématiques Pures et Appliquées*, **75** (1996) no.3, 273-297.

[129] **Ushakov V.I.**, Stabilization of solutions of the third mixed problem for a second order parabolic equation in a non-cylindric domain, (in Russian) *Matem. Sbornik*, **111** (1980) 95-115. Engl. transl. *Math. USSR Sb.*, **39** (1981) 87-105.

[130] **Varadhan S.R.S.**, On the behavior of the fundamental solution of the heat equation with variable coefficients, *Comm. Pure Appl. Math.*, **20** (1967) 431-455.

[131] **Varadhan S.R.S.**, Diffusion processes in a small time interval, *Comm. Pure Appl. Math.*, **20** (1967) 659-685.

[132] **Varopoulos N.Th.**, Brownian motion and random walks on manifolds, *Ann. Inst. Fourier*, **34** (1984) 243-269.

[133] **Varopoulos N.Th.**, Hardy-Littlewood theory for semigroups, *J. Funct. Anal.*, **63** (1985) no.2, 240-260.

[134] **Varopoulos N.Th.**, Small time Gaussian estimates of heat diffusion kernel. I. The semigroup technique, *Bull. Sci. Math.(2)*, **113** (1989) no.3, 253-277.

[135] **Varopoulos N.Th., Saloff-Coste L., Coulhon T.**, "Analysis and geometry on groups" Cambridge University Press, Cambridge, 1992.

[136] **Woess W.**, "Random walks on infinite graphs and groups"

[137] **Yafaev D.**, Spectral properties of the Schrödinger operator with positive slowly decreasing potential, (in Russian) *Funktsional. Anal. i Prilozhen.*, **16** (1982) 47–54. Engl. transl. *Functional Anal. Appl.*, **16** (1982) no.4, 280–285.

[138] **Yau S.-T.**, Harmonic functions on complete Riemannian manifolds, *Comm. Pure Appl. Math.*, **28** (1975) 201-228.

[139] **Yau S.-T.**, Isoperimetric constant and the first eigenvalue of a compact Riemannian manifold, *Ann. Sci. Ecole Norm. Sup., 4th serie*, **8** no.4, (1975) 487-507.

[140] **Yau S.-T.**, On the heat kernel of a complete Riemannian manifold, *J. Math. Pures Appl., ser. 9*, **57** (1978) 191-201.

[141] **Yau S.-T.**, Harnack inequality for non-self-adjoint evolution equations, *Math. Research Letters*, **2** (1995) 387-399.

Spectral theory of the Schrödinger operators on non-compact manifolds: qualitative results

MIKHAIL SHUBIN

ABSTRACT

The paper contains an expanded version of my lectures in the Edinburgh school on Spectral Theory and Geometry (Spring 1998). I tried to make the exposition as self-contained as possible.

The main object of this paper is a Schrödinger operator $H = -\Delta + V(x)$ on a non-compact Riemannian manifold M. We discuss two basic questions of the spectral theory for such operators: conditions of the essential self-adjointness (or quantum completeness), and conditions for the discreteness of the spectrum in terms of the potential V.

In the first part of the paper we provide a shorter and a more transparent proof of a remarkable result by I. Oleinik [81, 82, 83], which implies practically all previously known results about essential self-adjointness in absence of local singularities of the potential. This result gives a sufficient condition of the essential self-adjointness of a Schrödinger operator with a locally bounded potential in terms of the completeness of the dynamics for a related classical system. The simplification of the proof given by I. Oleinik is achieved by an explicit use of the Lipschitz analysis on the Riemannian manifold and also by additional geometrization arguments which include a use of a metric which is conformal to the original one with a factor depending on the minorant of the potential.

In the second part of the paper we consider the case when the potential V is semi-bounded below and the manifold M has bounded geometry. We provide a necessary and sufficient condition for the spectrum of H to be discrete in terms of V. It is formulated by use of the harmonic (Newtonian) capacity in geodesic coordinates on M. This result is due to V.A. Kondrat'ev and M. Shubin and it extends the famous result of A.M. Molchanov [78] where the case $M = \mathbb{R}^n$ was considered. We follow Molchanov's scheme of the proof but simplify and clarify some moments of this proof, at the same time generalizing it to manifolds of bounded geometry.

Somewhat shorter versions of the two parts of this paper can also be found in the papers [100], [58].

Table of Contents

Part I. Essential self-adjointness

1. Introduction

Let (M, g) be a Riemannian manifold (i.e. M is a C^∞-manifold, $g = (g_{ij})$ is a Riemannian metric on M), $\dim M = n$. We will always assume that M is connected. Let Δ denote the Laplace-Beltrami operator on scalar functions on M i.e.

$$\Delta u = \frac{1}{\sqrt{g}} \frac{\partial}{\partial x^i} (\sqrt{g} g^{ij} \frac{\partial u}{\partial x^j})$$

where $x^1, \ldots, x^n$ are local coordinates, (g^{ij}) is the inverse matrix to g_{ij}, $g = \det(g_{ij})$ and we use the usual summation convention.

The main object of our study will be the Schrödinger operator

$$H = -\Delta + V(x). \tag{1.1}$$

In the first part we will assume that the potential $V = V(x)$ is a real-valued measurable function which is in L^∞_{loc} i.e. V is locally bounded.

We will discuss conditions which guarantee that H is essentially self-adjoint in the Hilbert space $L^2(M) = L^2(M, d\mu)$, where $d\mu = \sqrt{g} dx^1 \ldots dx^n$ is the Riemannian volume element. This means that the closure of H from the original domain $C_c^\infty(M)$ is a self-adjoint operator. (Here $C_c^\infty(M)$ is the set of all C^∞ functions with compact support on M.)

The importance of the essential self-adjointness of H becomes clear if we turn to the quantum mechanics and try to use the differential expression (1.1) to produce a quantum observable (a Hamiltonian) associated with this expression: a self-adjoint operator in $L^2(M)$ which extends $H|_{C_c^\infty(M)}$. Such an extension always exists but essential self-adjointness means that this extension is unique. It is easy to see that this uniqueness is equivalent to the uniqueness of the solution of the following Cauchy problem for the evolutionary Schrödinger equation:

$$\frac{1}{i} \frac{\partial \psi(t)}{\partial t} = H\psi(t), \ \psi(0) = \psi_0 \in C_c^\infty(M), \ \psi(t) \in L^2(M) \text{ for all } t \in \mathbb{R}.$$

(See e.g. [4], Ch.VI, Sect.1.7.) Here H is applied to ψ in the sense of distributions and the derivative in t is taken in the norm sense.

In case when this uniqueness holds, it is natural to say that we have *quantum completeness* for the corresponding quantum system. (If the completeness does not hold, we need some extra data to construct a Hamiltonian, e.g. boundary conditions etc.)

Let us also consider the corresponding classical system: the Hamiltonian system with the Hamiltonian

$$h(p, x) = |p|^2 + V(x) \tag{1.2}$$

in the cotangent bundle T^*M (with the standard symplectic structure). Here p is considered as a cotangent vector at the point $x \in M$, $|p|$ means the length of p with respect to the metric induced by g on T^*M. In local coordinates $(x^1, \dots, x^n)$ we have

$$|p|^2 = g^{ij}(x)p_ip_j, \text{ where } p = p_j dx^j \in T_x^*M.$$

In the coordinates $(x^1, \dots, x^n, p_1, \dots, p_n)$ the hamiltonian system has the form

$$\frac{dx^i}{dt} = \frac{\partial h}{\partial p_i}, \ \frac{dp_i}{dt} = -\frac{\partial h}{\partial x^i}, \ i = 1, \dots, n. \tag{1.3}$$

Let us assume for a moment that $V \in C^2(M)$, so the local Hamiltonian flow associated with the classical Hamiltonian (1.2) is well defined. Let us say that the system is *classically complete* if all the hamiltonian trajectories, i.e. solutions of (1.3), with arbitrary initial conditions are defined for all values of t. Usually it is more natural to require that they are defined for almost all initial conditions (in the phase space T^*M), but this distinction will not play any role in our considerations, though it is relevant if we want to treat potentials with local singularities (e.g. Coulomb type potentials).

We refer to Reed and Simon [87] for a more detailed discussion about classical and quantum completeness.

In the future we will assume that

$$V(x) \geq -Q(x) \text{ for all } x \in M, \tag{1.4}$$

where Q is a real-valued function which is positive and somewhat more regular than V itself.

For any $x, y \in M$ denote by $d_g(x, y)$ the distance between x and y induced by the Riemannian metric g.

Now we can formulate the main result:

Theorem 1.1. (I. Oleinik [82], [83]) *Assume that V satisfies (1.4) where $Q(x) \geq 1$ for all $x \in M$ and the following conditions are satisfied:*

(a) The function $Q^{-1/2}$ is globally Lipschitz i.e.

$$|Q^{-1/2}(x) - Q^{-1/2}(y)| \leq C d_g(x, y), \ x, y \in M, \tag{1.5}$$

(b)
$$\int^\infty Q^{-1/2} ds = \infty,$$
where the integral is taken along any parametrized curve (with a parameter $t \in [a, \infty)$), such that it goes out to infinity (i.e. leaves any compact $K \subset M$ starting at some value of the parameter t), ds means the arc length element associated with the given metric g.

Then the operator H given by (1.1) is essentially self-adjoint.

Remark 1.2. The requirement (b) is related to the classical completeness of the system with the Hamiltonian $|p|^2 - Q(x)$ if we additionally assume that $Q \in$

$C^2(M)$. To illustrate this assume for simplicity that $M = \mathbb{R}^n$ and the metric g is the standard flat metric on $\mathbb{R}^n$. Now assume that (b) is satisfied. Then along the classical trajectory of the Hamiltonian $|p|^2 - Q(x)$ we have

$$|p|^2 - Q(x) = E = const.$$

It follows that

$$dt = \frac{ds}{|\dot{x}|} = \frac{ds}{2|p|} = \frac{ds}{2\sqrt{E + Q(x)}},$$

hence the classical completeness for the Hamiltonian $|p^2| - Q(x)$ follows from the condition (b).

Remark 1.3. If we assume that $Q \in C^2(M)$ then the condition (b) is equivalent to the geodesic completeness of the Riemannian metric $\tilde{g}$ given by $\tilde{g}_{ij} = Q^{-1}g_{ij}$ (so $\tilde{g}$ is conformal to the original metric g).

Note also that (b) implies that the original metric g is also complete because $Q \geq 1$.

Remark 1.4. The requirement (a) in the theorem does not impose any serious restrictions on the growth of Q at infinity, but rather restricts oscillations of Q. Indeed, we can equivalently rewrite (a) in the form of the following estimate:

$$|dQ| \leq 2CQ^{3/2},$$

where $|dQ|$ means the length of the cotangent vector dQ as above. Arbitrary tower of exponents

$$e^r, e^{e^r}, e^{e^{e^r}}, \ldots,$$

satisfies this estimate. (Here $r = r(x) = d_g(x, x_0)$ with a fixed $x_0 \in M$.)

Imposing appropriate conditions on V sometimes leads to the equivalence of the conditions of classical and quantum completeness. An example of such situation was provided by A. Wintner [111] in case $n = 1$, with the restrictions which mean that the derivatives of V are small compared with V itself. However some conditions are indeed necessary even in case $n = 1$. This was shown by J. Rauch and M. Reed [85] who refer to unpublished lectures of E. Nelson. Examples given in [85] show that the classical and quantum completeness conditions are independent if no additional restrictions on V are imposed.

Remark 1.5. Theorem 1.1 was extended to the Laplacian on forms by M.Braverman [10].

2. Preliminaries on the Lipschitz analysis on a Riemannian manifold

Let (M, g) be a Riemannian manifold. A function $f : M \to \mathbb{R}$ is called *a Lipschitz function with a Lipschitz constant* C if

$$|f(x) - f(x')| \leq C d_g(x, x'), \quad x, x' \in M. \tag{2.1}$$

It is well known that in this case f is differentiable almost everywhere and

$$|df| \leq C \tag{2.2}$$

with the same constant C. Here $|df|$ means the length of the cotangent vector df in the metric associated with g. The corresponding differential df, as well as the partial derivatives of the first order, coincide with the distributional derivatives. Vice versa if $df \in L^\infty(M)$, for the distributional differential $df = (\partial f/\partial x^j)dx^j$, then f can be modified on a set of measure 0 so that it becomes a Lipschitz function.

The estimate (2.2) can be also rewritten in the form

$$|\nabla f| \leq C, \tag{2.3}$$

(again with the same constant C), where ∇f means the gradient of f associated with g, i.e. the vector field which corresponds to dg and is given in local coordinates as

$$\nabla f = g^{ij} \frac{\partial f}{\partial x^i} \frac{\partial}{\partial x^j}.$$

In local form (in open subsets of $\mathbb{R}^n$) these facts are discussed e.g. in the book of V. Mazya [74], Sect.1.1. The correspondence between constants in (2.1), (2.2) and (2.3) is straightforward.

The Lipschitz vector fields, differential forms etc. are defined in an obvious way

We will need the Stokes formula, or rather the divergence formula for Lipschitz vector fields v on M in the following simplest form:

Proposition 2.1. *Let $v = v(x)$ be a Lipschitz vector field with a compact support on M. Then*

$$\int_M \operatorname{div} v \, d\mu = 0.$$

Here $\operatorname{div} v$ in local coordinates is given as

$$\operatorname{div} v = \frac{1}{\sqrt{g}} \frac{\partial}{\partial x^i} (\sqrt{g} v^i), \quad v = v^i \frac{\partial}{\partial x^i},$$

and $d\mu$ is the Riemannian volume element associated with g.

The proof of the Proposition can be easily reduced to the case when v is supported in a domain of local coordinates. After that we can use mollification (regularization) of v to approximate v by smooth vector fields. A more general statement can be found in [74], Sect. 6.2.

3. Proof of the main theorem

Let H_{min} and H_{max} be the minimal and maximal operators associated with the differential expression (1.1) in $L^2(M)$. Here H_{min} is the closure of H in $L^2(M)$ from the initial domain $C_c^\infty(M)$, $H_{max} = H_{min}^*$ (the adjoint operator to H_{min} in $L^2(M)$). Clearly

$$\mathrm{Dom}\,(H_{max}) = \{u \in L^2(M)|\ Hu \in L^2(M)\},$$

where Hu is understood in the sense of distributions.

It follows from the standard functional analysis arguments (see. e.g. [5], Appendix 1), that the essential self-adjointness of H is equivalent to the symmetry of H_{max} which means that

$$(3.1)\qquad (H_{max}u, v) = (u, H_{max}v), \quad u, v \in \mathrm{Dom}\,(H_{max}).$$

To establish the symmetry of H_{max} we need some information about Dom (H_{max}). This information is provided by the following

Lemma 3.1. *If $u \in \mathrm{Dom}\,(H_{max})$, then*

$$(3.2)\qquad \int_M Q^{-1}|\nabla u|^2 d\mu \le 2[(8C^2+1)\|u\|^2 + \|u\|\cdot\|Hu\|] < \infty.$$

Here $\|\cdot\|$ means the norm in $L^2(M)$, and C is the Lipschitz constant for $Q^{-1/2}$ from (1.5).

PROOF. Let us choose a Lipschitz function $\phi : M \to \mathbb{R}$, such that ϕ has a compact support and

$$(3.3)\qquad 0 \le \phi \le Q^{-1/2}.$$

Note that this implies that $\phi \le 1$.

Let us estimate the quantity $I \ge 0$ where

$$I^2 = \int_M \phi^2|\nabla u|^2 d\mu.$$

To this end note first that

$$\begin{aligned}\phi^2|\nabla u|^2 = &\ \mathrm{div}(\phi^2 u\nabla u) - 2\phi u\nabla\phi\cdot\nabla u - \phi^2 u\cdot\Delta u = \\ &\ \mathrm{div}(\phi^2 u\nabla u) - 2\phi u\nabla\phi\cdot\nabla u + \phi^2 u\cdot Hu - \phi^2u^2V \le \\ &\ \mathrm{div}(\phi^2 u\nabla u) - 2\phi u\nabla\phi\cdot\nabla u + \phi^2 u\cdot Hu + \phi^2u^2Q.\end{aligned}$$

Here $\nabla\phi\cdot\nabla u$ means the scalar g-product of the tangent vectors $\nabla\phi$ and ∇u. Let us integrate the inequality over M. By the Stokes formula (Proposition 2.1) the integral of the first term in the right hand side vanishes. Taking into account that $0 \le \phi \le 1$ and $\phi^2 Q \le 1$ due to (3.3), we can estimate the integral of the last two terms by $\|u\|(\|u\| + \|Hu\|)$. Now denote by $\tilde{C}$ the Lipschitz constant of ϕ, so that $|\nabla\phi| \le \tilde{C}$. Then we obtain by the Cauchy-Schwarz inequality

$$2\left|\int_M \phi u \nabla\phi \cdot \nabla u\right| =$$
$$2\left|\int_M (u\nabla\phi)\cdot(\phi\nabla u)\right| \le 2\tilde{C}I\|u\|.$$

Overall we obtain the inequality

$$I^2 \le 2\tilde{C}I\|u\| + \|u\|(\|u\| + \|Hu\|).$$

Estimating

$$2\tilde{C}I\|u\| \le \frac{1}{2}I^2 + 8\tilde{C}^2\|u\|^2,$$

we arrive at the estimate

$$I^2 \le 2[(8\tilde{C}^2 + 1)\|u\|^2 + \|u\|\cdot\|Hu\|]. \tag{3.4}$$

Now it is easy to construct a sequence of Lipschitz functions ϕ_k, $k = 1, 2, \ldots$, such that ϕ_k satisfies

$$0 \le \phi_k \le Q^{-1/2}, \quad |\nabla\phi_k| \le (C + \frac{1}{k}),$$

(3.3) for any k, $\phi_1 \le \phi_2 \le \ldots$, and

$$\lim_{k\to\infty} \phi_k(x) = Q^{-1/2}(x), \quad x \in M.$$

Indeed, take a function $\chi : \mathbb{R} \to \mathbb{R}$, such that $\chi \in C^\infty(\mathbb{R})$, $0 \le \chi \le 1$, $\chi(t) = 1$ if $t \le 1$, $\chi(t) = 0$ if $t \ge 3$, and $|\chi'| \le 1$. Then we can take

$$\phi_k(x) = \chi(k^{-1}d_g(x, x_0))Q^{-1/2}(x),$$

where $x_0 \in M$ is an arbitrary fixed point. The estimate (3.4) holds for ϕ_k with $\tilde{C} = C + \frac{1}{k}$. Taking the limit as $k \to \infty$, we obtain (3.2). □

Proof of Theorem 1.1. Let us introduce the metric $\tilde{g}_{ij} = Q^{-1}g_{ij}$ and denote the corresponding distance function by $\tilde{d}$. This means that

$$\tilde{d}(x, y) = \inf\left\{\int_\gamma Q^{-1/2}ds \mid \gamma : [0, 1] \to M,\ \gamma(0) = x, \gamma(1) = y\right\},$$

where $\gamma \in C^\infty$ and ds means the element of the arc length of γ associated with g. Denote also

$$P(x) = \tilde{d}(x, x_0),$$

where $x_0 \in M$ is fixed. The completeness condition (b) means exactly that

$$P(x) \to \infty \text{ as } x \to \infty,$$

or, equivalently, that the set $\{x|\ P(x) \leq t\} \subset M$ is compact for any $t \in \mathbb{R}$.

Clearly, $|dP|_{\tilde{g}} \leq 1$, which can be rewritten as $|dP|_g^2 \leq Q^{-1}$, or

$$|\nabla P| \leq Q^{-1/2}.$$

(Here, as above, ∇P means the gradient with respect to the original metric g, and $|\nabla P|$ means the length of the tangent vector ∇P with respect to g.)

Now for two real-valued functions $u, v \in \operatorname{Dom}(H_{max})$ consider the following integral:

$$I_t = \int_{\{x|\ P(x)\leq t\}} \left(1 - \frac{P(x)}{t}\right)(u \cdot Hv - v \cdot Hu)d\mu.$$

By the dominated convergence theorem we obviously have

$$I_t \to \int_M (u \cdot Hv - v \cdot Hu)d\mu = (u, Hv) - (Hu, v) \quad \text{as } t \to \infty. \tag{3.5}$$

(Here $(\cdot,\cdot)$ means the scalar product in $L^2(M)$.)

Now note that

$$u \cdot Hv - v \cdot Hu = v \cdot \Delta u - u \cdot \Delta v = \operatorname{div}(v\nabla u - u\nabla v),$$

and rewrite the integrand of I_t as

$$\operatorname{div}\left[\left(1 - \frac{P(x)}{t}\right)(v\nabla u - u\nabla v)\right] + \frac{1}{t}(\nabla P)(v\nabla u - u\nabla v).$$

The integral of the first term vanishes due to Proposition 2.1. Therefore using the Cauchy-Schwarz inequality we obtain

$$\begin{aligned}
|I_t| &= \left|\frac{1}{t}\int_{\{x|\ P(x)\leq t\}} (\nabla P)\cdot(v\nabla u - u\nabla v)d\mu\right| \\
&= \left|\frac{1}{t}\int_{\{x|\ P(x)\leq t\}} (Q^{1/2}\nabla P)\cdot Q^{-1/2}(v\nabla u - u\nabla v)d\mu\right| \\
&\leq \frac{1}{t}(\|v\|\|Q^{-1/2}\nabla u\| + \|u\|\|Q^{-1/2}\nabla v\|).
\end{aligned}$$

By Lemma 3.1 the right hand side is $O(1/t)$, so $I_t \to 0$ as $t \to \infty$. Due to (3.5) this proves that H_{max} is symmetric i.e. (3.1) holds. This ends the proof of Theorem 1.1. □

4. Examples and further comments

In this section we will provide several examples, further results and relevant bibliographical comments (by necessity incomplete).

We will start with some particular cases of Theorem 1.1.

Theorem 4.1. (M. Gaffney [40]) *Let (M, g) be a complete Riemannian manifold. Then the Laplace-Beltrami operator Δ is essentially self-adjoint.*

PROOF. Take $Q(x) \equiv 1$ and use Theorem 1.1. □

Note that in fact the proof of Theorem 1.1 uses some elements of the Gaffney's proof.

Theorem 4.2. *Let (M, g) be a complete Riemannian manifold, $V \in L^\infty_{loc}(M)$, and $V(x) \geq -C$, $x \in M$, with a constant C. Then the Schrödinger operator $H = -\Delta + V(x)$ is essentially self-adjoint.*

In case when $M = \mathbb{R}^n$ (with the standard metric) this result is due to T. Carleman [16], and the Carleman proof is reproduced in the book of I.M. Glazman [42], Theorem 34 in Sect.3. The requirement $V \in L^\infty_{loc}$ can be completely removed, i.e. replaced by $V \in L^2_{loc}(\mathbb{R}^n)$, as was shown by T. Kato [53] (see also [87], Sect. X.4). This can be done with the help of the Kato inequality

$$\Delta|u| \geq \mathrm{Re}[(\mathrm{sgn}\, u)\Delta u],$$

for any $u \in L^1_{loc}$ such that $\Delta u \in L^1_{loc}$. Some non-positive perturbations can be allowed as well. For example, it is sufficient to require that $V = V_1 + V_2$ where $V_1 \in L^2_{loc}$, $V_1 \geq 0$, and V_2 is bounded with respect to $-\Delta$ with the $-\Delta$-bound $a < 1$. In particular, it is sufficient to assume that

$$V_+ = \max(V, 0) \in L^2_{loc}, \quad V_- = \min(V, 0) \in L^p + L^\infty,$$

where $p = 2$ if $n \leq 3$; $p > 2$ if $n = 4$, and $p = n/2$ if $n \geq 5$. The work by T. Kato was partially motivated by the paper of B. Simon [101] who proved the essential self-adjointness under an additional restriction compared with [53]. The reader may consult Chapters X.4, X.5 in M. Reed and B. Simon [87] for more references, motivations and a review.

It is actually sufficient to require only that the operator H_{min} is semi-bounded below, as was suggested by I.M. Glazman and proved by A.Ya. Povzner [84]. Another proof was suggested by E. Wienholtz [110] and also reproduced in [42].

Though the completeness requirement looks natural in case of semi-bounded operators, sometimes it can be relaxed and incompleteness may be compensated by a specific behavior of the potential (see e.g. A.G. Brusentsev [15] and also the references there).

The following theorem in case $M = \mathbb{R}^n$ is due to D.B. Sears (see e.g. [98, 104, 5]), who followed an idea of an earlier paper by E.C. Titchmarsh.

Theorem 4.3. *Let us fix $x_0 \in M$ and denote $r = r(x) = d_g(x, x_0)$. Assume that $V(x) \geq -Q(r)$ where $Q(r) \geq 1$ for all $r \geq 0$,*

$$\int_0^\infty \frac{dr}{\sqrt{Q(r)}} = \infty, \tag{4.1}$$

and one of the following two conditions is satisfied:
(a) $Q^{-1/2}$ is globally Lipschitz, i.e.

$$|Q^{-1/2}(r) - Q^{-1/2}(r')| \leq C|r - r'|, \quad r, r' \in [0, \infty); \tag{4.2}$$

(b) Q is monotone increasing.
Then the operator (1.1) is essentially self-adjoint.

PROOF. Under condition (a) this theorem clearly follows from Theorem 1.1.
Now assume that (b) is satisfied. Then we can follow F.S. Rofe-Beketov [89] to reduce this to the case when in fact (a) is satisfied. It is enough to construct a new function $\tilde{Q}$, such that $\tilde{Q}(r) \geq Q(r)$ for all $r \geq 0$ and $\tilde{Q}$ satisfies both (4.1) and (a). To this end we can define $\tilde{Q}(n) = Q(n+1)$, $n = 0, 1, 2, \ldots$, and then extend $\tilde{Q}^{-1/2}$ to the semi-axis $[0, \infty)$ by linear interpolation, i.e. take

$$\tilde{Q}^{-1/2}(\alpha n + (1-\alpha)(n+1)) = \alpha \tilde{Q}^{-1/2}(n) + (1-\alpha)\tilde{Q}^{-1/2}(n+1),$$

where $0 \leq \alpha \leq 1$, $n = 0, 1, \ldots$. It is easy to see that $\tilde{Q}$ satisfies the desired conditions. □

Remark 4.4. F.S. Rofe-Beketov [90] proved (in case $M = \mathbb{R}^n$) that the local inequality $V(x) \geq -Q(x)$ can be replaced by an operator inequality

$$H \geq -\varepsilon\Delta - Q(x)$$

with a constant $\varepsilon > 0$. This allows in particular some potentials which are unbounded below. I. Oleinik [83] noticed that this result can be carried over to the case of manifolds as well.

Remark 4.5. F.S. Rofe-Beketov [89] noticed that if in Theorem 4.3 we have $Q(r) < \infty$ for all $r \geq 0$ and Q satisfies (4.2), then we can always replace Q by another function $Q_1 \in C^\infty$ such that Q_1 also satisfies all the conditions (including (a) with a possibly bigger Lipschitz constant).
Indeed, it suffices to construct a globally Lipschitz C^∞ function $Q_1 : [0, \infty) \to [1, \infty)$ so that $Q(r)/2 \leq Q_1(r) \leq 2Q(r)$ for all $r \geq 0$. To this end we can first

mollify $Q^{-1/2}$ on each of the overlapping intervals $[0,4], [2,6], [6,10], \ldots$, by convolution with a positive smooth probability measure supported in a small neighborood of 0. This neighborhood should depend on the chosen interval to insure the desired inequalities. Note that the convolution does not change the Lipschitz constant. Then we can use a partition of unity on $[0, \infty)$ such that it is subordinated to the covering of $[0, \infty)$ by the intervals above and consists of functions which have uniformly bounded derivatives of any fixed order (e.g. translations of an appropriately fixed C^∞ function). Using such partition of unity to glue locally mollified function $Q^{-1/2}$ we arrive to the desired approximation $Q_1^{-1/2}$.

Remark 4.6. A more general Sears-type result was obtained by T. Ikebe and T. Kato [46] where magnetic Schrödinger operators in $\mathbb{R}^n$ with possibly locally singular potentials were considered. The allowed local singularities are most naturally described by the Stummel type conditions first introduced by F. Stummel [103]; see also E. Wienholtz [110], E. Nelson [79], K. Jörgens [49], G. Hellwig [45], T. Kato [53], B. Simon [101], H. Kalf and F.S. Rofe-Beketov [51] and references there for other results on operators with singular potentials. In particular a recent paper by H. Kalf and F.S. Rofe-Beketov [51] contains most general results which provide the essential self-adjointness of a Schrödinger operator in $\mathbb{R}^n$ under the condition that the operator is locally self-adjoint and appropriate Sears type conditions at infinity are imposed.

Remark 4.7. B.M. Levitan [62] gave a new proof of Theorem 4.3 (in case $M = \mathbb{R}^n$ with the flat metric). His proof uses the wave equation and the finite propagation speed argument. Similar arguments were later used by A.A. Chumak [23], P. Chernoff [22] and T. Kato [54] to prove essential self-adjointness in a somewhat different context. A.A. Chumak considered semi-bounded Schrödinger operators on complete Riemannian manifolds. P. Chernoff proves in particular the essential self-adjointness for the powers of such operators as well as Dirac operators, whereas T. Kato extends the arguments and results to the powers H^m, $m = 1, 2, \ldots$, (in $\mathbb{R}^n$) under the condition that $H \geq -a - b|x|^2$ with some constants a, b.

Note however that the self-adjointness of the powers of the Laplacian on a complete Riemannian manifold was first established by H.O. Cordes [25] without finite propagation speed argument. (See also the book [26] for a variety of results on essential self-adjointness of semi-bounded Schrödinger-type operators on manifolds and their powers.)

There are many results on self-adjointness of more general higher order operators – see e.g. M. Schechter [96] for operators in $\mathbb{R}^n$ (and also for similar L^p results in $\mathbb{R}^n$) and also M. Shubin [99] for operators on manifolds of bounded geometry, as well as F.S. Rofe-Beketov [91] and references there.

Now we will formulate a result of I. Oleinik which shows that in fact it is sufficient to restrict the behavior of the potential V only on some sequence of layers or shells which eventually surround all the points in M. From the classical point of view this is obvious because the classical completeness can be guaranteed if the classical particle escaping to infinity spends infinite time already inside the layers. The first result of this kind in case $n = 1$ is due to P. Hartman [44], and further generalizations were obtained in one-dimensional case by R. Ismagilov [48] (higher order operators), and in case $M = \mathbb{R}^n$ by M.G. Gimadislamov [41], F.S. Rofe-Beketov [90], M.S.P. Eastham, W.D. Evans, J.B. McLeod [34] and A. Devinatz [30] (the last two references also include magnetic field terms).

Theorem 4.8. (I. Oleinik [83]) *Let $\{\Omega_k | \, k = 0, 1, \dots, \}$ be a sequence of open relatively compact subsets with smooth boundaries in M, $\overline{\Omega_k} \subset \Omega_{k+1}$, $\cup_k \Omega_k = M$. Denote $T_k = \Omega_{2k+1} \setminus \overline{\Omega_{2k}}$, and let h_k be the minimal thickness of the layer T_k, i.e. $h_k = \mathrm{dist}_g(\Omega_{2k}, M \setminus \Omega_{2k+1})$. Assume that*

$$V(x) \geq -C\gamma_k, \quad x \in T_k, \quad k = 0, 1, \dots, \tag{4.3}$$

where $C > 0$, $\gamma_k \geq 1$, and

$$\sum_{k=0}^{\infty} \min\{h_k^2, h_k \gamma_k^{-1/2}\} = \infty. \tag{4.4}$$

Then the operator (1.1) is essentially self-adjoint.

PROOF. Following F.S. Rofe-Beketov [90] and I. Oleinik [83] we will construct a minorant Q for the potential V, so that the conditions (a) and (b) in Theorem 1.1 are satisfied.

We will start by constructing for any $k = 0, 1, \dots,$ a function $Q_k \geq 0$ on M such that $Q_k = +\infty$ on $M \setminus T_k$, then assemble $Q^{-1/2}$ as a linear combination of the functions $Q_k^{-1/2}$.

Denote for any $x \in M$

$$\delta_{2k}(x) = \mathrm{dist}_g(x, \Omega_{2k}), \quad \delta_{2k+1}(x) = \mathrm{dist}_g(x, M \setminus \Omega_{2k+1}), \quad k = 0, 1, \dots.$$

For $p = 2k, 2k+1$ introduce sets

$$\Omega_p' = \{x | \, \delta_p(x) \leq h_k/4\}$$

and functions $\delta_p' : M \to [0, \infty)$,

$$\delta_p'(x) = \mathrm{dist}_g(x, M \setminus \Omega_p').$$

Now define

$$Q_k^{-1/2}(x) = h_k^{-1}, \quad x \in M \setminus (\Omega_{2k}' \cup \Omega_{2k+1}'),$$

and

$$Q_k^{-1/2}(x) = h_k^{-1} \delta_p(x)(\delta_p(x) + \delta_p'(x))^{-1}, \quad x \in \Omega_p',$$

where $p = 2k$ or $2k+1$. Clearly $0 \le Q_k^{-1/2}(x) \le h_k^{-1}$ on M and $Q_k^{-1/2}(x) = 0$ if $x \notin T_k$.

Let us evaluate the Lipschitz constant for $Q_k^{-1/2}$. To this end denote $f(s,t) = s/(s+t)$, and observe that the absolute values of both partial derivatives of f in s and t are bounded by $(s+t)^{-1}$ if $s,t \ge 0$, $s+t > 0$. Also both δ_p and δ_p' are Lipschitz with the Lipschitz constant 1. Now note that it is easily follows from the triangle inequality that

$$\delta_p(x) + \delta_p'(x) \ge h_k/4, \quad x \in M.$$

Hence by the chain rule we see that

$$|\nabla(Q_k^{-1/2})| \le 2h_k^{-1} \cdot 4h_k^{-1} = 8h_k^{-2}.$$

Hence the Lipschitz constant of $Q_k^{-1/2}$ does not exceed $8h_k^{-2}$.

Now let us define

$$Q^{-1/2}(x) = \sum_{k=0}^{\infty} a_k Q_k^{-1/2},$$

where we will adjust the coefficients $a_k \ge 0$ so that all the conditions are satisfied. Let us list these conditions turn by turn.

(a) We need the condition $V \ge -Q$ to be satisfied which will be guaranteed if $-C\gamma_k \ge -Q(x)$, $x \in T_k$. This is equivalent to $Q_k^{-1/2} \le (C\gamma_k)^{-1/2}$, $k = 0, 1, \ldots$, and will be guaranteed if $a_k h_k^{-1} \le (C\gamma_k)^{-1/2}$ or

$$a_k \le C^{-1/2} h_k \gamma_k^{-1/2}. \tag{4.5}$$

(b) The Lipschitz constant of $Q^{-1/2}$ is evaluated by $8\sup_k(a_k h_k^{-2})$, so for $Q^{-1/2}$ to be Lipschitz it is sufficient to have

$$a_k \le C_1 h_k^2 \tag{4.6}$$

with some constant $C_1 > 0$.

(c) At last we need the condition (b) of Theorem 1.1 to be satisfied. Note that the minimal thickness of the internal layer $T_k' = M \setminus (\Omega_{2k}' \cup \Omega_{2k+1}')$ is at least $h_k/2$, and $Q^{-1/2} = a_k h_k^{-1}$ in T_k'. It follows that the condition (b) in Theorem 1.1 will be satisfied if we require

$$\sum_{k=0}^{\infty} a_k = \infty. \tag{4.7}$$

Now taking $C_1 = C^{-1/2}$ we can choose

$$a_k = C^{-1/2} \min\{h_k^2, h_k \gamma_k^{-1/2}\},$$

so the conditions (4.5), (4.6) will be automatically satisfied. The condition (4.7) will be satisfied if we require the condition (4.4) to hold. □

Part II.
Discreteness of spectrum

5. Introduction

We will use notations from Sect.1. Let (M, g) be a Riemannian manifold (i.e. M is a C^∞-manifold, $g = (g_{ij})$ is a Riemannian metric on M), $\dim M = n$. For simplicity of formulations assume that M is connected. Let Δ denote the Laplace-Beltrami operator on scalar functions on M, $H = -\Delta + V(x)$ be the Schrödinger operator acting on scalar functions on M (as in (1.1)), where the potential $V = V(x)$ is a real-valued measurable function which is locally in L^2 and globally semi-bounded below i.e.

$$V(x) \geq -C, \quad x \in M, \tag{5.1}$$

with a constant $C \in \mathbb{R}$. We will consider H as a self-adjoint operator in the Hilbert space $L^2(M) = L^2(M, d\mu)$, where $d\mu = \sqrt{g}dx^1 \ldots dx^n$ is the Riemannian volume element. The operator H is determined by the closure of the quadratic form which is a priori defined on the space $C_c^\infty(M)$ of all C^∞ functions with compact support on M.

Let $B(x,r)$ denote the open ball in M with the radius r and the center at x, $\bar{B}(x,r)$ the closure of $B(x,r)$. We will also use the same notation for the balls in a tangent space to M.

In this paper we will assume that (M, g) is a manifold of bounded geometry, i.e. the following two conditions are satisfied:

(a) $r_{inj} > 0$ where r_{inj} is the radius of injectivity of M;

(b) $|\nabla^m R| \leq C_m$, where $\nabla^m R$ is the m-th covariant derivative of the curvature tensor.

The condition (a) is equivalent to the fact that for any $x \in M$ the geodesic exponential map $\exp_x : T_x M \to M$ is defined on a ball $B(0,r) \subset T_x M$ for any $r < r_{inj}$ and provides a diffeomorphism of this ball onto the ball $B(x,r) \subset M$. It follows that the manifold (M, g) is complete. There are normal (geodesic) coordinates in each ball $B(x,r)$ which are induced by the exponential map and some orthonormal coordinates in $T_x M$. The condition (b) is equivalent to the existence of $r_0 > 0$ such that each derivative of the transition function between two sets of the geodesic coordinates centered at two points $x, x' \in M$ and considered in the balls $B(x, r_0)$ and $B(x', r_0)$, is bounded by a constant which depends only on the order of the derivative ([88]). Throughout the whole paper we will fix a sufficiently small $r_0 \in (0, r_{inj})$, satisfying this condition.

It is easy to extend all the arguments of this paper to the case when V is locally in L^1 (instead of L^2). The corresponding operator should be defined then by the closure of corresponding quadratic form. We will discuss this in more detail in Sect.6.

We can also allow the operator H to be considered on $G = M\setminus S$ where $S \subset M$ is a closed subset (possibly of positive measure). In this case we impose the Dirichlet boundary conditions on ∂G. This means that we should start with the closure of the quadratic form which is defined on $C_c^\infty(G)$ and then take the corresponding operator. This will be discussed in more detail in Sect.10.

Let us say that H has a *discrete spectrum* if its spectrum $\sigma(H)$ consists of eigenvalues of finite multiplicity (with the only accumulation point $+\infty$). We will abbreviate this by writing $\sigma = \sigma_d$.

The main result of this part of the paper is the following

Theorem 5.1. (V.A. Kondrat'ev, M.A. Shubin) *Assume that (M, g) is a Riemannian manifold of bounded geometry, H is the Schrödinger operator (1.1) with the potential satisfying (5.1). There exists $c > 0$, depending only on (M, g), such that $\sigma = \sigma_d$ if and only if the following condition is fulfilled:*

(D) For any sequence $\{x_k \mid k = 1, 2, \dots\} \subset M$ such that $x_k \to \infty$ as $k \to \infty$, for any $r < r_0/2$ and any compact subsets $F_k \subset \bar{B}(x_k, r)$ such that $\operatorname{cap}(F_k) \le cr^{n-2}$ in case $n \ge 3$ and $\operatorname{cap}(F_k) \le c\left(\ln \frac{1}{r}\right)^{-1}$ in case $n = 2$,

$$\int_{B(x_k,r)\setminus F_k} V(x)\,d\mu(x) \to \infty \quad as \quad k \to \infty\,. \tag{5.2}$$

Here $\operatorname{cap}(F_k)$ for $n \ge 3$ means the harmonic (or Newtonian) capacity of the set F_k in the normal (geodesic) coordinates centered at x_k, and for $n = 2$ it means the same capacity with respect to a ball $B(0, R)$ of fixed radius $R < r_0$.

For the necessary definitions concerning capacity see Section 7.

Note that replacing the Riemannian volume element $d\mu$ in (5.2) by the Lebesgue measure dx in the local geodesic coordinates leads to an equivalent condition.

Remark 5.2. Let us clarify the status of the constant c in Theorem 5.1. It is clear that the condition (D) becomes weaker when c decreases. Therefore this condition is necessary for all sufficiently small $c > 0$. But it is seen from the proof of sufficiency (Sect.9) that it is also sufficient for arbitrarily small $c > 0$. Therefore in the Theorem 5.1 we could replace words

There exists $c > 0$, depending only on (M, g), such that $\sigma = \sigma_d$ if and only if the following condition is satisfied

by the words

Then $\sigma = \sigma_d$ if and only if for all sufficiently small $c > 0$ the following condition is satisfied

It is well known (and also follows from Theorem 5.1) that the condition

$$V(x) \to +\infty \quad \text{as} \quad x \to \infty \tag{5.3}$$

implies that $\sigma = \sigma_d$. (For the case $M = \mathbb{R}^n$ this result is due to K. Friedrichs [38].) On the other hand it follows from Theorem 5.1 that $\sigma = \sigma_d$ implies that for any fixed $r > 0$

$$\int_{B(x,r)} V(y)d\mu(y) \to +\infty \quad \text{as} \quad x \to \infty. \tag{5.4}$$

This condition is also sufficient (hence necessary and sufficient) for the spectrum of the one-dimensional Schrödinger operator $H = -d^2/dx^2 + V(x)$ in $L^2(\mathbb{R})$ to be discrete. This was proved by A.M. Molchanov [78] (see also [42]).

A.M. Molchanov [78] treated also the case $M = \mathbb{R}^n$ with $n \geq 3$ and in this context he established the result of Theorem 5.1. A.M. Molchanov claimed that the case $n = 2$ can be settled similarly but in fact it seems that the main necessity argument in [78] needs a substantial modification for $n = 2$.

In early 1960th V.G. Maz'ya discovered new powerful technique in embedding theorems for functional spaces (see short notes [67], [68], [69], [70] and also [71], [72], [73], [76] and especially [74], [75] for more details and further developments). This technique is based on use of isoperimetric inequalities applied on level sets of the functions under consideration. In particular, the paper [69] contains an inequality which implies the Cheeger's lower bound [19] for the smallest Dirichlet eigenvalue of the Laplacian in domains on Riemannian manifolds (see more details in Sect.3 of [75]). M.S.Birman established in [6] (see also [9]), that various spectral statements such as positivity, semiboundedness, discreteness or finiteness of the negative spectrum, discreteness of the whole spectrum can be formulated as some embedding theorems. Accordingly the above mentioned results of V.G. Maz'ya imply these spectral results for the operators of higher order on $\mathbb{R}^n$ (see [69], [71], [72] and especially Sections 2.5 and 12.5 in [74]), and in particular provide another proof of the Molchanov theorem for arbitrary $n \geq 2$ and also similar results for higher order operators (see Maz'ya's papers [71], [72] and his book [74]). A particular case of Maz'ya result for the higher order operators when the capacity is not needed (the case $2l > n$, where $2l$ is the order of the operator), was established earlier by M.S. Birman and B.S. Pavlov [7].

Later V.G. Maz'ya and M. Otelbaev [76] (see also Sect.12.5 in [74]) established even stronger result: 2-sided estimates for the bottom of the essential spectrum through so-called "inner diameter" which is defined in terms of an appropriate capacity.

We will provide more references and a brief review of related results in Sect.10.

We give a self-contained proof of Theorem 5.1 which does not rely on results from [78], though we used many Molchanov's arguments, improving them and correcting when this was needed. In particular we extend Molchanov's arguments to the case $n = 2$.

We choose to follow Molchanov's method because it seems to us simpler for a self-contained exposition.

In Section 6 we describe localization technique which reduces the problem to estimates of the bottom of the Dirichlet or Neumann spectrum of H in balls of a fixed small radius.

In Section 7 we discuss definitions and preliminaries on the harmonic (Newtonian) capacity.

In Section 8 we establish that the condition (D) is necessary for the spectrum of H to be discrete.

In Section 9 we prove that this condition is sufficient.

Finally in Section 10 we formulate and prove corollaries, generalizations and applications of the main theorem, and also give a brief review of most closely related results. We prove that replacing capacity by the Lebesgue measure in Theorem 5.1 gives a sufficient condition for the discreteness of the spectrum. We also formulate and sketch the proof of a generalization of Theorem 5.1 to the case when H is considered not on M but on an arbitrary open subset in M with the Dirichlet boundary condition. In particular this gives a necessary and sufficient condition for the Dirichlet spectrum of $-\Delta$ to be discrete on an open subset in M. (In case $M = \mathbb{R}^n$ these results again go back to A.M. Molchanov [78] and V.G. Maz'ya - see e.g. [73], [74].)

The main result of this part of the paper was announced in [57].

6. Localization

In this section we will show that the discreteness of the spectrum for the Schrödinger operator H is equivalent to some estimates of the Dirichlet or Neumann spectrum of H on the balls of a fixed sufficiently small radius, or to the discreteness of the Dirichlet or Neumann spectrum of H on any infinite disjoint union of such balls.

Let (M, g) be a complete Riemannian manifold. We will use the Riemannian norm $|\cdot| = |\cdot|_g$ induced by the metric g both on tangent and cotangent vectors. Namely, for a tangent vector $v = v^j\partial_j \in T_xM$, where $\partial_j = \partial/\partial x_j$, $(x^1, \ldots, x^n)$ are local coordinates, we have

$$|v|^2 = g_{ij}v^iv^j,$$

and for a cotangent vector $p = p_j dx^j$

$$|p|^2 = g^{ij}p_ip_j\ .$$

For any $u \in C^\infty(M)$ define its gradient ∇u as the vector field corresponding to du, i.e.

$$(\nabla u)^i = g^{ij}\frac{\partial u}{\partial x_j}\ .$$

Then $|\nabla u| = |du|$.

Let us consider a Schrödinger operator H of the form (1.1) with the potential V such that $V(x) \geq 1, x \in M$. Due to (5.1) we can always assume this without loss of generality.

Let us consider the corresponding quadratic form

$$Q(u,u) = \int_M (|\nabla u(x)|^2 + V(x)|u(x)|^2) d\mu(x). \tag{6.1}$$

The corresponding hermitian form is

$$Q(u,v) = \int_M (\nabla u(x) \cdot \overline{\nabla v(x)} + V(x)u(x)\overline{v(x)}) d\mu(x). \tag{6.2}$$

We refer to the books by M.S. Birman and M.Z. Solomyak [8], and M. Reed and B. Simon [87], VIII.6, X.3, for the basics on the theory of quadratic forms.

Let us start with $C_c^\infty(M)$ as the original domain of Q and then take the closure of Q in $L^2(M)$. It is well known that the closure exists. This means that if we take the completion $\mathcal{H}_1$ of $C_c^\infty(M)$ with the norm induced by the scalar product (6.2), then the natural map of $\mathcal{H}_1$ to $L^2(M)$ is injective. To prove this we can for example note first that $\mathcal{H}_1$ is naturally imbedded into the complete Hilbert space $\tilde{\mathcal{H}}_1$ of functions $u \in L^2_{loc}(M)$ such that $\nabla u \in L^2_{loc}(M)$ (here ∇u is understood in the sense of distributions) and

$$\int_M |\nabla u(x)|^2 d\mu(x) < \infty, \int_M V(x)|u(x)|^2 d\mu(x) < \infty,$$

and second, that $\tilde{\mathcal{H}}_1$ is naturally imbedded in $L^2(M)$. In fact we have the following well known

Lemma 6.1. *In the notations above $C_c^\infty(M)$ is dense in $\tilde{\mathcal{H}}_1$, so $\tilde{\mathcal{H}}_1 = \mathcal{H}_1$.*

SKETCH OF THE PROOF. Using appropriate smooth or Lipschitz cut-off functions we can prove that functions with compact support are dense in $\tilde{\mathcal{H}}_1$. Let u be such a function. We would like to approximate it by functions from $C_c^\infty(M)$. We can assume that u is real-valued. For any $N = 1, 2, \ldots$ define $u_N(x) = u(x)$ if $|u(x)| < N$, $u_N(x) = N$ if $u(x) \geq N$, $u_N(x) = -N$ if $u(x) < -N$. Then $u_N \to u$ in $\tilde{\mathcal{H}}_1$. Therefore it is sufficient to prove that bounded functions with compact support from $\tilde{\mathcal{H}}_1$ can be approximated by functions from $C_c^\infty(M)$. This can be done by using the standard mollifying and the dominated convergence theorem. □

Let us recall that the operator H associated with the quadratic form (6.1) can be obtained as follows. Its domain Dom (H) consists of all $u \in \mathcal{H}_1$ such that there exists $f \in L^2(M)$ so that

$$Q(u,v) = (f,v), \quad \text{for all } v \in \mathcal{H}_1, \tag{6.3}$$

and then $Hu = f$. Here $(\cdot,\cdot)$ means the scalar product in $L^2(M)$. It is clearly sufficient to require (6.3) to be true for all $v \in C_c^\infty(M)$. Then we can rewrite the equation there as $-\Delta u + Vu = f$ where Δ is applied in the sense of distributions. Now it is clear that we can also describe Dom(H) as the set of all $u \in \mathcal{H}_1$ such that $-\Delta u + Vu \in L^2(M)$.

Note also that $\mathcal{H}_1 = \text{Dom}(H^{1/2})$.

These considerations are not needed if we have $V \in L^2_{loc}(M)$. In this case we can define our operator on $C_c^\infty(M)$ and it will be essentially self-adjoint, which can be proved by the use of an appropriate modification of the Kato inequality (see [53] or X.4 in [87] where the case $M = \mathbb{R}^n$ is considered). In case when $V \in L^\infty_{loc}(M)$ more general results (for manifolds) can be found e.g. in [22], [23], [26], [81], [82], [99], [100].

Denote

$$(6.4)\qquad \begin{aligned}\mathcal{L} &= \{u \in C_c^\infty(M) | \int_M (|\nabla u|^2 + V|u|^2) d\mu \le 1\} \\ &= \{u \in C_c^\infty(M) | (Hu, u) \le 1\}.\end{aligned}$$

Lemma 6.2. *$\sigma = \sigma_d$ if and only if $\mathcal{L}$ is precompact in $L^2(M)$.*

Proof. Clearly the spectrum is discrete for H if and only if it is true for $H^{1/2}$, which in turn is equivalent to the compactness of the imbedding $\mathcal{H}_1 \subset L^2(M)$. Now note that it follows from Lemma 6.1 that $\mathcal{L}$ is dense in the unit ball of $\mathcal{H}_1$. Hence the precompactness of this unit ball in $L^2(M)$ is equivalent to the precompactness of $\mathcal{L}$ which proves the Lemma.

Note that in case $V \in L^2_{loc}(M)$ the result follows also from the essential self-adjointness of H. □

Lemma 6.3. *$\sigma = \sigma_d$ if and only if for any $\varepsilon > 0$ there exists $R > 0$ such that*

$$(6.5)\qquad \int_{M\setminus B(x_0,R)} |u|^2 d\mu < \varepsilon \textit{ for any } u \in \mathcal{L},$$

or, in other words,

$$\int_{M\setminus B(x_0,R)} |u|^2 d\mu \to 0 \textit{ as } R \to \infty, \textit{ uniformly in } u \in \mathcal{L}.$$

Here x_0 is an arbitrarily fixed point in M.

Proof. The set of restrictions of the functions $u \in \mathcal{L}$ to any fixed ball $B(x_0, R)$ is precompact in $L^2(B(x_0, R), d\mu)$ due to the Sobolev compactness of imbedding theorem. It remains to notice that precompactness of a set $\mathcal{L} \subset L^2(M)$ is equivalent to the precompactness of all such restriction sets together with the condition (6.5). □

Let Ω be an open subset in M. Define the following numbers depending on Ω:

$$\lambda(\Omega) = \inf\left\{\frac{\int_\Omega (|\nabla u|^2 + V|u|^2)d\mu}{\int_\Omega |u|^2 d\mu},\ u \in C_c^\infty(\Omega) \setminus 0\right\}, \tag{6.6}$$

$$\mu(\Omega) = \inf\left\{\frac{\int_\Omega (|\nabla u|^2 + V|u|^2)d\mu}{\int_\Omega |u|^2 d\mu},\ u \in C^\infty(\Omega) \setminus 0\right\}, \tag{6.7}$$

i.e. $\lambda(\Omega)$ and $\mu(\Omega)$ are bottoms of the Dirichlet and Neumann spectra respectively, in the usual variational understanding (see e.g. [27], [52]).

In both (6.6) and (6.7) we can restrict ourselves to real-valued functions u, so we will always consider real-valued functions unless specified otherwise.

Proposition 6.4.

(a) If $\sigma = \sigma_d$ then $\lambda(\Omega) \to \infty$ as $\Omega \to \infty$.

(Here $\Omega \to \infty$ means $dist(x_0, \Omega) \to \infty$.)

(b) If $\mu(M \setminus \bar{B}(x_0, R)) \to \infty$ as $R \to \infty$ then $\sigma = \sigma_d$.

PROOF. (a) Assume that $\sigma = \sigma_d$. Choose an arbitrarily small $\varepsilon > 0$ and assume that $\Omega \subset M \setminus \bar{B}(x_0, R)$ where $R = R(\varepsilon)$ corresponds to ε according to (6.5). Then for any $u \in C_c^\infty(\Omega)$ with $(Hu, u) \le 1$ we should have $(u, u) < \varepsilon$. Therefore

$$(u, u) \le \varepsilon(Hu, u), \quad u \in C_c^\infty(\Omega),$$

or

$$\frac{(Hu, u)}{(u, u)} \ge \frac{1}{\varepsilon}, \quad u \in C_c^\infty(\Omega) \setminus \{0\}.$$

It follows from (6.6) that $\lambda(\Omega) \ge 1/\varepsilon$. This proves that $\lambda(\Omega) \to \infty$ as $\Omega \to \infty$.

(b) Assume that $\mu(M \setminus \bar{B}(x_0, R)) \to \infty$ as $R \to \infty$. According to (6.7) this implies that for any $\varepsilon > 0$ there exists $R > 0$ such that

$$\int_{M\setminus \bar{B}(x_0,R)} (|\nabla u|^2 + V|u|^2)d\mu \ge \frac{1}{\varepsilon} \int_{M\setminus \bar{B}(x_0,R)} |u|^2 d\mu$$

for any $u \in \mathrm{Dom}\,(H)$. Recalling that $V \ge 1$, we deduce that

$$\int_{M\setminus \bar{B}(x_0,R)} |u|^2 d\mu \le \varepsilon, \quad u \in \mathcal{L}.$$

This implies $\sigma = \sigma_d$ due to Lemma 6.3. □

Our next goal will be to give conditions of the discreteness of the spectrum in terms of the behavior of Dirichlet and Neumann spectrum on small balls. The following well known geometric lemma will provide us with convenient coverings of M by balls.

Let $M = \cup_{j\in J} U_j$ be a covering of M by open sets U_j. We define the *multiplicity* of this covering as the maximum possible number N of different $j_1, \ldots, j_N \in J$ such that $U_{j_1} \cap \ldots \cap U_{j_N} \neq \emptyset$.

Lemma 6.5. *There exist $r_0 > 0$ and $N > 0$ such that for any $r \in (0, r_0)$ there exists a covering of M by balls $B(x_j, r)$ with the multiplicity of this covering not greater than N.*

PROOF. We will use an idea of Gromov [43] (see also [99] and [77]). Let us choose a maximal subset $S = \{x_1, x_2, \ldots\} \subset M$ such that $d(x_i, x_j) \geq r$ if $i \neq j$. (Here $d(x, y)$ means the Riemannian distance between x and y.) This can be done by the Zorn lemma. Then for every $x \in M$ there exists j such that $d(x, x_j) < r$ because otherwise we could add x to S which contradicts the maximality of S. This means that $M = \cup_j B(x_j, r)$.

Let us show that the multiplicity N of the covering of M by the balls $B(x_j, r)$ is bounded if $r < r_0$ where r_0 is sufficiently small. Assume that

$$x \in B(x_{j_1}, r) \cap \ldots \cap B(x_{j_N}, r).$$

Then $B(x, 2r) \supset B(x_{j_s}, r/2)$ for all $s = 1, \ldots, N$, and on the other hand the balls $B(x_{j_1}, r/2), \ldots, B(x_{j_N}, r/2)$ are disjoint by the triangle inequality. Therefore

$$N \leq \frac{\sup_{x\in M} \operatorname{vol}(B(x, 2r))}{\inf_{x\in M} \operatorname{vol}(B(x, r/2))}.$$

Here the right hand side is a constant (independent of r) in $\mathbb{R}^n$ with the standard metric. It is bounded in M if $r \in (0, r_0)$ and r_0 is sufficiently small. Indeed, the Jacobi matrices of the exponential maps and their inverses are uniformly bounded on small balls (see [88]), hence the volumes of small balls in M are estimated from both sides by the volumes of the corresponding balls in the tangent spaces (with the constants which are independent of r). □

Remark 6.6. The bounded geometry requirement is redundant for Lemma 6.5 to hold. In fact, an estimate from above for the ratio of the volumes of the balls holds if we only require that the Ricci curvature is bounded below (see e.g. [20] or [97], Theorem 1.3).

Now we will show that the domains Ω and $M \setminus \bar{B}(x_0, R)$ in Proposition 6.4 can be replaced by the balls $B(x, r)$ of arbitrarily fixed small radius r with $x \to \infty$.

Proposition 6.7.

(a) If $\sigma = \sigma_d$, then $\lambda(B(x, r)) \to \infty$ as $x \to \infty$ for any fixed $r \in (0, r_0/2)$.

(b) If $\mu(B(x, r)) \to \infty$ as $x \to \infty$ for some fixed $r \in (0, r_0/2)$, then $\sigma = \sigma_d$.

PROOF. (a) obviously follows from (a) in Proposition 6.4.

To prove (b) fix $r \in (0, r_0/2)$ and choose a covering of M by balls $B(x_k, r)$, $k = 1, 2, \ldots$, as in Lemma 6.5, so that the multiplicity of this covering is $\leq N$. Denote

$$\mu_R = \inf\{\mu(B(x_k, r))|\ B(x_k, r) \cap (M \setminus B(x_0, R)) \neq \emptyset\}\,.$$

Clearly the condition in (b) implies that $\mu_R \to \infty$ as $R \to \infty$.

Denote

$$I_R = \{k|\ B(x_k, r) \cap (M \setminus B(x_0, R + 2r)) \neq \emptyset\}\,.$$

For any $R > 0$ and any $u \in C^\infty(M \setminus B(x_0, R))$ we have

$$\int_{M\setminus B(x_0,R)} (|\nabla u|^2 + V|u|^2)d\mu \geq \frac{1}{N}\sum_{k\in I_R}\int_{B(x_k,r)} (|\nabla u|^2 + V|u|^2)d\mu \geq$$
$$\frac{1}{N}\mu_R \sum_{k\in I_R}\int_{B(x_k,r)} |u|^2 d\mu \geq \frac{1}{N}\mu_R \int_{M\setminus B(x_0,R+2r)} |u|^2 d\mu\,.$$

It follows that

$$\int_{M\setminus B(x_0,R+2r)} |u|^2 d\mu \leq N\mu_R^{-1} \quad \text{for any} \quad u \in \mathcal{L},$$

where $\mathcal{L}$ is defined by (6.4). Hence $\sigma = \sigma_d$ due to Lemma 6.3. □

To proceed further we need a well known Lemma which estimates the L^2-norm of a function in a ball $B = B(0, r) \subset \mathbb{R}^n$ (with the flat metric) by the L^2 norm of this function in a smaller homothetic ball provided we can control the Dirichlet integral of the function. (See e.g. [78] where parallelepipeds were considered instead of balls.) To formulate the Lemma let us introduce the following temporary short notations:

$$\|\psi\| = \|\psi\|_{L^2(B)} = \left(\int_B |\psi|^2 dx\right)^{1/2}, \quad \|\psi\|_t = \|\psi\|_{L^2(B_t)},$$

where $0 < t \leq 1$, dx is the Lebesgue measure on $\mathbb{R}^n$, $B_t = B(0, tr)$. Similarly define

$$\|\nabla\psi\| = \|\nabla\psi\|_{L^2(B)} = \left(\int_B |\nabla\psi|^2 dx\right)^{1/2},$$
$$\|\nabla\psi\|_t = \|\nabla\psi\|_{L^2(B_t)}.$$

Lemma 6.8. *The following estimates hold true:*

(6.8) $$\|\psi\| \leq t^{-n/2}\|\psi\|_t + 2^{n+1}r(1-t)\|\nabla\psi\|, \quad t \in [1/2, 1];$$

(6.9) $$\|\psi\|^2 \le 2t^{-n}\|\psi\|_t^2 + 2^{2n+3}r^2(1-t)^2\|\nabla\psi\|^2, \quad t \in [1/2, 1].$$

In these estimates we assume that ψ is a real-valued function from $C^\infty(B)$.

PROOF. To prove (6.8) denote $\psi_t(x) = \psi(tx)$ and consider the function

$$f(t) = \int_B \psi^2(tx)dx = t^{-n}\int_{B_t}\psi^2(x)dx = t^{-n}\|\psi\|_t^2,$$

where $0 < t \le 1$. Differentiating it we get for $t \in (0, 1)$

$$f'(t) = \int_B 2\psi_t(x)(x \cdot \nabla\psi(tx))dx = 2t^{-1}\int_B \psi_t(x \cdot \nabla\psi)_t dx.$$

Using the Cauchy-Schwarz inequality we obtain

$$|f'(t)|^2 \le 4t^{-2}\|\psi_t\|^2\|(x\cdot\nabla\psi)_t\|^2 = 4t^{-2n-2}\|\psi\|_t^2\|(x\cdot\nabla\psi)\|_t^2 \le 4t^{-2n-2}r^2\|\psi\|^2\|\nabla\psi\|^2.$$

Therefore,

$$|f(1) - f(t)| \le \int_t^1 |f'(\tau)|d\tau \le 2r\|\psi\|\|\nabla\psi\|\int_t^1 \tau^{-n-1}d\tau$$

Since

$$\int_t^1 \tau^{-n-1}d\tau = \frac{1-t^n}{nt^n} \le 2^n(1-t), \ t \in [1/2, 1],$$

we further obtain

$$|f(1) - f(t)| \le 2^{n+1}r(1-t)\|\psi\|\|\nabla\psi\|.$$

In particular,

$$f(1) \le f(t) + 2^{n+1}r(1-t)\|\psi\|\|\nabla\psi\|,$$

or

$$\|\psi\|^2 \le t^{-n}\|\psi\|_t^2 + 2^{n+1}r(1-t)\|\psi\|\|\nabla\psi\|.$$

Rewrite this inequality in the form

$$y^2 - 2ay - b^2 \le 0, \ y = \|\psi\|, \ a = 2^n r(1-t)\|\nabla\psi\|, \ b = t^{-n/2}\|\psi\|_t.$$

It follows that $(y-a)^2 \le a^2 + b^2$, hence $y - a \le \sqrt{a^2+b^2} \le a + b$ and $y \le 2a + b$ which is exactly (6.8)

Clearly (6.9) follows from (6.8) if we use the inequality $(a+b)^2 \le 2(a^2+b^2)$, $a, b \in \mathbb{R}$. □

Now we would like to compare $\lambda(B(x,r))$ and $\mu(B(x,r))$. This is done in the following

Lemma 6.9. *There exist $C_1, C_2 > 0$ depending only on (M, g) such that*

$$\mu(B(x,r)) \le \lambda(B(x,r)) \le C_1\mu(B(x,r)) + C_2 r^{-2}, \tag{6.10}$$

for any $x \in M$, $r \in (0, r_0/2)$.

PROOF. A. Note first that the inequality $\mu(\Omega) \le \lambda(\Omega)$ holds for any open set $\Omega \subset M$ due to the definitions of $\lambda(\Omega)$ and $\mu(\Omega)$ (see (6.6), (6.7)). Hence we need to prove only the second inequality in (6.10).

Note that due to the bounded geometry conditions replacing the gradient ∇ and the measure $d\mu$ by the Euclidean gradient and measure in normal coordinates in (6.6) and (6.7) leads to comparable ratios which differ from the ones in (6.6), (6.7) by a bounded factor which is also separated from 0 uniformly in u and $r \in (0, r_0/2)$. Therefore it is sufficient to prove (6.10) in the Euclidean case and for $x = 0$. This was done by A.M. Molchanov [78] (for parallelepipeds instead of balls which does not make any difference) but for the sake of completeness we will reproduce his arguments in our notations.

B. Now we will argue in the ball $B = B(0, r) \subset \mathbb{R}^n$ with the flat metric. Let us choose an arbitrary $\varepsilon > 0$ and according to (6.7) take a real-valued $\psi \in C^\infty(B)$ so that

$$\|\psi\|^2 = 1 \quad \text{and} \quad \int_B (|\nabla\psi|^2 + V|\psi|^2)dx \le \mu(B) + \varepsilon. \tag{6.11}$$

It follows in particular that

$$\int_B |\nabla\psi|^2 dx \le \mu(B) + \varepsilon. \tag{6.12}$$

Now let us make an appropriate choice of t, so as to be able to evaluate $\|\psi\|_t$ from below using (6.9). To this end we would like the estimate

$$2^{2n+3} r^2 (1-t)^2 \|\nabla\psi\|^2 \le \frac{1}{2}\|\psi\|^2 \tag{6.13}$$

to hold true. According to (6.12) it is sufficient to require that

$$2^{2n+4} r^2 (1-t)^2 (\mu(B) + \varepsilon) \le 1.$$

The minimal $t \ge 1/2$ satisfying this condition is

$$t = \max\left\{\frac{1}{2}, 1 - \frac{1}{2^{n+2} r\sqrt{\mu(B) + \varepsilon}}\right\}. \tag{6.14}$$

Let us fix this value of t in the subsequent arguments. Then (6.9) and (6.13) imply

$$\int_{B_t} |\psi|^2 dx = \|\psi\|_t^2 \ge \frac{1}{4} t^n \ge \frac{1}{2^{n+2}}. \tag{6.15}$$

C. Let us choose a cut-off function $\chi \in C_c^\infty(B)$ such that $\chi = 1$ on B_t, $0 \le \chi(x) \le 1$ and

$$|\nabla\chi(x)| \le \frac{2}{r(1-t)} \tag{6.16}$$

for all $x \in B$. Then take $\phi = \chi\psi$ to serve as a test function in evaluating $\lambda(B)$ from (6.6). We have then

$$\begin{aligned}|\nabla\phi|^2 + V\phi^2 &= |\chi\nabla\psi + \psi\nabla\chi|^2 + V\chi^2\psi^2\\ &\le |\nabla\psi|^2 + 2|\chi\psi\nabla\chi\cdot\nabla\psi| + \psi^2|\nabla\chi|^2 + V\psi^2\\ &\le 2(|\nabla\psi|^2 + V\psi^2 + \psi^2|\nabla\chi|^2).\end{aligned}$$

Integrating this over B and using (6.11) and (6.16) we obtain

$$\int_B (|\nabla\phi|^2 + V\phi^2)dx \le 2\left[\mu(B(0,r)) + \varepsilon + \frac{4}{r^2(1-t)^2}\right].$$

Due to (6.15) we also have

$$\int_B \phi^2 dx \ge \frac{1}{2^{n+2}}.$$

Therefore (6.6) implies

$$\lambda(B) \le 2^{n+3}\left[\mu(B) + \varepsilon + \frac{4}{r^2(1-t)^2}\right] \tag{6.17}$$

Note that (6.14) implies

$$\frac{1}{(1-t)^2} = \max\{4, 2^{2n+4}r^2(\mu(B)+\varepsilon)\} \le 4 + 2^{2n+4}r^2(\mu(B)+\varepsilon).$$

Therefore (6.17) gives

$$\lambda(B) \le 2^{n+3}(1+2^{2n+6})(\mu(B)+\varepsilon) + 2^{n+7}r^{-2}.$$

Since $\varepsilon > 0$ is arbitrary, this estimate also holds for $\varepsilon = 0$ i.e.

$$\lambda(B) \le 2^{n+3}(1+2^{2n+6})\mu(B) + 2^{n+7}r^{-2}.$$

This proves the desired estimate (6.10). □

Now we are ready to formulate the localization theorem which is the main result of this section.

Theorem 6.10. *Let (M, g) be a manifold of bounded geometry, and H is a Schrödinger operator (1.1) on M with the potential which is semi-bounded below. Then the following conditions are equivalent:*

(a) $\sigma = \sigma_d$ i.e. H has a discrete spectrum;

(b) $\lambda(B(x,r)) \to \infty$ *as* $x \to \infty$ *for any fixed* $r \in (0, r_0/2)$;
(c) $\mu(B(x,r)) \to \infty$ *as* $x \to \infty$ *for any fixed* $r \in (0, r_0/2)$;
(d) there exists $r \in (0, r_0/2)$ *such that* $\lambda(B(x,r)) \to \infty$ *as* $x \to \infty$;
(e) there exists $r \in (0, r_0/2)$ *such that* $\mu(B(x,r)) \to \infty$ *as* $x \to \infty$.

PROOF. 1) Obviously (b) implies (d), and (c) implies (e).

2) Due to Lemma 6.9, (b) is equivalent to (c), and (d) is equivalent to (e).

3) Due to Proposition 6.7, (a) implies (b), and (e) implies (a).

4) Considering the graph of all the implications listed in 1)–3), it is easy to conclude that all the conditions (a)–(e) are equivalent. □

7. Capacity

We will use the harmonic (or Newtonian) capacity for compact subsets $F \subset \mathbb{R}^n$ with respect to an open set $\Omega \subset \mathbb{R}^n$ which includes F, so $F \subset \Omega \subset \mathbb{R}^n$. We will always assume that Ω is bounded, connected and has a C^∞ boundary. In fact we only need $\Omega = B(0,R)$, where $B(0,R)$ is the open ball of a fixed radius $R > 0$ centered at 0.

Let us recall the necessary definitions. We will use the variational version as the main definition of the capacity.

Definition 7.1. For any compact set $F \subset \Omega$ define the *harmonic (or Newtonian) capacity* of F with respect to Ω as

$$\text{(7.1)}\quad \operatorname{cap}_\Omega(F) = \inf\left\{ \int_\Omega |\nabla u|^2 dx |\ u \in C_c^\infty(\Omega), u = 1 \text{ near } F,\ 0 \le u \le 1 \text{ in } \Omega \right\}.$$

Clearly $F_1 \subset F_2$ implies $\operatorname{cap}_\Omega F_1 \le \operatorname{cap}_\Omega F_2$ for any compact subsets $F_1, F_2 \subset \Omega$.

The following lemma allows to approximate the capacity of an arbitrary compact set $F \subset \Omega$ by the capacity of compacts of the form $\bar{U} \subset \Omega$ where U is an open set with a C^∞ boundary, $\bar{U}$ denotes its closure in $\mathbb{R}^n$.

Lemma 7.2. *In the notations of Definition 7.1*

$$\operatorname{cap}_\Omega(F) = \inf\{ \operatorname{cap}_\Omega(\bar{U}) |\ F \subset U \subset \bar{U} \subset \Omega,\ U \text{ has a } C^\infty \text{ boundary.}\}$$

PROOF. The proof follows immediately from the Definition 7.1. □

Let us list some equivalent versions of the Definition 7.1. Denote by $\operatorname{Lip}(\Omega)$ the set of Lipschitz functions in Ω, i.e. functions $f : \Omega \to \mathbb{R}$ such that

$$\text{(7.2)}\qquad |f(x) - f(y)| \le C|x - y|,\ x, y \in \Omega,$$

where $C \geq 0$ does not depend on x, y. Note that any function $f \in \mathrm{Lip}(\Omega)$ extends by continuity to $\bar{\Omega}$ (the closure of Ω in $\mathbb{R}^n$), and satisfies (7.2) on $\bar{\Omega}$ with the same constant C, i.e. $\mathrm{Lip}(\Omega) = \mathrm{Lip}(\bar{\Omega})$. It is well known that $f \in \mathrm{Lip}(\Omega)$ implies that all distributional derivatives $\partial f/\partial x_j$ are in $L^\infty(\Omega)$. Vice versa, if all first distributional derivatives of a distribution f are in $L^\infty(\Omega)$ then f as a distribution coincides with a (uniquely defined) Lipschitz function. (See e.g. [74], [102]). Denote by $\mathrm{Lip}_c(\Omega)$ the set of functions $f \in \mathrm{Lip}(\Omega)$ which have a compact support in Ω.

Proposition 7.3. *The conditions*

$$u \in C_c^\infty(\Omega),\ u = 1 \text{ near } F,\ 0 \leq u \leq 1 \text{ in } \Omega; \tag{7.3}$$

in (7.1) can be replaced (without changing the left hand side) by any of the following sets of conditions:

$$u \in \mathrm{Lip}_c(\Omega),\ u = 1 \text{ near } F,\ 0 \leq u \leq 1 \text{ in } \Omega; \tag{7.4}$$

$$u \in \mathrm{Lip}(\Omega),\ u = 1 \text{ near } F,\ u|_{\partial\Omega} = 0,\ 0 \leq u \leq 1 \text{ in } \Omega; \tag{7.5}$$

$$u \in \mathrm{Lip}(\Omega),\ u = 1 \text{ near } F,\ u|_{\partial\Omega} = 0; \tag{7.6}$$

$$u \in \mathrm{Lip}(\Omega),\ u = 1 \text{ on } F,\ u|_{\partial\Omega} = 0,\ 0 \leq u \leq 1 \text{ in } \Omega; \tag{7.7}$$

$$u \in \mathrm{Lip}(\Omega),\ u \geq 1 \text{ near } F,\ u|_{\partial\Omega} = 0; \tag{7.8}$$

$$u \in \mathrm{Lip}(\Omega),\ u = 1 \text{ on } F,\ u|_{\partial\Omega} = 0; \tag{7.9}$$

$$u \in \mathrm{Lip}(\Omega),\ u = 1 \text{ near } F,\ u|_{\partial\Omega} = 0,\ \Delta u = 0 \text{ in } \Omega \setminus u^{-1}(\{1\}), \tag{7.10}$$

where $u^{-1}(\{1\}) = \{x|\ x \in \Omega,\ u(x) = 1\}$.

PROOF. Equivalence of (7.3) and (7.4) can be easily obtained by the standard mollifying procedure which leads to an approximation of any function $u \in \mathrm{Lip}_c(\Omega)$ by functions $u_k \in C_c^\infty(\Omega)$ so that for the approximating functions we still have $0 \leq u_k \leq 1$ and $u_k = 1$ near F, and also

$$\int_\Omega |\nabla u_k|^2 dx \to \int_\Omega |\nabla u|^2 dx$$

as $k \to \infty$.

Similarly using an appropriate sequence of cut-off functions we come to the conclusion that (7.5) is equivalent to (7.4).

To prove the equivalence of (7.5) and (7.6) we should replace a function u satisfying (7.6) first by $\max(u,0)$ and then by $\min(u,1)$ using the fact that the Dirichlet integral decreases under these operations.

The equivalence of (7.6) and (7.7) can be obtained if we approximate u from (7.7) by functions u_k satisfying (7.5). The functions u_k can be obtained e.g. by a procedure suggested by V. G. Mazya in [74], Sect.2.2.1, i.e. as $u_k(x) = \lambda_k(u(x))$. where

$$0 \le \lambda_k'(t) \le 1 + \frac{1}{k}, \ \lambda(t) = 0 \text{ near } (-\infty, 0]. \text{ and } 1 \text{ near } [1, \infty), \ 0 \le \lambda \le 1 .$$

The same approximation leads to the equivalence of (7.8) and (7.9) to the previous conditions.

Finally the equivalence of (7.10) and (7.6) can be obtained if we replace a function u satisfying (7.6) by a Lipschitz function v which satisfies the same conditions as u and besides is harmonic in $\Omega \setminus \bar{U}$ where U is a small neighborhood of F, $U \subset u^{-1}(\{1\})$, such that U has a C^∞ boundary. (This can be done by solving the appropriate Dirichlet problem.) By the well known variational property of the Dirichlet problem we will have then

$$\int_\Omega |\nabla v|^2 dx \le \int_\Omega |\nabla u|^2 dx,$$

which proves the equivalence of (7.10) and (7.6). □

Corollary 7.4. *If $F = \bar{U} \subset \Omega$ where U has a C^∞ boundary, then*

$$\operatorname{cap}_\Omega(F) = \int_{\Omega \setminus F} |\nabla u|^2 dx = \int_{\partial U} \frac{\partial u}{\partial \nu} d\sigma, \tag{7.11}$$

where u is the solution of the following Dirichlet problem

$$\Delta u = 0 \text{ in } \Omega \setminus F, \ u|_{\partial U} = 1, u|_{\partial \Omega} = 0, \tag{7.12}$$

ν is the exterior unit normal vector to $\partial(\Omega \setminus F)$, $d\sigma$ is the (euclidean) area element on the boundary, the derivative $\partial u / \partial \nu$ is understood as $-\partial u / \partial \nu_{in}$ where $\nu_{in} = -\nu$ is the interior unit normal vector.

PROOF. The second equality in (7.11) follows from the divergence formula

$$\int_{\Omega \setminus F} |\nabla u|^2 dx = \int_{\Omega \setminus F} [\operatorname{div}(u \nabla u) - u \Delta u] dx = \int_{\Omega \setminus F} \operatorname{div}(u \nabla u) dx = \int_{\partial(\Omega \setminus F)} u \frac{\partial u}{\partial \nu} d\sigma$$

and the boundary conditions in (7.12). The first equality then follows by an easy limit transition over a neighborhoods of F with C^∞ boundary as in (7.10). □

This corollary allows in particular to calculate the capacity of a ball F with respect to a bigger ball with the same center. Namely, for $U = B(0,r)$ and $\Omega = B(0,R)$ with $r < R$ the solution of the Dirichlet problem (7.12) has the form

$$u(x) = \frac{|x|^{2-n} - R^{2-n}}{r^{2-n} - R^{2-n}}, \qquad n \geq 3,$$
$$u(x) = \left(\ln \tfrac{R}{r}\right)^{-1} \ln \tfrac{R}{|x|}, \quad n = 2.$$

Using these formulas to calculate the right hand side of (7.11) we easily obtain

$$\text{(7.13)} \qquad \operatorname{cap}_{B(0,R)}(\bar{B}(0,r)) = (n-2)\omega_n r^{n-2}(1 - r^{n-2}/R^{n-2})^{-1}, \quad n \geq 3,$$

$$\text{(7.14)} \qquad \operatorname{cap}_{B(0,R)}(\bar{B}(0,r)) = 2\pi \left(\ln \tfrac{R}{r}\right)^{-1}, \qquad n = 2,$$

where ω_n is the area of the unit sphere in $\mathbb{R}^n$.

Note that when R is fixed and $r \to 0$, the capacity $\operatorname{cap}_{B(0,R)} \bar{B}(0,r)$ is equivalent to $(n-2)\omega_n r^{n-2}$ for $n \geq 3$ and to $2\pi \left(\ln \frac{1}{r}\right)^{-1}$ for $n = 2$. In case $n \geq 3$ we also have

$$\lim_{R \to \infty} \operatorname{cap}_{B(0,R)}(\bar{B}(0,r)) = (n-2)\omega_n r^{n-2}.$$

Generally for $n \geq 3$ and any compact $F \subset \mathbb{R}^n$ using the maximum principle it is easy to prove the existence of the limit

$$\operatorname{cap}(F) = \lim_{R \to \infty} \operatorname{cap}_{B(0,R)}(F),$$

which is usually called the (absolute) harmonic (or Newtonian) capacity. This absolute capacity can be used instead of the relative one because asymptotically for small r it gives the same result. In case $n = 2$ the $R \to \infty$ limit vanishes, so we really have to use the relative capacity with a fixed $R > 0$.

Remark 7.5. The absolute capacity of a compact $F \subset \mathbb{R}^n$ for $n \geq 3$ can be also defined as follows:

$$\text{(7.15)} \quad \operatorname{cap}(F) = \sup\{m(F) | \frac{1}{(n-2)\omega_n} \int_F \frac{dm(y)}{|x-y|^{n-2}} \leq 1,\ x \in \mathbb{R}^n \setminus F\},$$

where m is a finite positive Radon measure on F (a "charge").

This definition corresponds to the physical notion of capacity (in electrostatics), which is the maximal charge which you can "load" on the capacitor so that the potential of this charge is ≤ 1 everywhere.

In case $F = \bar{U}$ where U is a bounded open set in $\mathbb{R}^n$ with a C^∞ boundary, the optimal measure m in (7.15) can be determined as follows. Let us take $u \in \operatorname{Lip}(\mathbb{R}^n)$, such that $u = 1$ on F and u is the solution of the Dirichlet problem similar to (7.12):

$$\Delta u = 0 \text{ in } \mathbb{R}^n \setminus F,\ u|_{\partial U} = 1,\ u(x) \to 0 \text{ as } x \to \infty.$$

Then $m = -\Delta u$ where Δ is understood in the sense of distributions. This means that in fact m is supported on ∂U and has there a density $(\partial u/\partial \nu)d\sigma$.

The corresponding variational definition of the absolute capacity is

$$\operatorname{cap}(F) = \inf\left\{\int_{\mathbb{R}^n} |\nabla u|^2 dx \,|\, u \in C_c^\infty(\mathbb{R}^n), u \geq 1 \text{ on } F\right\},$$

where u is assumed to be real-valued.

The capacity of a set as a mathematical notion was first introduced by N. Wiener [107] who used it in [108] to establish his famous criterion for regularity of a boundary point with respect to the Dirichlet problem. The most important early developments are due to C.J. de La Vallée-Poussin [105] and O. Frostman [39].

For more details about the capacity we refer to the monographs by D. Adams and L. Hedberg [1], L. Carleson [17], E.M. Landis [59], N.S. Landkof [60], V.G. Maz'ya [74] and J. Wermer [106], as well as to the papers by M.V. Keldysh [55] and V.G. Maz'ya [75].

8. Necessary condition

In this section we will establish that the condition (D) in Theorem 5.1 is necessary for the spectrum of H to be discrete. Let us fix $R > 0$ and denote $\Omega = B(0, R) \subset \mathbb{R}^n$. For simplicity of notations in this section we will write $\operatorname{cap}(F)$ instead of $\operatorname{cap}_\Omega(F)$ for any compact set $F \subset \Omega$. We will consider the Schrödinger operator (1.1) in the ball Ω (with the usual Euclidean Laplacian).

Our main result in this section will be the following

Theorem 8.1. *If $F \subset \bar{B} \subset \mathbb{R}^n$, where $B = B(0, r)$, $r \leq R/2$, and F is compact, then*

$$\mu(B) \leq 2^{n+4} r^{-2} [\operatorname{cap}(\bar{B})]^{-1} \left(\int_{B\setminus F} V(x)dx + \operatorname{cap}(F)\right), \tag{8.1}$$

provided

$$\operatorname{cap}(F) < 2^{-2n-6} \operatorname{cap}(\bar{B}). \tag{8.2}$$

Here $\operatorname{cap}(\bar{B})$ can be found by the formulas (7.13), (7.14).

PROOF. A. Let us choose an arbitrary $\varepsilon > 0$ and take a function $u \in C_c^\infty(\Omega)$ such that $u = 1$ in a neighborhood of F, $0 \leq u(x) \leq 1$ for all $x \in \Omega$, and

$$\int_\Omega |\nabla u|^2 dx \leq \operatorname{cap}(F) + \varepsilon$$

Let us use the function $\psi = 1 - u$ as a test function instead of u in (6.7), so

$$\mu(B) \le \frac{\int_B (|\nabla\psi|^2 + V\psi^2)dx}{\int_B \psi^2 dx}. \tag{8.3}$$

Note that

$$\int_B |\nabla\psi|^2 dx = \int_B |\nabla u|^2 dx = \int_{B\setminus F} |\nabla u|^2 dx \le \int_{\Omega\setminus F} |\nabla u|^2 dx \le \operatorname{cap}(F) + \varepsilon.$$

Since $\psi = 0$ in a neighborhood of F and $0 \le \psi \le 1$ everywhere on Ω, we have then

$$\int_B (|\nabla\psi|^2 + V\psi^2)dx \le \operatorname{cap}(F) + \varepsilon + \int_{B\setminus F} V dx. \tag{8.4}$$

This gives an estimate of the numerator in the right hand side of (8.3). So it remains to estimate the denominator from below.

B. We will start by an estimate from below for $\|\psi\|^2_{L^2(B_2)}$, where $B_2 = B(0, 2r)$. Take the following cut-off function $\theta \in \operatorname{Lip}(\Omega)$:

$$\theta(x) = 0 \text{ on } B, \quad \theta(x) = \frac{|x|}{r} - 1 \text{ on } B_2 \setminus B, \quad \theta(x) = 1 \text{ on } \Omega \setminus B_2.$$

It follows that $|\nabla\theta| \le r^{-1}$. Now consider $\tilde\psi = \theta\psi \in \operatorname{Lip}(\Omega)$. Clearly,

$$\tilde\psi = 0 \text{ on } B, \qquad \tilde\psi = \psi \text{ on } \Omega \setminus B_2.$$

We have

$$\begin{aligned} \textstyle\int_\Omega |\nabla\tilde\psi|^2 dx = \int_\Omega |\nabla(\theta\psi)|^2 dx \ &\le 2\int_\Omega |\nabla\psi|^2 dx + 2r^{-2}\int_{B_2\setminus B} \psi^2 dx \\ &\le 2(\operatorname{cap}(F) + \varepsilon) + 2r^{-2}\int_{B_2} \psi^2 dx. \end{aligned}$$

On the other hand $\tilde\psi = 1$ near $\partial\Omega$, so if we take $\tilde u = 1 - \tilde\psi$, then $\tilde u = 1$ on B and 0 near $\partial\Omega$, so $\tilde u$ can serve as a test function for estimating $\operatorname{cap}(\bar B)$. Therefore

$$\int_\Omega |\nabla\tilde\psi|^2 dx \ge \operatorname{cap}(\bar B).$$

Combining this inequality with the previous estimate we obtain

$$\operatorname{cap}(\bar B) \le 2(\operatorname{cap}(F) + \varepsilon) + 2r^{-2}\int_{B_2} \psi^2 dx,$$

hence

$$\int_{B_2} \psi^2 dx \ge \frac{1}{2} r^2 \operatorname{cap}(\bar B) - r^2(\operatorname{cap}(F) + \varepsilon).$$

Now assuming that

$$\text{(8.5)} \qquad \operatorname{cap}(F) \le \frac{1}{4}\operatorname{cap}\bar{B} - \varepsilon,$$

we obtain

$$\text{(8.6)} \qquad \int_{B_2} \psi^2 dx \ge \frac{1}{4} r^2 \operatorname{cap}(\bar{B}).$$

C. Now we should estimate the norm $\|\psi\|^2_{L^2(B)}$ from below. To this end let us use the estimate (6.9) with $t = 1/2$ and with r replaced by $2r$, i.e.

$$\|\psi\|^2_{L^2(B_2)} \le 2^{n+1}\|\psi\|^2_{L^2(B)} + 2^{2n+3}r^2\|\nabla\psi\|^2_{L^2(B_2)}.$$

Using the same ψ as above, we obtain

$$\|\nabla\psi\|^2_{L^2(B_2)} \le \|\nabla\psi\|^2_{L^2(\Omega)} \le \operatorname{cap}(F) + \varepsilon,$$

hence

$$\|\psi\|^2_{L^2(B)} \ge 2^{-n-1}\|\psi\|^2_{L^2(B_2)} - 2^{n+2}r^2\|\nabla\psi\|^2_{L^2(B_2)}$$
$$\ge 2^{-n-1}\|\psi\|^2_{L^2(B_2)} - 2^{n+2}r^2\|\nabla\psi\|^2_{L^2(\Omega)} \ge 2^{-n-1}\|\psi\|^2_{L^2(B_2)} - 2^{n+2}r^2(\operatorname{cap}(F) + \varepsilon).$$

Together with (8.6) this gives

$$\|\psi\|^2_{L^2(B)} \ge 2^{-n-3}r^2 \operatorname{cap}(\bar{B}) - 2^{n+2}r^2(\operatorname{cap}(F) + \varepsilon),$$

provided (8.5) is satisfied.

Now imposing the condition

$$\operatorname{cap}(F) \le 2^{-2n-6}\operatorname{cap}\bar{B} - \varepsilon$$

which is obviously stronger than (8.5), we arrive to the estimate

$$\int_B \psi^2 dx \ge 2^{-n-4}r^2 \operatorname{cap}(\bar{B}).$$

Together with the inequality (8.4) this gives the estimate

$$\mu(B) \le 2^{n+4}r^{-2}[\operatorname{cap}(\bar{B})]^{-1}\left(\int_{B\setminus F} V(x)dx + \operatorname{cap}(F) + \varepsilon\right).$$

This estimate holds for any compact $F \subset B$ satisfying (8.2), with an arbitrary $\varepsilon > 0$ such that

$$\varepsilon < 2^{-2n-6}\operatorname{cap}(\bar{B}) - \operatorname{cap}(F).$$

Taking the limit as $\varepsilon \to 0$, we arrive to (8.1). □

Corollary 8.2. *There exists $c > 0$ such that the condition (D) in Theorem 5.1 is necessary i.e. $\sigma = \sigma_d$ implies (D).*

PROOF. If (D) is not satisfied whatever $c > 0$, then Theorem 8.1 implies that $\mu(B(x,r))$ does not go to ∞ as $x \to \infty$. But then due to the localization theorem (Theorem 6.10) the spectrum of H is not discrete. □

9. Sufficient condition

In this section we will prove that the condition (D) in Theorem 5.1 is sufficient for the spectrum of the Schrödinger operator (1.1) to be discrete, i.e. (D) implies that $\sigma = \sigma_d$. As in Sect.6 we will also assume that $V \geq 1$. Using the localization theorem 6.10 and the bounded geometry conditions we could again argue in an Euclidean ball $B = B(0,r) \subset \mathbb{R}^n$ of sufficiently small radius $r > 0$. But it proves to be more convenient to use cubes instead of balls. Namely, we will use *geodesic cubic cells* which we will call simply *cells* i.e. sets which are given in geodesic coordinates as follows:

$$C(a,d) = \{x = (x^1,\ldots,x^n) | \, |x^j - a^j| < d/2, \ j = 1,\ldots,n\}.$$

Here $d > 0$ will be called the *size* of the cell, the point $a \in M$ is called the *center* of the cell, and all the points $x = (x^1,\ldots,x^n) \in M$ are supposed to belong to a ball $B(x_0, R)$ where the geodesic coordinates $x^1,\ldots,x^n$ are defined. We restrict ourselves to balls of sufficiently small radius $R < r_0$ with r_0 as in Sect.5. So by definition the cell $C(a,d)$ belongs to such a ball and working with this cell we will assume that the geodesic coordinates are fixed.

The closure of the cell $C(a,d)$ will be denoted $\bar{C}(a,d)$.

The advantage of using cells (instead of balls) is that we can subdivide cells into smaller cells up to sets of measure 0.

In the future we will mostly work in a fixed small cell $C = C(0,d) \subset \mathbb{R}^n$, i.e. the cell of the size d centered at the origin in $\mathbb{R}^n$. Here $\mathbb{R}^n$ is considered with the standard metric. The closure of this cell will be denoted $\bar{C}$.

Let us start with the following well known

Lemma 9.1. (Poincaré inequality)
Let $\bar{v}$ denote the mean value of a real-valued function $v \in \mathrm{Lip}(C)$ i.e.

$$\bar{v} = \frac{1}{\mathrm{vol}(C)} \int_C v(x)dx = \frac{1}{d^n} \int_C v(x)dx.$$

Then

$$\int_C (v - \bar{v})^2 dx \leq \pi^{-2} d^2 \int_C |\nabla v|^2 dx.$$

PROOF. Note that by separation of variables the first non-zero Neumann eigenvalue of $-\Delta$ in the cell $C = C(0,d) \subset \mathbb{R}^n$ is $\mu_1 = \pi^2 d^{-2}$.

Now the desired inequality follows from the standard variational formula for μ_1:

$$\mu_1 = \inf\left\{ \frac{\int_C |\nabla u|^2 dx}{\int_C |u|^2 dx} \,\middle|\, u \in \mathrm{Lip}\,(C), \int_C u dx = 0 \right\}.$$

□

Let us introduce a class of *negligible* (or, more precisely, *c-negligible*) subsets in an open set $G \subset M$, such that G lies in a domain of geodesic coordinates, more precisely $G \subset B(x,R)$ for some $x \in M$ and $R < r_0/2$ where r_0 is chosen as in Sect.5. Using geodesic coordinates we can identify G with a subset in $\mathbb{R}^n$. Then we define

$$\mathcal{N}_c(G) = \{F \mid F \text{ is a compact subset in } \bar{G} \text{ and } \mathrm{cap}\,(F) \le c\,\mathrm{cap}\,(\bar{G})\}.$$

Here $c > 0$ is a constant, and cap means cap_Ω where $\Omega = B(0,R) \subset \mathbb{R}^n$ as in the previous section, $R \in (0, r_0/2)$ is fixed. The class $\mathcal{N}_c(G)$ depends on G and also on the choice of the constant c which will be eventually chosen sufficiently small.

For the future we will only use G which is a small ball or a small cell i.e. $G = B$ or $G = C$, so we will assume that the boundary of G is piecewise smooth.

Note that it only makes sense to consider $\mathcal{N}_c(G)$ with $c < 1$ because otherwise this class consists of *all* compact subsets in G.

We will also need another class of negligible sets which depends on $\mu(G)$:

$$\mathcal{N}_c'(G) = \mathcal{N}_{c\mu(G)}(G).$$

Note that $\mu(G) \ge 1$ because $V \ge 1$. Therefore $\mathcal{N}_c(G) \subset \mathcal{N}_c'(G)$.

The main necessity Theorem 8.1 can be reformulated as the following estimate:

$$\mu(B) \le A(r) \inf_{F \in \mathcal{N}_c(B)} \left(\int_{B \setminus F} V(x) dx + \mathrm{cap}\,(F) \right),$$

where $A(r) = 2^{n+5} r^{-2} [\mathrm{cap}\,(\bar{B})]^{-1}$, $c = 2^{-2n-7}$.

Our main goal in this section will be the proof of a somewhat opposite estimate which is formulated in the following

Theorem 9.2. *For any $c > 0$ there exists $d_1 > 0$ such that*

$$\inf_{F \in \mathcal{N}_c'(C)} \int_{C \setminus F} V(x) dx \le 4 d^n \mu(C), \tag{9.1}$$

provided $0 < d < d_1$.

In fact instead of (9.1) we will prove a stronger estimate which makes use of a more restrictive notion of negligibility. For any $A > 0$ define

$$\tilde{\mathcal{N}}_A(C) = \{F \mid F \text{ is a compact subset in } \bar{C} \text{ and } \operatorname{cap}(F) \leq Ad^n\}.$$

Similarly introduce also the corresponding class depending on $\mu(C)$:

$$\tilde{\mathcal{N}}'_A(C) = \tilde{\mathcal{N}}_{A\mu(C)}(C).$$

Again we have $\tilde{\mathcal{N}}_A(C) \subset \tilde{\mathcal{N}}'_A(C)$.

The following inclusions of subsets in $\mathbb{R}^n$

$$\bar{B}(0, d/2) \subset \bar{C}(0, d) \subset \bar{B}(0, d\sqrt{n}/2)$$

imply that

$$\operatorname{cap}(\bar{B}(0, d/2)) \leq \operatorname{cap}(\bar{C}(0, d)) \leq \operatorname{cap}(\bar{B}(0, d\sqrt{n}/2)).$$

Hence due to (7.13), (7.14) we see that

$$\operatorname{cap}(\bar{C}(0, d)) \asymp \operatorname{cap}(\bar{B}(0, d)). \tag{9.2}$$

(Let us recall that for two real-valued functions f_1, f_2 defined on the same set S, the relation $f_1 \asymp f_2$ means that there exists a constant $C > 0$ such that $C^{-1} f_1(x) \leq f_2(x) \leq C f_1(x)$ for all $x \in S$.)

Using (9.2) together with (7.13), (7.14), we easily conclude that for any $A > 0$ and $c > 0$ there exists $d_1 > 0$ such that

$$\tilde{\mathcal{N}}_A(C) \subset \mathcal{N}_c(C) \text{ and } \tilde{\mathcal{N}}'_A(C) \subset \mathcal{N}'_c(C) \text{ if } 0 < d < d_1,$$

because $d^n = o(\operatorname{cap}(\bar{C}))$ as $d \to 0$.

Proposition 9.3. *There exists $A = A_n > 0$ such that*

$$\inf_{F \in \tilde{\mathcal{N}}'_A(C)} \int_{C \setminus F} V(x)\,dx \leq 4d^n \mu(C). \tag{9.3}$$

Since for small d the left-hand side of (9.3) is not smaller than the left-hand side of (9.1), Proposition 9.3 implies Theorem 9.2.

PROOF OF PROPOSITION 9.3. A. Let us choose an arbitrary $\varepsilon > 0$ and take a real-valued function $\psi \in C^\infty(\bar{C})$ which nearly minimizes the fraction in (6.7), i.e. such that

$$\int_C (|\nabla\psi|^2 + V\psi^2)\,dx \leq (1 + \varepsilon)\mu(C) \int_C \psi^2\,dx.$$

Let us normalize ψ so that $\overline{\psi^2} = 1$, i.e. $\|\psi\|^2_{L^2(C)} = \mathrm{vol}\,(C) = d^n$. We will have then

(9.4)
$$\int_C (|\nabla\psi|^2 + V\psi^2)dx \le (1+\varepsilon)d^n\mu(C).$$

Now let us take

(9.5)
$$F = \{x|\ x \in \bar{C},\ |\psi(x)| \le 1/2\}.$$

Then

$$\int_{C\setminus F} V dx \le 4\int_{C\setminus F} V\psi^2 dx \le 4\int_C (|\nabla\psi|^2 + V\psi^2)dx \le 4d^n(1+\varepsilon)\mu(C).$$

After taking infimum over all $F \in \mathcal{N}'_A(C)$ in the left-hand side here, we can replace ε by 0 in the right-hand side provided we can choose $A = A_n$ which does not depend on ε (and on d). Then the desired estimate (9.3) will follow.

B. Now we have to estimate the capacity of the set F defined by (9.5). Using the same function ψ as above, let us consider $\tilde{\psi} \in \mathrm{Lip}\,(C)$ defined as follows:

$$\tilde{\psi}(x) = \max(|\psi(x)| - \frac{1}{2}, 0).$$

Clearly $|\tilde{\psi}| \le |\psi| \le |\tilde{\psi}| + 1/2$ on C and $\tilde{\psi} = 0$ on F. Also $|\nabla\tilde{\psi}(x)| \le |\nabla\psi(x)|$ for almost all $x \in C$, hence

(9.6)
$$\int_C |\nabla\tilde{\psi}|^2 dx \le \int_C |\nabla\psi|^2 dx.$$

We also have

$$d^n = \int_C \psi^2 dx \le \int_C \left(|\tilde{\psi}| + \frac{1}{2}\right)^2 dx \le 2\int_C |\tilde{\psi}|^2 dx + \frac{1}{2}d^n.$$

Therefore

(9.7)
$$\int_C \tilde{\psi}^2 dx \ge \frac{1}{4}d^n.$$

Normalize $\tilde{\psi}$ in the same way as ψ i.e. take $\phi = a\tilde{\psi}$ with $a > 0$ so that $\overline{\phi^2} = 1$. Then it follows from (9.7) that we can take $1 \le a \le 4$, hence (9.6) and (9.4) imply that

(9.8)
$$\int_C |\nabla\phi|^2 dx \le 4(1+\varepsilon)d^n\mu(C).$$

Note that $\phi = 0$ on F.

We will also assume (replacing ϕ by $-\phi$ if necessary) that $\bar{\phi} \ge 0$.

C. Now define $v = 1 - \phi$, so $v = 1$ on F. Let us extend v to $C_2 = C(0, 2d)$ by reflections in hyperplanes constituting the faces of the cube $C = C(0, d)$. Namely, we may e.g. first use reflections in the planes $x^1 = \pm d/2$ to extend v to the set

$(-d,d) \times (-d/2,d/2)^{n-1}$, then use reflections in the planes $x^2 = \pm d/2$ to extend the resulting function to the set $(-d,d)^2 \times (-d/2,d/2)^{n-2}$ etc. The result will be a function $\hat{v} \in \mathrm{Lip}\,(C_2)$ such that

$$\int_{C_2} \hat{v}^2 dx = 2^n \int_C v^2 dx, \tag{9.9}$$

and

$$\int_{C_2} |\nabla \hat{v}|^2 dx = 2^n \int_C |\nabla v|^2 dx.$$

Since $|\nabla v| = |\nabla \phi|$ on C, we obtain also according to (9.8):

$$\int_{C_2} |\nabla \hat{v}|^2 dx \leq 2^{n+2} d^n (1+\varepsilon)\mu(C). \tag{9.10}$$

For the simplicity of notations we will write further v instead of $\hat{v}$ which obviously does not lead to a confusion.

D. To get a test function for estimating the capacity of F, we need to cut-off v to obtain a Lipschitz function in Ω such that it vanishes at $\partial\Omega$. In fact we will even make it vanish outside C_2. To this end let us take a cut-off function $\chi \in \mathrm{Lip}\,(C_2)$ such that $\chi = 1$ on C, $\chi = 0$ on ∂C_2, $0 \leq \chi \leq 1$ everywhere and $|\nabla\chi| \leq 2\sqrt{n}d^{-1}$. (We can take e.g. $\chi(x) = \chi_0(x^1)\chi_0(x^2)\ldots\chi_0(x^n)$ where $\chi_0(t) = 1$ on $[-d/2,d/2]$, 0 on $\mathbb{R} \setminus (-d,d)$ and χ_0 is linear on each of the intervals $[-d,-d/2]$ and $[d/2,d]$.)

Define $u = \chi v \in \mathrm{Lip}\,(C_2)$, so $u = 1$ on F and $u = 0$ on ∂C_2. We can now extend u by 0 to $\Omega \setminus \bar{C}_2$.

Let us estimate the Dirichlet integral for u. We have

$$\int_{C_2} |\nabla u|^2 dx = \int_{C_2} |\chi\nabla v + v\nabla\chi|^2 dx \leq 2\int_{C_2} |\nabla v|^2 dx + 8nd^{-2}\int_{C_2} v^2 dx. \tag{9.11}$$

E. Our next task will be to prove that $\|v\|^2_{L^2(C_2)}$ is small, so that we may in fact omit the last term from the estimate (9.11), installing an additional constant factor in front of $\|\nabla v\|^2_{L^2(C_2)}$ instead. The idea is that the L^2-norm of $\nabla\phi$ is small, therefore ϕ is "almost constant", hence close to 1. We will prove that this is indeed true in the L^2 sense. (The proof makes use of the Poincaré inequality.) It will follow that $v = 1 - \phi$ is close to 0 in the same sense.

Note that (9.9) shows that it is sufficient to evaluate $\|v\|^2_{L^2(C)}$.

Our first step will be to show that $\bar{v} = 1 - \bar{\phi}$ is small, more precisely:

$$|\bar{\phi} - 1| \leq 2\pi^{-1}[(1+\varepsilon)\mu(C)]^{1/2} d. \tag{9.12}$$

Using the triangle inequality and Lemma 9.1 we obtain

$$d^{n/2} = \|\phi\| \leq \|\phi - \bar{\phi}\| + \bar{\phi}d^{n/2} \leq \pi^{-1}d\|\nabla\phi\| + \bar{\phi}d^{n/2},$$

where $\|\cdot\|$ means the norm in $L^2(C)$. Now using (9.8) we obtain

$$d^{n/2} \leq 2\pi^{-1}[(1+\varepsilon)\mu(C)]^{1/2} d^{n/2+1} + \bar{\phi} d^{n/2},$$

or

$$\bar{\phi} \geq 1 - 2\pi^{-1}[(1+\varepsilon)\mu(C)]^{1/2} d. \tag{9.13}$$

Similarly we have

$$\begin{aligned}\bar{\phi} d^{n/2} &= \|\bar{\phi}\| \leq \|\phi - \bar{\phi}\| + \|\phi\| \leq \pi^{-1} d \|\nabla \phi\| + d^{n/2} \\ &\leq 2\pi^{-1}[(1+\varepsilon)\mu(C)]^{1/2} d^{n/2+1} + d^{n/2},\end{aligned}$$

hence

$$\bar{\phi} \leq 1 + 2\pi^{-1}[(1+\varepsilon)\mu(C)]^{1/2} d. \tag{9.14}$$

Combining (9.13) and (9.14) we obtain (9.12).

F. Clearly (9.12) can be rewritten as

$$|\bar{v}| \leq 2\pi^{-1}[(1+\varepsilon)\mu(C)]^{1/2} d.$$

Now arguing as in the proof of (9.12) we can write

$$\begin{aligned}\|v\| &\leq \|\bar{v}\| + \|v - \bar{v}\| \leq 2\pi^{-1}[(1+\varepsilon)\mu(C)]^{1/2} d^{n/2+1} + \pi^{-1} d \|\nabla v\| \\ &\leq 2\pi^{-1}[(1+\varepsilon)\mu(C)]^{1/2} d^{n/2+1} + 2\pi^{-1}[(1+\varepsilon)\mu(C)]^{1/2} d^{n/2+1} \\ &= 4\pi^{-1}[(1+\varepsilon)\mu(C)]^{1/2} d^{n/2+1},\end{aligned}$$

so we finally obtain the desired estimate

$$\int_C v^2 dx \leq 16\pi^{-2}(1+\varepsilon)\mu(C) d^{n+2},$$

which implies due to (9.9):

$$\int_{C_2} v^2 dx \leq 2^{n+4}\pi^{-2}(1+\varepsilon)\mu(C) d^{n+2}. \tag{9.15}$$

G. Now let us return to estimating u. We can use (9.11), (9.10) and (9.15) to conclude that

$$\operatorname{cap}(F) \leq A_n \mu(C) d^n,$$

where $A_n = 2^{n+3}(1+\varepsilon)(1+16n\pi^{-2})$. This concludes the proof of Proposition 9.3. □

Proof of Theorem 5.1. A. We already proved in Sect.8 that the condition (D) is necessary. To prove that it is also sufficient, let us assume the opposite i.e. that (D) is satisfied but $\sigma \neq \sigma_d$. It follows due to the localization theorem 6.10 that there exist small $r > 0$ and a sequence of balls $B(x_k, r)$, $k = 1, 2, \ldots$, such that $x_k \to \infty$ as $k \to \infty$ but $\lambda(B(x_k, r)) \le L$ where $L > 0$ does not depend on k. Using the geodesic coordinates centered at x_k and the monotonicity of $\lambda(G)$ with respect to the ordering of G by inclusion we deduce that $\lambda(C(x_k, 2r)) \le L$ with the same constant L because $B(x_k, r) \subset C(x_k, 2r)$. It follows that

$$\mu(C(x_k, d)) \le L, \tag{9.16}$$

where $d = 2r$.

B. Fixing k for a moment and choosing an arbitrary integer $N > 0$ we can in a standard way subdivide the cell $C = C(x_k, d)$ into N^n subcells

$$C_j = C(x'_{kj}, d/N), \qquad j = 1, \ldots, N^n,$$

so that the closures of the subcells C_j cover exactly the closure of C and the interiors of these subcells do not intersect. We claim that then there exists $j \in \{1, \ldots, N^n\}$ such that

$$\mu(C_j) \le L + 2,$$

with the same L as in (9.16).

Indeed, assuming opposite, we conclude that for any j the inequality

$$\int_{C_j} (|\nabla\psi|^2 + V|\psi|^2) d\mu \ge (L+1) \int_{C_j} |\psi|^2 d\mu$$

holds for any function $\psi \in C^\infty(C_j)$. But then we can take an arbitrary $\psi \in C^\infty(C)$ and sum up the inequalities above, applying them to the restrictions of ψ to the subcells C_j. In this way we see that the inequality

$$\int_C (|\nabla\psi|^2 + V|\psi|^2) d\mu \ge (L+1) \int_C |\psi|^2 d\mu$$

holds for any $\psi \in C^\infty(C)$. This means that $\mu(C) \ge L + 1$, which contradicts (9.16).

We conclude therefore that for any integer $N > 0$ there exists a sequence of points $\{x_k^{(N)} |\ k = 1, 2, \ldots\}$ in M such that $x_k^{(N)} \to \infty$ as $k \to \infty$ for any fixed N, but $\mu(C(x_k^{(N)}, d/N)) \le L + 1$ with a constant L which is independent of N.

C. Let us denote temporarily $\tilde{C} = C(x_k^{(N)}, d/N)$. We can use Theorem 9.2 to deduce that for any $c > 0$ there exists $N > 0$ such that in geodesic coordinates

$$\inf_{F \in \mathcal{N}_c(\tilde{C})} \int_{\tilde{C} \setminus F} V(x) dx \le 4\tilde{L} N^{-n} d^n,$$

where $\tilde{L} > 0$ does not depend on k and N. Let us show that the same will be true if we replace the cell $\tilde{C}$ by the ball $\tilde{B} = B(x_k^{(N)}, d/2N) \subset \tilde{C}$, which clearly leads to a contradiction with the condition (D) and ends the proof of Theorem 5.1 (we should take $c > 0$ sufficiently small so that the condition (D) is also necessary).

D. To be able to replace $\tilde{C}$ by $\tilde{B}$ it is sufficient to establish that for any $c > 0$ there exists $b > 0$ such that

$$\inf_{F_1 \in \mathcal{N}_c(\tilde{C})} \int_{\tilde{C}\setminus F_1} V(x)dx \geq \inf_{F_2 \in \mathcal{N}_b(\tilde{B})} \int_{\tilde{B}\setminus F_2} V(x)dx. \tag{9.17}$$

To this end note first that for any compact set $F_1 \subset \tilde{C}$ we can take $F_2 = F_1 \cap \bar{\tilde{B}}$ to get the inequality

$$\int_{\tilde{C}\setminus F_1} V(x)dx \geq \int_{\tilde{B}\setminus F_2} V(x)dx.$$

This would be enough to establish (9.17) if F_2 would be a compact subset in $\tilde{B}$. But we can now replace F_2 by an intersection of F_2 with a smaller ball, e.g. by $F_2^{(m)} = F_2 \cap B(x_k^{(N)}, d/2N - 1/m)$ and use the obvious relation

$$\int_{\tilde{B}\setminus F_2} V(x)dx = \lim_{m\to\infty} \int_{\tilde{B}\setminus F_2^{(m)}} V(x)dx.$$

This leads to (9.17) and ends the proof of Theorem 5.1. □

Remark 9.4. The arguments given in the proof of Theorem 5.1 show that the condition (D) for balls is equivalent to the similar condition for geodesic cubic cells.

Another strange corollary of the arguments above is the fact that the negligibility restriction $\operatorname{cap}(F) \leq c \operatorname{cap}(\bar{B})$ can be replaced by a formally much stronger restriction $\operatorname{cap}(F) \leq Ar^n$ (here we should take arbitrary $A > 0$ but then introduce restriction $r \in (0, r_1)$ where $r_1 = r_1(A)$). The resulting version $(D)_s$ of the condition (D) is equivalent to the original condition. Indeed, we actually proved that $(D)_s$ implies that $\sigma = \sigma_d$ which in turn implies (D), but also obviously (D) implies $(D)_s$, hence (D) and $(D)_s$ are equivalent.

10. Applications, examples and further results

10.1. Measure conditions

We will start by proving that replacing the capacity by the Lebesgue measure in the condition (D) leads to a sufficient condition of the discreteness of the spectrum.

Theorem 10.1. *1) Assume that (M, g) is a Riemannian manifold of bounded geometry, H is the Schrödinger operator (1.1) with the potential satisfying (5.1).*

Let us assume that there exists $c > 0$ such that the following condition is satisfied:

(D_L) *For any sequence $\{x_k \mid k = 1, 2, \ldots\} \subset M$ such that $x_k \to \infty$ as $k \to \infty$, for any $r < r_0/2$ and any compact subsets $F_k \subset B(x_k, r)$ with $\operatorname{mes} F_k \le cr^n$ we have*

$$\int_{B(x_k,r)\setminus F_k} V(x)d\mu(x) \to \infty \quad as \quad k \to \infty \,. \tag{10.1}$$

Then $\sigma = \sigma_d$.

Here $\operatorname{mes} F_k$ means the Riemannian measure of the set F_k or its Lebesgue measure in the normal (geodesic) coordinates centered at x_k.

2) If $n = 2$, then a stronger statement also holds. Namely, assume that there exist $N > 0$, $c > 0$ and $r_1 > 0$ such that the condition (10.1) holds provided $r \in (0, r_1)$ and $\operatorname{mes} F_k \le cr^N$. Then $\sigma = \sigma_d$.

PROOF. We will use the following inequalities comparing the capacity with the Lebesgue measure in $\mathbb{R}^n$ ([74], Sect.2.2.3):

$$\operatorname{cap}_\Omega(F) \ge \omega_n^{2/n} n^{(n-2)/n}(n-2)(\operatorname{mes} F)^{(n-2)/n}, \qquad n \ge 3, \tag{10.2}$$

and

$$\operatorname{cap}_\Omega(F) \ge 4\pi \left[\log\left(\frac{\operatorname{mes}\Omega}{\operatorname{mes} F}\right)\right]^{-1}, \qquad n = 2. \tag{10.3}$$

Here $\Omega = B(0, R) \subset \mathbb{R}^n$ and $R > 0$ is fixed, F is a compact subset in Ω.

Now we will prove that (D_L) implies (D), possibly with a different constant c. For simplicity of notations we will write cap instead of $\operatorname{cap}_\Omega$.

Assume first that $n \ge 3$ and (D_L) is satisfied. This means that (10.1) is satisfied provided

$$\operatorname{mes} F_k \le cr^n \tag{10.4}$$

with a constant $c > 0$. We need to prove that (10.1) is true under the condition

$$\operatorname{cap}(F_k) \le \tilde{c}r^{n-2} \tag{10.5}$$

with another constant $\tilde{c} > 0$. To this end it is sufficient to prove that (10.5) with an appropriate choice of the constant $\tilde{c}$ implies (10.4). But this is obvious because $\operatorname{mes} F_k \le A_n[\operatorname{cap}(F_k)]^{n/(n-2)}$ due to (10.2).

Arguing similarly in the case $n = 2$, we see that it is sufficient to prove that

$$\operatorname{cap}(F_k) \le \tilde{c}\left(\log\frac{1}{r}\right)^{-1} \tag{10.6}$$

implies

$$\operatorname{mes} F_k \le cr^N,$$

provided $r \in (0, r_1)$ and the constants $r_1, \tilde{c}$ are appropriately chosen. To establish this we can use (10.3) and (10.6) to conclude that

$$\begin{aligned} \text{mes}\, F_k &\leq \text{mes}\, \Omega \cdot \exp(-4\pi(\,\text{cap}\,(F_k))^{-1}) \leq \text{mes}\, \Omega \cdot \exp(-4\pi\tilde{c}^{-1} \log \frac{1}{r}) \\ &= \text{mes}\, \Omega \cdot r^{4\pi\tilde{c}^{-1}} \leq cr^N, \end{aligned}$$

provided $r \in (0, r_1)$ and both $\tilde{c}, r_1$ are sufficiently small. □

Corollary 10.2. *Under the conditions of Theorem 5.1 assume that for any $A > 0$ and any $r \in (0, r_0/2)$*

$$\text{mes}\,\{y|\; y \in B(x,r), V(y) \leq A\} \to 0 \quad \text{as} \quad x \to \infty. \tag{10.7}$$

Then $\sigma = \sigma_d$.

Proof. Let mes denote the Lebesgue measure in geodesic coordinates. Let us prove that (10.7) implies the condition (D_L) in Theorem 10.1. Indeed, we can obviously take c to be arbitrarily small. Then we will have

$$\text{mes}\,(B(x_k, r)) \setminus F_k) \geq \frac{1}{2}\,\text{mes}\, B(x_k, r).$$

It follows that for any $A > 0$ and large $k \geq k_A$ we have

$$\text{mes}\,(\{y|\, V(y) \geq A\} \cap (B(x_k, r) \setminus F_k)) \geq \frac{1}{4}\,\text{mes}\, B(x_k, r),$$

hence

$$\int_{B(x_k,r)\setminus F_k} V(x)dx \geq \frac{1}{4} A\,\text{mes}\, B(x_k, r),$$

which implies (10.1). □

Example 10.3. For $n = 2$ in $\mathbb{R}^2$ with the standard metric consider the Schrödinger operator $H = -\Delta + x^2y^2$ where $x = x^1, y = x^2$. It is easy to see that the condition (10.7) is satisfied. (The reason is that the set $\{(x, y)|\, x^2y^2 \leq A\}$ becomes narrow at infinity, so its intersections with a ball of a fixed radius r have the measure which tends to 0 as the ball goes to infinity.) Therefore $\sigma = \sigma_d$.

Similarly we can use Corollary 10.2 to establish that the operator

$$H = -\Delta + x^2y^2 + x^2z^2 + y^2z^2$$

in $\mathbb{R}^3 = \{(x, y, z)\}$ has discrete spectrum.

10.2. Dirichlet spectrum in open subsets

Let us start with some generalities. Let (M, g) be a Riemannian manifold with bounded geometry. Let us choose an open subset $G \subset M$ and consider the Schrödinger operator (1.1) with the potential $V \in L^1_{loc}(G)$, $V \geq 0$, and with the Dirichlet boundary condition. It is defined as the self-adjoint operator H_G which is canonically associated with the closure of the quadratic form (6.1) defined on $C_c^\infty(G)$. (See the beginning of Sect.6 for the definitions.)

It is easy to see that the form Q indeed admits the closure (see the arguments at the beginning of Sect.6).

A particular case is the Laplacian with the Dirichlet boundary condition (the case $V \equiv 0$). The corresponding operator will be denoted $-\Delta_G$. In this case the quadratic form is the Dirichlet integral

$$\int_M |\nabla u|^2 d\mu = \int_M g^{ij} \frac{\partial u}{\partial x^i} \frac{\partial u}{\partial x^j} \sqrt{g} dx$$

defined a priori on $C_c^\infty(G)$.

We can consider H_G as a particular case (or the limit case) of the Schrödinger operator $H = -\Delta + V(x)$ with $V = +\infty$ on $M \setminus G$ but a mechanical application of the formulations taken from the case $V \in L^1_{loc}(M)$ in fact gives wrong result. Some arguments should be modified, though not very much, so basically the proof of Theorem 5.1 works in this situation. The following theorem is an appropriate reformulation of Theorem 5.1 for this situation. In case $M = \mathbb{R}^n$ it is due to V.G. Maz'ya [74], Sect.12.5.

Theorem 10.4. *There exists $c = c(M, g) > 0$ such that the spectrum of H_G is discrete if and only if the following condition is fulfilled:*

(D_G) *Assume that we are given a sequence $\{x_k |\ k = 1, 2, \ldots\} \subset M$ such that $x_k \to \infty$ as $k \to \infty$, a number $r \in (0, r_0/2)$ and compact subsets $F_k \subset \bar{B}(x_k, r)$ satisfying the conditions*

$$F_k \supset (M \setminus G) \cap \bar{B}(x_k, r),$$

$\operatorname{cap}(F_k) \leq cr^{n-2}$ *in case* $n \geq 3$ *and* $\operatorname{cap}(F_k) \leq c\left(\log \frac{1}{r}\right)^{-1}$ *in case* $n = 2$*. Then*

$$\int_{B(x_k, r)\setminus F_k} V(x) d\mu(x) \to \infty \quad \text{as} \quad k \to \infty.$$

An equivalent condition is obtained if we require that (D_G) is true for all sufficiently small $c > 0$.

It can happen that there is no $\{x_k\}$, r and $\{F_k\}$ satisfying the conditions in (D_G). Then it is understood that the condition (D_G) is satisfied, hence $\sigma = \sigma_d$ for

H_G for any potential $V \in L^1_{loc}(G)$, $V \geq 0$. This will be true if and only if $\sigma = \sigma_d$ for $-\Delta_G$. Let us formulate this result as the following

Corollary 10.5. *There exists $c = c(M, g) > 0$ such that the spectrum of $-\Delta_G$ is discrete if and only if for any $r \in (0, r_0/2)$*

$$\liminf_{x\to\infty} \operatorname{cap}((M \setminus G) \cap \bar{B}(x,r)) \geq cr^{n-2}, \quad \text{if} \quad n \geq 3,$$

and

$$\liminf_{x\to\infty} \operatorname{cap}((M \setminus G) \cap \bar{B}(x,r)) \geq c\left(\log \frac{1}{r}\right)^{-1}, \quad \text{if} \quad n = 2.$$

An equivalent condition is obtained if we require that these conditions hold for all sufficiently small $c > 0$.

This Corollary in case $M = \mathbb{R}^n$ was first mentioned by A.M. Molchanov (see [78], Theorem 8.1).

SKETCH OF PROOF OF THEOREM 10.4. A. We can assume without loss of generality that $V(x) \geq 1$ for all $x \in G$. Let us introduce the set

$$\mathcal{L}_G = \{u \in C_c^\infty(G) | \int_G (|\nabla u|^2 + V|u^2|)d\mu \leq 1\}.$$

Then $\sigma = \sigma_d$ for H_G if and only if $\mathcal{L}_G$ is precompact in $L^2(G)$. Clearly Lemmas 6.2 and 6.3 hold in this situation.

Now for any open set $\Omega \subset M$ we should introduce the numbers

$$\lambda_G(\Omega) = \inf\left\{\frac{\int_\Omega(|\nabla u|^2 + V|u|^2)d\mu}{\int_\Omega |u|^2 d\mu}, \; u \in C_c^\infty(\Omega \cap G) \setminus 0\right\}, \tag{10.8}$$

$$\mu_G(\Omega) = \inf\left\{\frac{\int_\Omega(|\nabla u|^2 + V|u|^2)d\mu}{\int_\Omega |u|^2 d\mu}, \; u \in C_c^\infty(G) \setminus 0\right\}. \tag{10.9}$$

Clearly $\lambda_G(\Omega)$ is the bottom of the Dirichlet spectrum of H in $\Omega \cap G$, and $\mu_G(\Omega)$ is the bottom of the spectrum of H in $\Omega \cap G$ with the Dirichlet condition on $\partial G \cap \partial(\Omega \cap G)$ and the Neumann condition on the remaining part of ∂G (which is contained in $\partial\Omega$). Then it is easy to establish the localization result which is obtained from Theorem 6.10 by replacing λ and μ by λ_G and μ_G respectively. To this end we can use Lemma 6.8 and note that all arguments from the proofs of Proposition 6.4, Proposition 6.7, Lemma 6.9 and Theorem 6.10 work without any changes if we restrict ourselves to functions which vanish near ∂G.

B. We can extend Theorem 8.1 to evaluate $\mu_G(B)$ where G is an open subset in $\mathbb{R}^n$, if we require that $F \supset (\mathbb{R}^n \setminus G) \cap \bar{B}$. (Of course it is possible that we will not be able to find such a set F which satisfies the capacity restriction (8.2) but otherwise the proof works without any changes and provides the necessity of the condition D_G.

C. The arguments of Sect.9 work as well towards the proof of sufficiency of (D_G). Again we should work with $\mu_G(C)$ instead of $\mu(C)$ and start with a function $\psi \in C_c^\infty(G)$ which nearly minimizes the ratio in (10.9). Then for the set F defined by (9.5) we have

$$F \supset (M \setminus G) \cap \bar{C},$$

and the rest of the proof of sufficiency does not require any changes. □

Remark 10.6. Note that if we try to apply formally Theorem 5.1 in the situation of Theorem 10.4, by extending V by $+\infty$ on $M \setminus G$, then we arrive to a condition which is not equivalent to (D_G) even for $V \equiv 0$, hence we get wrong result even for $-\Delta_G$.

Again we can use the comparison between the capacity and the Lebesgue measure to obtain a sufficient condition which is similar to Theorem 10.1:

Theorem 10.7. *1) Let us assume that there exists $c > 0$ such that the following condition is satisfied:*

(D_{GL}) For any sequence $\{x_k \mid k = 1, 2, \ldots\} \subset M$ such that $x_k \to \infty$ as $k \to \infty$, for any $r < r_0/2$ and any compact subsets $F_k \subset \bar{B}(x_k, r)$ with $F_k \supset (M \setminus G) \cap \bar{B}(x_k, r)$ and $\operatorname{mes} F_k \leq cr^n$ we have

$$\int_{B(x_k,r)\setminus F_k} V(x)d\mu(x) \to \infty \quad as \quad k \to \infty . \tag{10.10}$$

Then H_G has a discrete spectrum.

2) If $n = 2$, then a stronger statement also holds. Namely, assume that there exist $N > 0$, $c > 0$ and $r_1 > 0$ such that the condition (10.10) holds provided $r \in (0, r_1)$, $F_k \supset (M \setminus G) \cap \bar{B}(x_k, r)$ and $\operatorname{mes} F_k \leq cr^N$. Then H_G has a discrete spectrum.

For the particular case $V \equiv 0$ we get the following

Corollary 10.8. *1) Assume that there exist $c > 0$ and $r_1 > 0$ such that for any $r \in (0, r_1)$*

$$\liminf_{x\to\infty} \operatorname{mes}((M \setminus G) \cap \bar{B}(x, r)) \geq cr^n.$$

Then $-\Delta_G$ has a discrete spectrum.

2) Assume that $n = 2$ and there exist $c > 0$, $N > 0$ and $r_1 > 0$ such that for any $r \in (0, r_1)$

$$\liminf_{x\to\infty} \operatorname{mes}((M \setminus G) \cap \bar{B}(x, r)) \geq cr^N.$$

Then $-\Delta_G$ has a discrete spectrum.

In particular, we have

Corollary 10.9. *Assume that for any fixed* $r \in (0, r_1)$

$$\operatorname{mes}(G \cap B(x,r)) \to 0 \quad \text{as} \quad x \to \infty.$$

Then $-\Delta_G$ *has a discrete spectrum.*

This in turns implies

Corollary 10.10. *If* $\operatorname{vol}(G) < \infty$, *then* $-\Delta_G$ *has a discrete spectrum.*

The condition that G is an open set in a manifold of bounded geometry plays an important role here. For example, it is well known that non-compact hyperbolic surfaces of finite volume never have a discrete spectrum. Also it is well known that even in a bounded open set $G \subset \mathbb{R}^n$ the Laplacian $-\Delta$ with the Neumann boundary condition may have non-discrete spectrum.

Of course there are plenty of open sets G with infinite volume and with a discrete spectrum of $-\Delta_G$. The following interesting example is well known.

Example 10.11. We will describe now "the spiny urchin" which is a connected open subset $G \subset \mathbb{R}^2$ introduced by C. Clark [24]. It is obtained by removing countably many disjoint rays from $\mathbb{R}^2$. Namely, let us identify $\mathbb{R}^2 = \mathbb{C}$ and take

$$G = \mathbb{R}^2 \setminus \bigcup_{k=1}^{\infty} \{z | \arg z = n\pi 2^{-k} \text{ for some integer } n, \ |z| \geq k\}.$$

Any circle $\{z|\ |z| = a\}$ with $k-1 < a < k$ is divided by the rays into 2^k equal parts. Therefore it is clear that if we take a closed disc $\bar{B}(z,r) \subset \mathbb{C}$ with a fixed $r > 0$ and $z \to \infty$, then for large $|z|$ this disc will be uniformly densely pierced by a number of rays which goes to infinity as $|z| \to \infty$. It can be deduced from this geometric picture that $\operatorname{cap}(F \cap \bar{B}(z,r)) \to \operatorname{cap}(\bar{B}(z,r))$ as $z \to \infty$. (Here we can assume that $r < 1$ and define the capacity of $\operatorname{cap}(F \cap \bar{B}(z,r))$ as the capacity with respect to $B(z,1)$.) Therefore Corollary 10.5 (but not Corollary 10.8) allows us to conclude that for the open set G described above the Laplacian $-\Delta_G$ has a discrete spectrum. as was first established in [24].

C. Clark [24] in fact established the discreteness of the spectrum for $-\Delta_G$ by elementary methods. We can also deduce it directly from the G-version of the localization theorem 6.10 (see the proof of Theorem 10.4). Indeed, $\lambda_G(B(z,r)) = \lambda(G \cap B(z,r))$ is equal to the minimum Dirichlet eigenvalue for all connected components of $G \cap B(z,r)$. But these components become more and more narrow as $z \to \infty$ i.e. their maximal width tends to 0 as $z \to \infty$. It easily follows that $\lambda(G \cap B(z,r)) \to \infty$ as $z \to \infty$, hence the spectrum of $-\Delta_G$ is discrete.

Of course more can be said about the spectrum of this particular domain. For example, J. Fleckinger ([36], [37]) obtained an asymptotic of the eigenvalues distribution function for the spiny urchin.

The following theorem generalizes Molchanov's Theorem 8.3 [78] and follows from the results by H. Donnelly [32] who only required lower bound for the Ricci curvature instead of bounded geometry.

Theorem 10.12. *Let (M, g) be a Riemannian manifold of bounded geometry, $G \subset M$ is an open set such that ∂G has bounded coefficients of the second quadratic form in M ("bounded curvature"). Then $-\Delta_G$ has a discrete spectrum if and only if for any fixed $r > 0$ the set G does not contain any infinite set of disjoint balls with the radius r.*

SKETCH OF THE PROOF. Clearly the condition is necessary i.e. if there exists $r > 0$ such that G contains an infinite set of disjoint balls with the radius r, then $\sigma \neq \sigma_d$ (for $-\Delta_G$). Let us prove the inverse statement. Assume that $\sigma \neq \sigma_d$. Then by the Corollary 10.5 there exists an infinite set of disjoint balls $\bar{B}(x_k, r) \subset M$ which have "negligible" intersections with $M \setminus G$ i.e.

$$\text{cap}\,((M \setminus G) \cap \bar{B}(x_k, r)) \leq c\,\text{cap}\,(\bar{B}(x_k, r)) \tag{10.11}$$

with a small $c > 0$. The "bounded curvature" restriction on ∂G implies that there exists $r_1 > 0$ such that for any $x \in M$ and $r \in (0, r_1)$ the connected components of $\partial G \cap \bar{B}(x, r)$ either belong to a boundary layer $\bar{B}(x, r) \setminus B(x, \frac{1}{3}r)$ or are transversal to the sphere $\partial B(x, r)$ and behave as hyperplane sections of $\bar{B}(x, r)$ in geodesic coordinates. In particular if such a component intersects $B(x, \frac{1}{3}r)$ then it has the Riemannian $(n-1)$-measure which is comparable with r^{n-1}, hence due to an isoperimetric inequality (see e.g. Sect.2.3.3 in [74]) it has the capacity which is comparable with the capacity of the ball $\bar{B}(x, r)$.

So we have the following alternative: every connected component of the intersection $((M \setminus G) \cap \bar{B}(x, r))$ either is disjoint with the smaller ball $B(x, r/3)$ or has a capacity which is comparable with the capacity of the ball $B(x, r)$.

Applying this to the balls $B(x_k, r)$ chosen above (we can assume of course that $r < r_1$), we see that the second possibility is forbidden by (10.11), hence $B(x_k, r/3) \subset G$ for all k. □

In case $n = 2$ the "bounded curvature" restriction can be replaced by the requirement of the connectedness of $M \setminus G$ as it is seen in the following result generalizing another Molchanov's result (Theorem 8.2 in [78]):

Theorem 10.13. *Assume that (M, g) is a manifold of bounded geometry with $\dim M = 2$, $G \subset M$ is open and $M \setminus G$ is connected. Then the spectrum of $-\Delta_G$ is discrete if and only if for any fixed $r > 0$ the domain G does not contains any infinite set of disjoint discs with the radius r.*

SKETCH OF THE PROOF. The proof is similar to the proof of Theorem 10.12. Note that the connectedness of $M \setminus G$ implies that either $B(x, r/3) \subset G$ or the diameter of a connected component of $(M \setminus G) \cap \bar{B}(x, r)$ is at least $\frac{2}{3}r$.

In the last case it follows that $\text{cap}\,((M \setminus G) \cap \bar{B}(x,r))$ is comparable with the capacity of the ball $\bar{B}(x,r)$ (see [74], Proposition 2 in Sect. 9.1.2). Hence we can again apply Corollary 10.5 to end the proof. □

Remark 10.14. Results similar to Theorems 10.12, 10.13 for the Neumann boundary condition must include some regularity conditions for the boundary of G, because it is well known that even bounded domains in $\mathbb{R}^n$ may have non-discrete Neumann spectrum (see e.g. [2], and Sect.XIII.14 in [87]). It is easy to see that if the Neumann spectrum of $-\Delta$ in a domain $G \subset \mathbb{R}^n$ is discrete, then $\text{vol}\,(G) < \infty$. V.G. Maz'ya ([74], Sect.4.10) gave a necessary and sufficient condition (in terms of capacity) for the discreteness of the spectrum in this situation.

10.3. Review of some related results

The literature about the discreteness of spectrum starts with a paper by H. Weyl [109] who proved that the spectrum of the one-dimensional Schrödinger operator $H_1 = -d^2/dt^2 + V(t)$ in $L^2([0,+\infty))$ is discrete if $V(x)$ is monotone and $V(t) \to +\infty$ as $t \to +\infty$. The monotonicity requirement can be removed (see e.g. K. Friedrichs [38], and also books by E.C. Titchmarsh [104], I.M. Glazman [42], M. Reed and B. Simon [87], F.A. Berezin and M.A. Shubin [5]).

After the Molchanov's result [78] many improvements were made for the one-dimensional Schrödinger operator H_1 in $L^2(\mathbb{R})$. In particular, the semi-boundedness requirement was relaxed by I. Brinck [11] who proved that if we replace (5.1) by requiring only that

$$\int_J V(t)dx \geq -C \tag{10.12}$$

for any interval $J \subset \mathbb{R}$ of length ≤ 1, then Molchanov's condition (5.4) is still necessary and sufficient for the discreteness of the spectrum. Further improvements are due to R.S. Ismagilov [47] and L.B. Zelenko [113].

Another non-trivial one-dimensional result is due to I.S. Kac and M.G. Krein [50] who gave an explicit necessary and sufficient condition for the discreteness of spectrum of a singular string.

The one-dimensional Molchanov theorem and its above mentioned improvements were extended by several authors to the operators H_1 with operator-valued potentials V whose values are self-adjoint (not necessarily bounded) operators in a Hilbert space such that $\text{Dom}\,(V(t))$ does not depend of t. One of the first results of this kind is due to B.M. Levitan and G.A. Suvorchenkova [63] who proved that if V is bounded below in operator sense, then the Molchanov condition (5.4) is sufficient for the discreteness of the spectrum. In the same situation V.P. Maslov [66] was able to provide a beautiful necessary and sufficient condition which generalizes the one-dimensional Molchanov theorem and is easy to formulate but not so easy to check for an operator-valued potential because it includes an operator valued energy-type estimate. R. Kleine [56] extended Maslov's result to the case when the semiboundedness condition on V is replaced by the Brinck condition (10.12). Finally J. Brüning [14] introduced a most general "coerciveness" assumption on V which is weaker than all the conditions above (in particular the weakest Zelenko condition in the scalar case) but still allows to establish a necessary and sufficient Maslov type condition of the discreteness of the spectrum. J. Brüning obtained applications of his abstract result to the Laplacian on manifolds with nice ends (with the conditions of the discreteness of the spectrum formulated e.g. in terms of Ricci curvature and mean curvature).

The first mathematical result about the discreteness of spectrum for the multidimensional Schrödinger operator (in $\mathbb{R}^n$) seems to be due to K. Friedrichs [38] who proved that the condition $V(x) \to +\infty$ as $x \to \infty$ is sufficient for $\sigma = \sigma_d$. (This proof, which is now considered elementary, can be also found e.g. in the books [104], [42], [87], [5].)

Molchanov's multidimensional result from [78] and further work by V.G. Maz'ya [71], [72], [74] revealed relevance of capacity in spectral theory. Further developments in the description of the spectrum under the Molchanov condition (in $\mathbb{R}^n$) are due to G.V. Rozenblum who established two-sided estimates for the eigenvalues distribution function and (under some mild additional restrictions) even the asymptotic for the Laplace and polyharmonic operators [92], as well as for the Schrödinger type operators [93]. The estimates themselves are formulated in terms of capacity, but simple (though strong) sufficient measure conditions follow.

Perturbation arguments allow to obtain some sufficient conditions of the discreteness of the spectrum in case when the potential V is allowed to be not semibounded below (see e.g. [87], Sect.XIII.14).

Other examples of usefulness of capacity in spectral theory were found by J. Rauch and M. Taylor [86] and G. Courtois [28]. In particular, G. Courtois established perturbation type estimates of eigenvalues of the Laplacian $-\Delta$ on $X \setminus A$ (here X is a closed compact Riemannian manifold, A is its compact subset) in terms of capacity of A, which is defined as follows:

$$\operatorname{cap}(A) = \inf\left\{\int_X |\nabla u|^2 d\mu \;\middle|\; u \in C^\infty(X),\ \int_X u\,d\mu = 0,\ u = 1 \text{ near } A\right\}.$$

For example, he proved that

$$C^{-1}\operatorname{cap}(A) \le \lambda_1(X \setminus A) \le C\operatorname{cap}(A),$$

$$0 \le \lambda_k(X \setminus A) - \lambda_k \le C_k(\operatorname{cap}(A))^{1/2}.$$

(Here λ_k is the kth eigenvalue of $-\Delta$ on X, in particular $\lambda_1 = 0$, $\lambda_k(X \setminus A)$ is the kth Dirichlet eigenvalue of $X \setminus A$, multiplicities counted.) The convergence $\lambda_k(X \setminus A) \to \lambda_k$ as $\operatorname{cap}(A) \to 0$ (without estimate of remainder) follows already from the results of [86] where also other applications of capacity are given.

Some conditions of discreteness of the spectrum for the Laplacian on Riemannian manifolds were established in the papers in [3], [12], [13], [31], [32], [33].

In particular, A. Baider [3] considered some special manifolds which are warped products and used separation of variables which sometimes gives necessary and sufficient condition for the discreteness of the spectrum.

R. Brooks [12], [13] established lower and upper estimates for the bottom of the essential spectrum of the Laplacian in geometric terms, namely through the volume growth and Cheeger's isoperimetric constant. These estimates give in particular some necessary and some sufficient conditions for the spectrum to be discrete. For example, the spectrum is never discrete for complete Riemannian manifolds with infinite volume and with polynomial growth of the volumes of the balls. Some of the Brooks results were recently generalized by L. Notarantonio [80] to operators generated by regular Dirichlet forms.

H. Donnelly and P. Li [33] proved that the spectrum of the Laplacian on a complete simply connected negatively curved manifold M is discrete provided the sectional curvature tends to $-\infty$ at infinity. If $\dim M = 2$ then instead of the simply connectedness it is sufficient to require that the fundamental group $\pi_1(M)$ is finitely generated. The proof is based on a comparison result and isoperimetric inequalities technique by J. Cheeger [19] and S.T. Yau [112] (see also I. Chavel [18], Sect.IV.3). H. Donnelly in [31] established somewhat inverse results, proving existence of the essential spectrum of the Laplacian with some estimates of this spectrum. For example he proved that if M is non-compact, complete and has Ricci curvature bounded below by $-(n-1)c$, where $n = \dim M$, $c \geq 0$, then the essential spectrum intersects the interval $[0, (n-1)^2c/4]$. It is interesting that no restriction on the injectivity radius is needed for this result. The proof relies on a comparison theorem by S.Y. Cheng [21] and some other similar comparison theorems.

The Rozenblum-Cwikel-Lieb bound (see e.g. [87], Sect.XIII.12) provides an estimate for the number of eigenvalues of a Schrödinger operator below some level E, whereas some essential spectrum above E can be present. In particular, it gives a condition that the spectrum below E is discrete. Recently D. Levin and M. Solomyak [61] proved that this bound holds in a very general context of Markov generators, in particular for the Schrödinger operators on manifolds where Ricci curvature is bounded below. The proof relies on the global Sobolev estimate (see e.g. [18], Sect.IV.3) which was proved in this context by L. Saloff-Coste [95], based on the ideas of P. Li and S.T. Yau [64]. Even more general result was obtained by G.V. Rozenblum and M.Z. Solomyak [94] in terms of the heat kernel behavior at 0 and ∞.

Note that the V.G.Maz'ya and M. Otelbaev result [76] (see also [74], Sect.12.5) which gives a two-sided estimate (in terms of capacity) for the bottom of the essential spectrum for the Schrödinger operator (and more general ones) in $\mathbb{R}^n$, implies in particular conditions for the spectrum below some level to be discrete.

Let us recall that discreteness of the spectrum for a self-adjoint operator is equivalent to the compactness of its resolvent. Hence a possibility to make the statement $\sigma = \sigma_d$ more precise is to establish that the resolvent belongs to a Schatten class $\mathcal{S}_p$ of compact operators, $1 \le p < \infty$. For the Dirichlet spectrum of the Laplacian in a domain $G \subset \mathbb{R}^n$ this was done by E.B. Davies [29] in terms of convergence of some integrals, namely

$$\int_G [\operatorname{dist}(x, \partial G)]^\alpha dx < \infty.$$

where α depends on p. (Some related results for manifolds were obtained by M. Lianantonakis [65].)

Acknowledgements

This work was partially supported by NSF grant DMS-9706038, Humboldt University (Berlin), Weizmann Institute of Science (Israel) and Institut des Hautes Ètudes Scientifiques (France). I am also very grateful to M.S. Birman, J. Brüning, E.B. Davies, A.A. Grigoryan, V.G. Maz'ya, N.N. Nadirashvili, Yu. Netrusov, F.S. Rofe-Beketov, M.Z. Solomyak, V. Turbiner and M. Zworski for useful discussions and references.

References

[1] D. Adams, L. Hedberg, *Function spaces and potential theory*, Springer-Verlag, Berlin, 1996

[2] S. Agmon, *Lectures on elliptic boundary value problems*, Van Nostrand, Princeton, N.J., 1965

[3] A. Baider, *Noncompact Riemannian manifolds with discrete spectra*, J. Differ. Geom., **14** (1979), 41–57

[4] Yu.M. Berezanski, *Expansions in eigenfunctions of self-adjoint operators,* Amer. Math. Soc. Translation of Math. Monographs, Providence, RI, 1968

[5] F.A. Berezin, M.A. Shubin, *The Schrödinger equation.* Kluwer Academic Publishers, Dordrecht e.a., 1991

[6] M.S. Birman, *On the spectrum of singular boundary-value problem*, Mat. Sbornik (N.S.), **55 (97)** (1961), 125–174 (Russian)

[7] M. S. Birman, B. S. Pavlov, *On compactness of some imbedding operators*, Vestnik Leningr. Univ., no. 1 (1961), 51–74 (Russian)

[8] M.S. Birman, M.Z. Solomyak, *Spectral theory of self-adjoint operators in Hilbert space*, D. Reidel Publishing Co., Dordrecht, 1987

[9] ———, *Schrödinger operator. Estimates for number of bound states as function-theoretical problem*, Amer. Math. Transl. (2), **150** (1992), 1–54

[10] M. Braverman, *On self-adjointness of a Schrödinger operator on differential forms,* Proc. Amer. Math. Soc., **126** (1998), 617–624

[11] I. Brinck, *Self-adjointness and spectra of Sturm-Liouville operators*, Math. Scand., **7** (1959), 219–239

[12] R. Brooks, *A relation between growth and spectrum of the Laplacian*, Math. Z., **178** (1981), 501–508

[13] ———, *On the spectrum of non-compact manifolds with finite volume*, Math. Z., **187** (1984), 425–432

[14] J. Brüning, *On Schrödinger operators with discrete spectrum*, J. Funct. Anal., **85** (1989), 117–150

[15] A.G. Brusentsev, *On essential self-adjointness of semi-bounded second order elliptic operators without completeness of the Riemannian manifold*, Math. Physics, Analysis, Geometry (Kharkov), **2** (1995), no. 2, 152–167 (in Russian)

[16] T. Carleman, *Sur la théorie mathématique de l'équation de Schrödinger*, Ark. Mat. Astr. Fys., **24B**, no. 11 (1934), 1–7

[17] L. Carleson, *Selected problems on exceptional sets*, Van Nostrand, Princeton, N.J., 1967

[18] I. Chavel, *Eigenvalues in Riemannian geometry*, Academic Press, Orlando e.a., 1984

[19] J. Cheeger, *A lower bound for the smallest eigenvalue of the Laplacian*, In: "Problems in Anaysis: A Symposium in honor of Salomon Bochner", Princeton Univ. Press, Princeton, N.J., 1970, 195–199

[20] J. Cheeger, M. Gromov, M. Taylor, *Finite propagation speed, kernel estimates for functions of the Laplace operator, and the geometry of complete Riemannian manifolds*, J. Diff. Geom., **17** (1982), 15–53

[21] S.Y. Cheng, *Eigenvalue comparison theorems and its geometric applications*, Math. Z., **143** (1975), 289–297

[22] P. Chernoff, *Essential self-adjointness of powers of generators of hyperbolic equations*, J. Funct. Analysis, **12** (1973), 401–414

[23] A.A. Chumak, *Self-adjointness of the Beltrami-Laplace operator on a complete paracompact manifold without boundary*, Ukrainian Math. Journal, **25** (1973), no. 6, 784-791 (in Russian)

[24] C. Clark, *Rellich's embedding theorem for a "spiny urchin"*, Canad. Math. Bull., **10** (1967), 731–734

[25] H.O. Cordes, *Self-adjointness of powers of elliptic operators on non-compact manifolds*, Math. Annalen, **195** (1972), 257-272

[26] ———, *Spectral theory of linear differential operators and comparison algebras*, London Math. Soc., Lecture Notes Series, **76**, Cambridge Univ. Press, 1987

[27] R. Courant, D. Hilbert, *Methods of mathematical physics*, Interscience Publishers, New York, 1953

[28] G. Courtois, *Spectrum of manifolds with holes*, J. Funct. Anal., **134** (1995), 194–221

[29] E.B. Davies, *Trace properties of the Dirichlet Laplacian*, Math. Zeitschrift, **188** (1985), 245–251

[30] A. Devinatz, *Essential self-adjointnessof Schrödinger-type operators*, J. Funct. Analysis, **25** (1977), 58–69

[31] H. Donnelly, *On the essential spectrum of a complete Riemannian manifold*, Topology, **20** (1981), 1–14

[32] ———, *Spectrum of domains in Riemannian manifolds*, Illinois J. Math., **31** (1987), 692–698

[33] H. Donnelly, P. Li, *Pure point spectrum and negative curvature for non-compact manifolds*, Duke Math. J., **46** (1979), 497–503

[34] M.S.P. Eastham, W.D. Evans, J.B. McLeod, *The essential self-adjointness of Schrödinger type operators*, Arch. Rat. Mech. Anal., **60** (1975/76), no. 2, 185–204

[35] W. Faris, R. Lavine, *Commutators and self-adjointness of Hamiltonian operators*. Commun. Math. Phys., **35** (1974), 39–48

[36] J. Fleckinger, *Répartition des valeurs propres d'opérateurs elliptiques sur des ouverts non bornés*, C. R. Acad. Sc. Paris, Sér.A, **286** (1978), 149–152

[37] ———, *Comportement des valeurs propres d'opérateurs elliptiques sur des ouverts non bornés*, Publication CNRS 7704, Bordeaux, 1977

[38] K. Friedrichs, *Spektraltheorie halbbeschränkter Operatoren und Anwendung auf die Spektralzerlegung von Differentialoperatoren*, Math. Ann., **109** (1934), 465–487, 685–713

[39] O. Frostman, *Potentiel d'équilibre et capacité des ensembles avec quelquels applications à la théorie des fonctions*, Medd. Lunds Univ. Mat. Sem., **3** (1935), 1–118

[40] M. Gaffney, *A special Stokes's theorem for complete Riemannian manifolds*, Ann. of Math., **60** (1954). 140–145

[41] M.G. Gimadislamov, *Sufficient conditions of coincidence of minimal and maximal partial differential operators and discreteness of their spectrum*, Math. Notes, **4**, no. 3 (1968), 301–317 (in Russian)

[42] I.M. Glazman, *Direct methods of qualitative spectral analysis of singular differential operators*, Israel Program for Scientific Translation, Jerusalem, 1965

[43] M. Gromov, *Curvature, diameter and Betti numbers*, Comment. Math. Helv., **56** (1981), 179–195

[44] P. Hartman, *The number of L^2-solutions of $x'' + q(t)x = 0$*, Amer. J. Math., **73** (1951), 635–645

[45] G. Hellwig, *Differential operators of mathematical physics. An introduction*. Addison-Wesley, 1964

[46] T. Ikebe, T. Kato, *Uniqueness of the self-adjoint extension of singular elliptic differential operators*, Arch. for Rat. Mech. and Anal., **9** (1962), 77–92

[47] R.S. Ismagilov, *Conditions for the semiboundedness and discreteness of the spectrum for one-dimensional differential equations*, Soviet Math. Dokl., **2** (1961), 1137–1140

[48] ———, *Conditions for self-adjointness of differential operators of higher order*, Dokl. Akad. Nauk SSSR, **142** (1962), 1239–1242. English translation: Soviet Math. Doklady, **3** (1962), 279–283

[49] K. Jörgens, *Wesentliche Selbstadjungiertheit singulärer elliptischer Differentialoperatoren zweiter Ordnung in $C_0^\infty(G)$*, Math. Scand., **15** (1964), 5–17

[50] I.S. Kac, M.G. Krein, *A criterion for the discreteness of the spectrum of a singular string*, Izvestiya vys. ucheb. zavedenii, no.2(3) (1958), 136–153 (Russian)

[51] H. Kalf, F.S. Rofe-Beketov, *On the essential self-adjointness of Schrödinger operators with locally integrable potentials*, Proc. Royal Soc. Edinburgh, **128A** (1998), 95–106

[52] T. Kato, *Perturbation theory for linear operators*, Spriger-Verlag, 1966

[53] T. Kato, *Schrödinger operators with singular potentials*, Israel J. Math., **13** (1972), 135–148

[54] T. Kato, *A remark to the preceding paper by Chernoff*, J. Funct. Analysis, **12** (1973), 415–417

[55] M.V. Keldysh, *On solvability and stability for the Dirichlet problem*. Uspekhi Matem. Nauk, no. VIII (1940), 171–231 (Russian)

[56] R. Kleine, *Discreteness condition for the Laplacian on complete non-compact Riemannian manifolds*, Dissertation, Duisburg, 1986

[57] V.A. Kondrat'ev, M.A. Shubin, *Conditions for the discreteness of the spectrum for Schrödinger operators on manifolds*, Funct. Anal. and Appl., **33** (1999)

[58] V.A. Kondrat'ev, M.A. Shubin, *Discreteness of spectrum for the Schrödinger operators on manifolds of bounded geometry*. To appear in Proceedings of the conference "Functional Analysis, Partial Differential Equations and Applications" dedicated to the V.G.Maz'ya 60th birthday (Rostock, August 1998).

[59] E.M. Landis, *Second order equations of elliptic and parabolic type*, Amer. Math. Soc., Providence, RI, 1998

[60] N.S. Landkof, *Foundations of modern potential theory*, Springer-Verlag, New York - Heidelberg, 1972

[61] D. Levin, M. Solomyak, *Rozenblum - Lieb - Cwikel inequality for Markov generators*. Journal d'Analyse Mathématique, **71** (1997), 173–193

[62] B.M. Levitan, *On a theorem by Titchmarsh and Sears*, Uspekhi Matem. Nauk, **16**, no.4 (1961), 175–178 (in Russian)

[63] B.M. Levitan, G.A. Suvorchenkova, *Sufficient conditions for a discrete spectrum in the case of a Sturm-Liouville equation with operator coefficients*, Funct. Anal. Appl., **2** (1968), 147–152

[64] P. Li, S.T. Yau, *On the Schrödinger equation and the eigenvalue problem*, Commun. Math. Phys., **88** (1983), 309-318

[65] M. Lianantonakis, *Nonclassical spectral asymptotics for weighted Laplace-Beltrami operators*, International Math. Research Notes (1993), no. 8, 241–244

[66] V.P. Maslov, *A criterion for discreteness of the spectrum of a Sturm-Liouville equation with an operator coefficient*, Funct. Anal. Appl., **2** (1968), 153–157

[67] V.G. Maz'ya, *Classes of regions and imbedding theorems for function spaces*, Soviet. Math. Dokl., **1** (1960), 882–885

[68] ———, *The p-conductivity and theorems of imbedding certain function spaces into a C-space*, Soviet. Math. Dokl., **2** (1961), 1200–1203

[69] ———, *The negative spectrum of the n-dimensional Schrödinger operator*, Soviet Math. Dokl., **3** (1962), 808–810

[70] ———, *The Dirichlet problem for elliptic equations of arbitrary order in unbounded regions*, Soviet Math. Dokl., **4** (1963), 860–863

[71] ———, *On the theory of the multidimensional Schrödinger operator*, Izv. Akad. Nauk SSSR, **28** (1964), 1145–1172 (Russian)

[72] ———, *On (p, l)-capacity, imbedding theorems and the spectrum of a selfadjoint elliptic operator*, Math. USSR-Izv., **7** (1973), 357–387

[73] ———, *On the connection between two kinds of capacity*, Vestnik Leningrad Univ. Math., **7** (1974), 135–145

[74] ———, *Sobolev spaces*, Springer Verlag, Berlin, 1985

[75] ——— *Classes of domains, measures and capacities in the theory of differentiable functions*, Encyclopaedia of Math. Sciences, **26**, Analysis III, Springer-Verlag, 1991, 141–211

[76] V.G. Maz'ya, M. Otelbaev, *Imbedding theorems and the spectrum of a pseudodifferential operator*, Siberian Math. Journal, **18** (1977), 758–769

[77] G.A. Meladze, M.A. Shubin *Properly supported uniform pseudo-differential operators on unimodular Lie groups*, Trudy Sem. Petrovsk., **11** (1986), 74–97 (Russian)

[78] A.M. Molchanov, *On the discreteness of the spectrum conditions for self-adjoint differential equations of the second order*, Trudy Mosk. Matem. Obshchestva (Proc. Moscow Math. Society), **2** (1953), 169–199 (Russian)

[79] E. Nelson, *Feynman integrals and the Schrödinger operators,* J. Math. Phys., **5** (1964), 332–343

[80] L. Notarantonio, *Growth and spectrum of diffusions.* Preprint, 1998

[81] I.M. Oleinik, *On the essential self-adjointness of the Schrödinger operator on a complete Riemannian manifold,* Mathematical Notes, **54** (1993), 934–939

[82] ———, *On the connection of the classical and quantum mechanical completeness of a potential at infinity on complete Riemannian manifolds,* Mathematical Notes, **55** (1994), 380–386

[83] ———, *On the essential self-adjointness of the Schrödinger-type operators on complete Riemannian manifolds,* PhD thesis, Northeastern University, May 1997

[84] A.Ya. Povzner, *On expansions of arbitrary functions in eigenfunctions of the operator $\Delta u + cu$,* Matem. Sbornik, **32** (74), no. 1 (1953), 109–156 (in Russian)

[85] J. Rauch, M. Reed, *Two examples illustrating the differences between classical and quantum mechanics,* Commun. Math. Phys., **29** (1973), 105–111

[86] J. Rauch, M. Taylor, *Potential and scattering theory on wildly perturbed domains,* J. Funct. Anal., **18** (1975), 27–59

[87] M. Reed, B. Simon, *Methods of modern mathematical physics, I: Functional analysis, II: Fourier analysis, self-adjointness, IV: Analysis of operators*, Academic Press, New York e.a., 1972, 1975, 1978

[88] J. Roe, *An index theorem on open manifolds. I*, J. Differ. Geom., **27** (1988), 87–113

[89] F.S. Rofe-Beketov, *On non-semibounded differential operators*, Theory of Functions, Functional Analysis and Applications (Teoriya funktsii, funkts. analyz i ikh prilozh.), no. 2, Kharkov (1966), 178–184 (in Russian)

[90] ———, *Conditions for the self-adjointness of the Schrödinger operator*, Mathematical Notes, **8** (1970), 888–894

[91] ———, *Self-adjointness of elliptic operators of higher order and energy estimates in* $\mathbb{R}^n$, Theory of Functions, Functional Analysis and Applications (Teoriya funktsii, funkts. analyz i ikh prilozh.), no. 56, Kharkov (1991), 35–46 (in Russian)

[92] G.V. Rozenblum, *On eigenvalues of the first boundary problem in unbounded domains*, Mat. Sbornik, **89**, no.2 (1972), 234–247

[93] ———, *On estimates of the spectrum of the Schrödinger operator*, In: Problems in Mathematical Analysis, Leningrad, **5** (1975), 152-165 (Russian)

[94] G.V. Rozenblum, M.Z. Solomyak, *CLR-estimate for the generators of positivity preserving and positively dominated semigroups*, Algebra i Analiz, **9**, no.6 (1997), 214–236 (Russian)

[95] L. Saloff-Coste, *Uniformly elliptic operators on Riemannian manifolds*, J. Differ. Geom., **36** (1992), 417–450

[96] M. Schechter, *Spectra of partial differential operators*, North-Holland, 1971

[97] R. Schoen, S.T. Yau, *Lectures on differential geometry*, International Press, Boston, 1994

[98] D.B. Sears, *Note on the uniqueness of Green's functions associated with certain differential equations*, Canad. J. Math., **2** (1950), 314–325

[99] M.A. Shubin, *Spectral theory of elliptic operators on non-compact manifolds*, Astérisque, **207**, (1992), 37–108

[100] ———, *Classical and quantum completeness for the Schrödinger operators on non-compact manifolds*, Preprint (1998). To appear in Proceedings of the minisymposium "Spectral Invariants, Heat Kernel Approach", Roskilde, Denmark, September 1998

[101] B. Simon, *Essential self-adjointness of Schrödinger operators with positive potentials*, Math. Annalen, **201** (1973), 211–220

[102] I.M. Stein, Singular integrals and differentiability properties of functions, Princeton Univ. Press, Princeton, N.J., 1970

[103] F. Stummel, *Singuläre elliptische Differentialoperatoren in Hilbertschen Räumen*, Math. Annalen, **132** (1956), 150–176

[104] E.C. Titchmarsh, *Eigenfunction expansions associated with second-order differential equations*, Clarendon Press, Oxford, Part I, 1946, Part II, 1958

[105] C.J. de La Vallée-Poussin, *L'extension de la méthode du balayage*, Annal. Inst. H. Poincaré, **2**, no.3 (1932), 169–232

[106] J. Wermer, *Potential theory*, Lecture Notes in Math., **408**, Springer-Verlag, Berlin, 1974

[107] N. Wiener, *Certain notions in potential theory*, J. Math. and Phys. Mass. Inst. Techn., **3**(1924), 24–51

[108] ———, *The Dirichlet problem*, J. Math. and Phys. Mass. Inst. Techn., **3**(1924), 127–146

[109] H. Weyl, *Über gewönliche Differentialgleichungen mit Singularitäten und die zugehörige Entwicklungen willkürlicher Funktionen*, Math. Ann., **68** (1910), 220–269

[110] E. Wienholtz, *Halbbeschränkte partielle Differentialoperatoren zweiter ordnung vom elliptischen Typus*, Math. Ann., **135** (1958), 50–80

[111] A. Wintner, *On the normalization of characteristic differentials in continuous spectra*, Phys. Rev., **72** (1947), 516–517

[112] S.T. Yau, *Isoperimetric constants and the first eigenvalue of a compact Riemannian manifold*, Ann. Scient. École Normale Sup., **8** (1975), 487–507

[113] L.B. Zelenko, *Conditions for semiboundedness and discrete spectrum of the Sturm-Liouville operator on the half-line*, Izv. Vyssh. Uchebn. Zaved. Mat., **9** (1967), 31–40 (Russian)

Department of Mathematics, Northeastern University, Boston, MA 02115, USA

1991 Mathematics Subject Classification: Primary 35P05, 35J10, 58G25; Secondary 35P15, 47B25, 47F05, 81Q10

Submitted: 19 February 1999

LECTURES ON WAVE INVARIANTS

STEVE ZELDITCH

INTRODUCTION

These are notes of some lectures on wave invariants and quantum normal form invariants of the Laplacian Δ at a closed geodesic γ of a compact boundaryless Riemannian manifold (M, g). Our purpose in the lectures was to give a survey of some recent developments involving these invariants and their applications to inverse spectral theory, mainly following the references [G.1] [G.2] [Z.1][Z.2][Z.3]. Originally, the notes were intended to mirror the lectures but in the intervening time we wrote another expository account on this topic [Z.4] and also extended the methods and applications to certain plane domains which were outside the scope of the original lectures [Z.5]. These events seemed to render the original notes obsolete. In their place, we have included some related but more elementary material on wave invariants and normal forms which do not seem to have been published before and which seem to us to have some pedagogical value. This material consists first of the calculation of wave invariants on manifolds without conjugate points using a global Hadamard-Riesz parametrix. Readers who are more familiar with heat kernels than wave kernels may find this calculation an easy-to-read entree into wave invariants. A short section on normal forms leads the reader into this more sophisticated – and more useful – approach to wave invariants. We illustrate this approach by putting a Sturm-Liouville operator on a finite interval (with Dirichlet boundary conditions) into normal form. The results are equivalent to the classical expansions of the eigenvalues and eigenfunctions as described in Marchenko [M] and Levitan-Sargsjan [LS], but the approach is quite different and possibly new. The main point is that it generalizes to higher dimensions. We hope the reader will find it stimulating to compare the (well-understood) one-dimensional case to the (still murky) higher dimensional ones. To highlight the murkiness we end these notes with a few open problems.

Date: May 7, 1999.

Research partially supported by NSF grant #DMS-9703775.

CONTENTS

1. Wave kernel of a compact Riemannian manifold

We begin by recalling some basic notions and notations from geometry and from the wave equation on a Riemannian manifold. For further background we suggest [Be] and [HoI-IV].

1.1. Geometric Preliminaries. We assume the reader is familiar with basic notions of Riemannian geometry. We now quickly run through the definitions and notations which we will need in analysing wave invariants. All the notions are discussed in detail in [Kl]. We also refer to [Su] for background in a context closely related to that of this article. Let:

(i)	$\Theta(x,y)$	denote the volume density in normal coordinates at x;
(ii)	$r = r(x,y)$	denote the distance function of M;
(iii)	$\Lambda(M)$	denote the H^1 loopspace of M;
(iv)	$\mathcal{G}(M)$	denote the subset of closed geodesics in $\Lambda(M)$;
(v)	$\mathcal{G}_{[\gamma]}$	denote the set of closed geodesics in $\mathcal{G}(M)$ whose free homotopy class is $[\gamma]$;
(vi)	$\lvert\xi\rvert_g : T^*M - 0 \to \mathbb{R}^+$	denote the length of a (co)-vector.
(vii)	$S^*M = \{\lvert\xi\rvert_g = 1\}$	denote the unit cosphere bundle;
(viii)	$G^t : T^*M - 0 \to T^*M - 0$	denote the geodesic flow, i.e. the Hamilton flow of $\lvert\xi\rvert_g$.
(ix)	γ	denote a closed geodesic, i.e. a closed orbit of G^t in S^*M.
(x)	$\operatorname{inj}(M,g)$	denote the injectivity radius.

We call a metric g is *non-degenerate* if the energy functional E on $\Lambda(M)$ is a Bott-Morse function, i.e. $\mathcal{G}(M)$ is a smooth submanifold of $\Lambda(M)$, and $T_c\mathcal{G}(M) = \ker J_c$ where $c \in \mathcal{G}(M)$ and J_c is the Jacobi operator (= index form) on $T_c\Lambda(M)$, i.e. $J_c = \nabla^2 + R(\dot{c},\cdot)\dot{c}$; We also say that g is *bumpy* if, for every $c \in \mathcal{G}(M)$, the orbit $S^1(c)$ of c under the S^1-action of constant reparametrization $c(t+s)$ of $c(t)$, is a non-degenerate critical manifold of E. For the sake of simplicity we will assume throughout this article that (M,g) is a bumpy Riemannian manifold.

We let $\mathcal{J}_\gamma^\perp \otimes \mathbb{C}$ denote the space of complex normal Jacobi fields along γ, a symplectic vector space of (complex) dimension $2n$ ($n = \dim M - 1$) with respect to the Wronskian

$$\omega(X,Y) = g(X, \frac{D}{ds}Y) - g(\frac{D}{ds}X, Y).$$

The linear Poincaré map P_γ is then the linear symplectic map on $\mathcal{J}_\gamma^\perp \otimes \mathbb{C}$ defined by $P_\gamma Y(t) = Y(t + L_\gamma)$.

Recall that, since P_γ is symplectic, its eigenvalues ρ_j come in three types: (i) pairs $\rho, \bar{\rho}$ of conjugate eigenvalues of modulus 1; (ii) pairs ρ, ρ^{-1} of inverse real eigenvalues; and (iii) 4-tuplets $\rho, \bar{\rho}, \rho^{-1}\bar{\rho}^{-1}$ of complex eigenvalues. We will often write them in the forms: (i) $e^{\pm i\alpha_j}$, (ii)$e^{\pm\lambda_j}$, (iii) $e^{\pm\mu_j \pm i\nu_j}$ respectively (with $\alpha_j, \lambda_j, \mu_j, \nu_j \in \mathbb{R}$), although a pair of inverse real eigenvalues $\{-e^{\pm\lambda}\}$ could be negative. Here, and throughout, we make the assumption that P_γ is *non-degenerate* in the sense that $det(I - P_\gamma) \neq 0$. In constructing the normal form we assume the stronger non-degeneracy assumption that

$$\Pi_{i=1}^{2n} \rho_i^{m_i} \neq 1, \qquad (\forall \rho_i \in \sigma(P_\gamma), \quad (m_1, \ldots, m_{2n}) \in \mathbf{N}^{2n}).$$

A closed geodesic is called *elliptic* if all of its eigenvalues are of modulus one, *hyperbolic* if they are all real, and *loxodromic* if they all come in quadruples as above.

1.2. **Wave group.** The wave group of a Riemannian manifold is the unitary group $U(t) = e^{it\sqrt{\Delta}}$ where $\Delta = -\frac{1}{\sqrt{g}} \sum_{i,j=1}^n \frac{\partial}{\partial x_i} g^{ij} g \frac{\partial}{\partial x_j}$ is the Laplacian of (M, g). Here, $g_{ij} = g(\frac{\partial}{\partial x_i}, \frac{\partial}{\partial x_j})$, $[g^{ij}]$ is the inverse matrix to $[g_{ij}]$ and $g = \det[g_{ij}]$. On a compact manifold, Δ has a discrete spectrum

$$(1) \qquad \Delta\varphi_j = \lambda_j^2 \varphi_j, \qquad \langle \varphi_j, \varphi_k \rangle = \delta_{jk}$$

of eigenvalues and eigenfunctions. The (Schwartz) kernel of the wave group can be constructed in two very different ways: in terms of the spectral data

$$(2) \qquad U(t)(x,y) = \sum_j e^{it\lambda_j} \varphi_j(x)\varphi_j(y)$$

or in terms of geometric data

$$(3) \quad U(t)(x,y) = \int_o^\infty e^{i\theta(r^2(x,y)-t^2)} \sum_{k=0}^\infty W_k(x,y)\theta^{\frac{d-3}{2}-k} d\theta \qquad (t < \mathrm{inj}(M,g))$$

where $U_o(x,y) = \Theta^{-\frac{1}{2}}(x,y)$ is the volume 1/2-density, where the higher coefficients are determined by transport equations, and where θ^r is regularized at 0 (see below). This formula is only valid for times $t < inj(M,g)$ but using the group property of $U(t)$ it determines the wave kernel for all times. It shows that for fixed (x,t) the kernel $U(t)(x,y)$ is singular along the distance sphere $S_t(x)$ of radius t centered at x, with singularities propagating along geodesics.

Closely related but somewhat simpler is the even part of the wave kernel, $\cos t\sqrt{\Delta}$ which solves the initial value problem

$$(4) \qquad \begin{cases} (\frac{\partial}{\partial t}^2 - \Delta)u = 0 \\ u|_{t=0} = f \qquad \frac{\partial}{\partial t} u|_{t=0} = 0 \end{cases}$$

Similar, the odd part of the wave kernel, $\frac{\sin t\sqrt{\Delta}}{\sqrt{\Delta}}$ is the operator solving

$$(5) \qquad \begin{cases} (\frac{\partial}{\partial t}^2 - \Delta)u = 0 \\ u|_{t=0} = 0 \qquad \frac{\partial}{\partial t} u|_{t=0} = g \end{cases}$$

These kernels only really involve Δ and may be constructed by the Hadamard-Riesz parametrix method. As above they have the form

$$(6)\qquad \int_0^\infty e^{i\theta(r^2-t^2)} \sum_{j=0}^{\infty} W_j(x,y)\theta_{reg}^{\frac{n-1}{2}-j} d\theta \mod C^\infty$$

where W_j are the Hadamard-Riesz coefficients determined inductively by the transport equations

$$(7)\qquad \begin{array}{l} \frac{\Theta'}{2\Theta}W_0 + \frac{\partial W_0}{\partial r} = 0 \\ 4ir(x,y)\{(\frac{k+1}{r(x,y)} + \frac{\Theta'}{2\Theta})W_{k+1} + \frac{\partial W_{k+1}}{\partial r}\} = \Delta_y W_k. \end{array}$$

The solutions are given by:

$$(8)\qquad \begin{array}{l} W_0(x,y) = \Theta^{-\frac{1}{2}}(x,y) \\ W_{j+1}(x,y) = \Theta^{-\frac{1}{2}}(x,y)\int_0^1 s^k \Theta(x,x_s)^{\frac{1}{2}}\Delta_2 W_j(x,x_s)ds \end{array}$$

where x_s is the geodesic from x to y parametrized proportionately to arc-length and where Δ_2 operates in the second variable.

Performing the integrals, one finds that

$$(9)\qquad \cos t\sqrt{\Delta}(x,y) = C_o|t|\sum_{j=0}^{\infty}(-1)^j w_j(x,y)\frac{(r^2-t^2)_-^{j-\frac{d-3}{2}-2}}{4^j\Gamma(j-\frac{d-3}{2}-1)} \mod C^\infty$$

where C_o is a universal constant and where $W_j = \tilde{C}_o e^{-ij\frac{\pi}{2}}4^{-j}w_j(x,y),$. Similarly

$$(10)\qquad \frac{\sin t\sqrt{\Delta}}{\sqrt{\Delta}}(x,y) = C_o sgn(t)\sum_{j=0}^{\infty}(-1)^j w_j(x,y)\frac{(r^2-t^2)_-^{j-\frac{d-3}{2}-1}}{4^j\Gamma(j-\frac{d-3}{2})} \mod C^\infty$$

Here and above t^{-n} is the distribution defined by $t^{-n} = Re(t+i0)^{-n}$ (see [Be], [G.Sh., p.52,60].) We recall that $(t+i0)^{-n} = e^{-i\pi\frac{n}{2}}\frac{1}{\Gamma(n)}\int_o^\infty e^{itx}x^{n-1}dx$ and also that $\int_o^\infty e^{itx}x^{n-1}dx$ has precisely the same singularity at $t=0$ as the sum $\sum_{k=0}^\infty e^{it(k+\frac{\beta}{4})}(k+\frac{\beta}{4})^{n-1}$.

1.3. **Wave invariants on a bumpy manifold.** Wave invariants are spectral invariants of Δ obtained by forming the (distribution) trace $TrU(t) = \sum_{\lambda_j \in Sp(\sqrt{\Delta})} e^{it\lambda_j}$ of the wave group. As above,

$$Sp(\sqrt{\Delta}) = \{\lambda_0 = 0 < \lambda_1 \le \lambda_2 \le \dots\}$$

is the spectrum of $\sqrt{\Delta}$.

The singular points of $TrU(t)$ occur when t lies in the length spectrum $Lsp(M,g)$, i.e. the set of lengths of closed geodesics. We denote the length of a closed geodesic γ by L_γ. For each $L = L_\gamma \in Lsp(M,g)$ there are at least two closed geodesics of that length, namely γ and γ^{-1} (its time reversal). The singularities due to these lengths are identical so one often considers the even part of $TrU(t)$ i.e. $TrE(t)$ where $E(t) = \cos(t\sqrt{\Delta})$. For background on the wave trace we refer to [D.G].

The trace of the wave group on a compact, bumpy riemannian manifold (M, g) has the singularity expansion

$$\text{(11)} \qquad \operatorname{Tr} U(t) = e_0(t) + \sum_{L \in \operatorname{Lsp}(M,g)} e_L(t)$$

with

$$e_0(t) = a_{0,-n}(t+iO)^{-n} + a_{0,-n+1}(t+iO)^{-n+1} + \cdots$$

$$\text{(12)} \qquad \begin{aligned} e_L(t) &= a_{L,-1}(t - L + iO)^{-1} + a_{L,0} \log(t - (L + iO)) \\ &+ a_{L,1}(t - L + iO) \log(t - (L + iO)) + \cdots \ , \end{aligned}$$

where $\cdots$ refers to homogeneous terms of ever higher integral degrees ([DG]). The wave coefficients $a_{0,k}$ at $t = 0$ are given by integrals over M of $\int_M P_j(R, \nabla R, ...)$dvol of homogeneous curvature polynomials. The principal wave invariant at $t = L$ in the case of a non-degenerate closed geodesic is given by

$$a_{L,-1} = \sum_{\gamma: L_\gamma = L} \frac{e^{\frac{i\pi}{4}} L_\gamma^\#}{|\det(I - P_\gamma)|^{\frac{1}{2}}}$$

where $\{\gamma\}$ runs over the set of closed geodesics, and where L_γ, $L_\gamma^\#$, m_γ, resp. P_γ are the length, primitive length, Maslov index and linear Poincaré map of γ. The problem we address here is to similarly determine the sub-principal wave invariants for non-zero $L \in Lsp(M, g)$.

1.4. **Wave invariants as non-commutative residues.** We begin the calculation of the wave invariants by characterizing them as non-commutative residues of $U(t)$ and its time derivatives. This is the method of [G.1][G.2] [Z.1]–[Z.5]. For the sake of completeness we recall the basic facts about residues.

To say that the wave invariants are non-commutative residues of $U(t)$ and its time derivatives at $t = L_\gamma$ is just to say that they are the residues of the zeta function

$$\text{(13)} \qquad \zeta(s,t) := \sum_{j=o}^{\infty} e^{it\lambda_j} \lambda_j^{-s} = TrU(t)\Delta^{-s/2} \qquad (Res >> 0)$$

and its t-derivatives at the simple poles $s = 1, 0, -1, ...$, with $\{\lambda_j\}$ the eigenvalues of $\sqrt{\Delta_g}$. For short we write

$$\text{(14)} \qquad a_{\gamma k} = resD_t^k U(t)|_{t=L_\gamma}$$

with $D_t = \frac{\partial}{\partial t}$. Of course, if there are several closed geodesics of the same length L, the residue is the sum of $a_{\gamma k}$ over all γ of length L. Also, we are assuming in the description of the poles that γ is an isolated, non-degenerate closed goedesic. To justify the definition we need to show that the zeta function in (13) has a meromorphic continuation to $\mathbb{C}$.

This definiton of the non-commutative residue is the natural extension to Fourier Integral operators of the non-commutative residue of a ψDO (pseudodifferential operator) A on a compact manifold M , defined by

$$\text{(0.3)} \qquad \text{res}(A) = 2\,\text{Res}_{s=0}\,\zeta(s, A)$$

where

$$\text{(15)} \qquad \zeta(s, A) = \text{ Tr } A\Delta^{-s/2} \ \ (\text{Re } s >> 0)\,.$$

Let us just summarize the results in the case of the wave group. We denote by $G^t : T^*M - 0 \to T^*M - 0$ the geodesic flow, i.e. the Hamilton flow of the metric Hamilton $|\xi|_g$. We will assume throughout that the fixed point set of G^t is clean for each t (see [D.G] for the definition). Then $\zeta(\cdot, t)$ has a meromorphic continuation to $\mathbb{C}$, with simple poles at $s = 1 + \frac{\dim s \text{ Fix}(G^t)-1}{2} - j$ ($j = 0, 1, \ldots, 2$). If the metric is bumpy (so that $\dim S$ Fix $G^t = 1$ if S Fix $G^t \neq \varphi$), then the poles occur at $s = 1 - j$ ($j = 0, 1, 2, \ldots$). Here, S Fix(G^t) is the set of unit vectors in the fixed point set of G^t.

It therefore makes sense to make the:

Definition: $\text{res}(A) := \text{Res}_{s=0}\,\zeta(s, A)$,

As in the pseudododifferential case, $res(A)$ is independent of the choice of Δ, is tracial ($res(AB) = res(BA)$) and there is a local formula for res(A). Indeed, the latter is what we wish to calculate for the wave group.

Let us observe that the residue of the wave group is the sub-principal wave invariant, i.e.

$$\text{res } U(L) = \begin{cases} 0 & L \notin \text{Lsp}(M, g) \\ a_{L,0} & L \in \text{Lsp}(M, g)\end{cases}.$$

By replacing $U(t)$ by $\Delta^{n/2}U(t)$, we can express all wave invariants as non-commutative residues.

1.4.1. *Non-commutative residues of pseudodifferential operators.* As preparation for the calculation of the residue of the wave group, let us briefly review the relations between the singularities of Tr $Ae^{\text{it}\sqrt{\Delta}}$ and the poles/residues of Tr $A\Delta^{-s/2}$ for $A \in \Psi^*(M)$. Here $\Psi^*(M)$ denotes the algebra of ψDO's on M; $\Psi^m(M)$ will denote the m-th order ones. Also, Δ will denote the Laplacian of some fixed metric g on M, whose normalized eigenfunctions/eigenvalues will be denoted φ_j/λ_j.

PROPOSITION 1.1. *For $A \in \Psi^m(M)$, let $N(\lambda, A) = \sum_{\sqrt{\lambda_j}\leq\lambda}(A\varphi_j, \varphi_j)$. Then for any $\rho \in \zeta(\mathbb{R})$ with $\hat{\rho} \in C_0^\infty(\mathbb{R})$, Supp $\hat{\rho} \cap$ Lsp$(M, g) = \{0\}$ and with $\hat{\rho} \equiv 1$*

in some interval around 0, we have:

$$\rho * dN(\lambda, A) \sim \sum_{k=0}^{\infty} \alpha_k \lambda^{n+m-k-1} \quad (\lambda \to +\infty)$$

where:

$$\begin{aligned} n &= \dim M \\ \alpha_0 &= \int_{S^*M} \sigma_A d\lambda \\ \alpha_k &= \int_{S^*M} \omega_k d\lambda \end{aligned}$$

where ω_k is determined from the k-jet of the complete symbol a of A.

Proof. This is essentially Proposition 2.1 of [D.G], except that they only consider the case $A = I$. For completeness, we will extend their argument to the general case.

Using the action of a ψDO on an exponential, the kernel $AU(t,x,y)$ of $AU(t)$ is easily seen to have the form

$$AU(t,x,y) = (2\pi)^{-n} \int_{\mathbb{R}^n} \alpha(t,x,y,\eta) e^{i\varphi(t,x,y,\eta)} d\eta$$

where α is a classical symbol of order m and where $\varphi(t,x,y,\eta) = \psi(x,y,\eta) - t|\eta|$, with $\psi(x,y,\eta) = 0$ if $\langle x-y, \eta\rangle = 0$. Hence,

$$\begin{aligned} \rho * dN(\lambda, A) &= \int_{\mathbb{R}} \int_M e^{i\lambda t} \hat{\rho}(t)\, AU(t,y,y) dt\, dy \\ &= (2\pi)^{-n} \int_M \int_{\mathbb{R}} \int_{\mathbb{R}^n} e^{i\lambda t} \hat{\rho}(t) e^{-it|\eta|} \alpha(t,y,y,\eta) dy\, d\eta\, dt \ . \end{aligned}$$

Now proceed as in [D.G]. ∎

Corollary **1.2.** *$\zeta(s,A)$ has a meromorphic continuation to $\mathbb{C}$, having only simple poles at $s = m+n-k$ ($k = 0,1,2,\dots$) with residues $(2\pi)^{-n}\alpha_k$.*

Proof. Essentially [[D.G], Corollary 2.2]. Since we will generalize the proof for the case of the wave group, we review it briefly.

Let $\sqrt{\lambda_1}$ be the lowest eigenvalue of $\sqrt{\Delta}$. Choose $\chi \in C^\infty(\mathbb{R})$, $\chi(\lambda) \equiv 0$ for $\lambda \le \varepsilon < \lambda_1$, $\chi \equiv 1$ for $\lambda \ge \lambda_1$ and let $\chi_s(\lambda) = \chi(\lambda)\lambda^{-s}$. Then

$$\begin{aligned} \zeta(s,A) &= \langle \chi_s(\lambda), dN(\lambda, A)\rangle \\ &= \langle \chi_s, \rho * dN(\lambda, A)\rangle + \langle \chi_s^\vee, (1-\hat{\rho}) dN(\cdot, A)^\wedge\rangle \ , \end{aligned}$$

where ρ is as in Proposition 1. By [DG, p. 49], $(1-\hat{\rho})\chi_s^\vee : \mathbb{C} \to \zeta(\mathbb{R})$ is entire, so

$$\langle \chi_s^\vee (1-\hat{\rho}) \ , \ \ dN(\cdot, A)^\vee\rangle$$

is entire (and of polynomial growth in any halfplane Re$s \geq C$). Hence, $\zeta(s, A)$ has a meromorphic continuation if and only if $\langle \chi_s, \rho * dN(\lambda, A)\rangle$ does, and the poles/residues are the same. But we obviously have

$$\begin{aligned}\langle \chi_s, \rho * dN(\lambda, A)\rangle &\sim \sum_{k=0}^{\infty} \alpha_k \int_{\mathbb{R}} \chi(\lambda)\lambda^{n+m-k-1-s} d\lambda \\ &\sim \psi(s) \sum_{k=0}^{\infty} \frac{\alpha_k}{s-(n+m-k)}\end{aligned}$$

where $\psi(s) = \int \lambda^{n+m-k-s} \frac{d\chi}{d\lambda} d\lambda$. Also, $\sim$ means that the error in taking $\sum_{k=0}^{N}$ on the right side is regular in Re$s \geq n+m-N$. We observe (following [D.G]) that ψ is entire and $\psi(n+m-k) = 1$. The proposition follows. ∎

COROLLARY **1.3.** $res(A) = \alpha_{n+m}$.

Let us make the relation to the singularity expansion of $S(t, A) = \mathrm{Tr}\, AU(t)$ more explicit. First, Proposition (1.1) is equivalent to the statement

$$S(t, A) \sim \sum_{k=0}^{\infty} \alpha_k \chi_{k-(n+m)}(t) \tag{16}$$

where χ_r is homogeneous distribution on $\mathbb{R}$ of degree $-r$ with $WF(\chi_r) = \{(0, \tau) : \tau \in \mathbb{R}^+\}$. The space of such distributions is one-dimensional, and is spanned by $\int_0^\infty e^{it\tau}\tau^{r-1}d\tau$ (regularized at $\tau = 0$ as in [[G.Sh], p. 342], see also [[G.Sh], p. 360]). For $r \neq 0, -1, -2, \ldots$, this integral is a multiple of $(t+i0)^{-r}$ (loc. cit.). For $r = 0, -1, -2$, this integral equals $a_0^{(r)} t^{-r} - a_{-1}^{(r)} t^{-r} ln(t+i0)$ for certain constants $a_0^{(r)}, a_{-1}^{(r)}$ [[G.Sh], p. 177]. We may therefore let $\chi_r(t) = (t+i0)^{-r}$ for $r \neq 0, -1, -2$ and $\chi(t) = t^n ln(t+i0)$ for $r = -n = 0, -1, -2, \ldots$

In particular, we see that res(A) is the coefficient of the $\chi_0 = ln(t+i0)$ term.

Finally, a word on the local formula for res(A). The key point is the independence of res(A) of the choice of Δ. Since we can localize the calculation of res(A), we may assume that A is essentially supported in a coordinate chart x, and let Δ be the flat Laplacian in the local coordinates. Then,

$$A\Delta_0^{-s/2}(x, y) \equiv \int_{|\xi|>1} |\xi|^{-s} a(x, \xi) e^{i\langle \xi, x-y\rangle} d\xi \pmod{C^\infty},$$

hence

$$\begin{aligned}\mathrm{Tr}\, A\Delta_0^{-s/2} &\equiv \int_M \int_{|\xi|>1} |\xi|^{-s} a(x, \xi) dx\, d\xi \\ &\equiv \sum_j \left(\int_{S^*M} a_j d\lambda\right) \int_1^\infty r^{n-1-s+j} dr \\ &\equiv \sum_j \frac{\alpha_j}{(n-s+j)}\end{aligned}$$

(modulo holomorphic functions). In particular,

$$\text{res}(A) = \int_{S^*M} a_{-n} d\lambda \, .$$

1.4.2. *res* $A : A \in I^*(M \times M, \Lambda)$. Let us now generalize the results to FIO's. We assume $A \in I^m(M \times M, \Lambda)$ for some canonical relation Λ which cleanly intersects the diagonal in $T^*(M \times M)$. In fact, let us assume Λ cleanly intersects the whole canonical relation C of the wave group.

Under this condition, $S(t, A) = \text{Tr}\, AU(t)$ is a Lagrangean distribution on $\mathbb{R}$ with singularities at the set of *sojourn times*,

$$\mathcal{ST} = \{T : \exists (x, \xi) \in S^*M : (x, \xi, G^T(x, \xi)) \in \Lambda\} \tag{17}$$

For a given sojourn time T, the corresponding set

$$W_T = \{(x, \xi) : (x, \xi, G^t(x, \xi)) \in \Lambda\}$$

of sojourn rays fills out a submanifold of S^*M of some dimension e_T. Letting $N(\lambda, A) = \sum_{\sqrt{\lambda_j} \leq \lambda} (A\varphi_j, \varphi_j)$ as before, we have a generalization of Proposition (1.1)

PROPOSITION **1.4.** *If* $\rho \in \zeta(\mathbb{R})$ *with* $\hat{\rho} \in C_0^\infty(\mathbb{R})$, $\text{Supp}\,(\hat{\rho}) \cap \mathcal{ST} = \{0\}$ *and* $\hat{\rho} \equiv 1$ *near 0, then*

$$\rho * dN(\lambda, A) \sim C_n \lambda^{m + \frac{e_0 - 1}{2}} \sum \alpha_j \lambda^{-j} \, ,$$

where C_n *is a universal constant. The coefficients have the form,*

$$\alpha_j = \int_{\Lambda_\Delta} \omega_j d\lambda_\Delta$$

where $\Lambda_\Delta = \Lambda \cap$ *diag*$(S^*M \times S^*M)$, $e_0 = \dim \Lambda_\Delta$, $d\lambda_\Delta$ *is a canonical density on* Λ_Δ *and the functions* ω_j *are determined by the* j*-jet of the (local) complete symbols of* AU *along* Λ_Δ.

Proof. This is essentially Proposition 1.10 (see also Lemma 3.1) of [Z.4] ∎

We now prove that $\zeta(s, t)$ has an analytic continuation to $\mathbb{C}$ with at most simple poles and that the residues are given by local invariants. We have:

$$\begin{aligned} \text{Tr}\, A\Delta^{-s/2} &= \langle \chi_s, dN(\bullet, A) \rangle \\ &= \langle \chi_s, \rho * dN(\bullet, A) \rangle + \text{ entire} \end{aligned}$$

(where ρ has the same meaning as above).

Hence,

$$\begin{aligned}\zeta(s,A) &= C_n \sum_{j=0}^{\infty} \alpha_j \int_1^{\infty} \lambda^{\frac{e_0-1}{2}+m-s-j} d\lambda + \text{entire} \\ &= C_n \sum_{j=0}^{\infty} \frac{\alpha_j}{m+1+\frac{e_0-1}{2}-(s+j)} + \text{entire}\,.\end{aligned}$$

In particular, if $A = U(L)$, then $\{0\}$ is a sojourn time if and only if $L \in \mathrm{Lsp}(M,g)$. If $L \notin \mathrm{Lsp}(M,g)$, $\zeta(s,L)$ is regular at 0. If $L \in \mathrm{Lsp}(M,g)$ and the metric is bumpy, then $m = 0$ and $e_0 = 1$ and the formula $\mathrm{res}\, U(L) = a_{L,0}$ follows.

Next, we verify that res(A) has the remarkable properties of the ψDO case. The proofs are essentially the same as in that case and follow the lines of [K].

Proposition 1.5. *Let $A_i \in I^{m_i}(M \times M, \Lambda_i)$ $(i = 1, 2)$ and assume that all compositions to follow are clean.*

Then

(a) res(A_j) is independent of the choice of Δ;

(b) If either Λ_1 or Λ_2 is a local canonical graph, then

$$\mathrm{res}(A_1A_2) = \mathrm{res}(A_2A_1)\,.$$

Proof. (a) [G.3] Let Δ_1, Δ_2 be two Laplacians.

Evidently,

$$\Delta_1^{-s}A = \Delta_2^{-s}Q(s)A$$

with $Q(s) = \Delta_1^{s}\Delta_2^{-s}$. $Q(s)$ is a holomorphic family of ψDO's of order 0, and

$$Q(s) = I + sQ'(0) + s^2Q^{(2)}(s)$$

where $Q^{(2)}(s)$ is holomorphic. Hence, $\mathrm{Tr}\,\Delta_1^{-s}A = \mathrm{Tr}\,\Delta_2^{-s}A$ is regular at 0 (since both traces have only simple poles at 0).

(b) ([K]) Suppose Λ_2 is a local canonical graph. Since the proposition is local, we may assume that Λ_2 is actually the graph of a canonical transformation χ.

Suppose temporarily that A_2 is actually invertible, so that

$$A_2^{-1} \in I^{-m}(M \times M, C_2^{-1})\,.$$

Then $A_2^{-1}\Delta A_2$ is a positive elliptic ψDO of order 2. By (a) (which only requires Δ_i to be positive elliptic),

$$\begin{aligned}\mathrm{Res}_{s=0}\,\mathrm{Tr}\,A_1A_2\Delta^{-s} &= \mathrm{Res}_{s=0}\,\mathrm{Tr}\,A_1A_2(A_2^{-1}\Delta A_2)^{-s} \\ &= \mathrm{Res}_{s=0}\,\mathrm{Tr}\,A_1A_2(A_2^{-1}\Delta^{-s}A_2) \\ &= \mathrm{Res}_{s=0}\,\mathrm{Tr}\,A_2A_1\Delta^{-s}\,,\end{aligned}$$

hence $\mathrm{res}(A_1A_2) = \mathrm{res}(A_2A_1)$.

Now let us drop the assumption that A_2 is invertible. Since C_2 is a graph, there must be an invertible element, say B, in $I^0(M \times M, C_2)$. For z sufficiently large, $A_2 + zB$ will be invertible, hence

$$\text{res } A_1(A_2 + zB) = \text{ res}(A_2 + zB)(A_1) .$$

Comparing coefficients, we get (b). ∎

COROLLARY **1.6.**

$$a_{L,k} = res(D_t^k U(t)|_{t=L})$$

$(D_t = \frac{1}{i}\frac{d}{dt}$.

Proof. Since $D_t^k U(t)|_{t=L} = \Delta^{k/2} U(L)$, the right side equals

$$\begin{aligned} \text{Res}_{s=-k}\zeta(s, L) &= C_n \alpha_{k+1} \\ &= a_{L,k} \ , \end{aligned}$$

as follows from the relation between the expansions above. ∎

1.5. **Calculating wave invariants.** The wave invariants can be calculated in a straightforward way by the residue method once one has constructed a parametrix for the wave kernel near γ. The results are of course only as explicit as is the parametrix. We outline the general calculation and then turn to the most favorable case of manifolds without conjugate points.

First, we assume given a microlocal parametrix for $U(t)$ in a neighborhood of $\gamma \times \gamma$, i.e. a kernel

$$F_{\gamma,N}(t,x,y) = \int_{\mathbb{R}^m} a_{\gamma,N}(t,x,y,\theta) e^{i\varphi_\gamma(t,x,y,\theta)} d\theta \tag{18}$$

with the property that

$$U(t,x,y) - F_{\gamma,N}(t,x,y) \in H_N^{\text{loc}} \tag{19}$$

at $T^*_{L_\gamma}\mathbb{R} \times \mathbb{R}^+\gamma \times \mathbb{R}^+\gamma$ (cf. [Hö III, 18.1.30] for the relevant defintion). Then

$$a_{\gamma,k} = \text{ res}(D_t^k F_{\gamma,N}(t)|_{t=L_\gamma}) \ . \tag{20}$$

Second, we choose an elliptic operator Δ (or, more generally, a gauging) which best simplifies the calculation of the right side. An obvious choice is to use the following gauging of $F_{\gamma,N}(t,x,y)$:

$$F_{\gamma,N}(t,x,y,s) = \int_{\mathbb{R}^m} a_{\gamma,N}(t,x,y,\theta)|\theta|^{-s} e^{i\varphi_\gamma} d\theta. \tag{21}$$

Since the residue trace is independent of the gauging, we have

$$a_{\gamma,k} = \text{ Res}_{s=0} \int_M \int_{\mathbb{R}^m} a_{\gamma,n}^{(k)}(L,x,x,\theta)|\theta|^{-s} e^{i\varphi_\gamma(L,x,x,\theta)} d\theta dV \ , \tag{22}$$

where $a^{(k)}_{\gamma,N}$ is the amplitude one gets after applying D^k_t to $F_{\gamma,N}$, and where dV is the volume form. Next, we change to polar coordinates $\theta = r\omega$ with $r = |\theta|$ and $\omega \in S^{m-1}$, and expand the amplitude in powers of r:

$$a^{(k)}_{\gamma,N}(L,x,x,r\omega) \sim \sum_{p=d}^{-\infty} a^{(k)}_{\gamma,N,j}(L,x,x,\omega)r^j \,, \tag{23}$$

leading to

(24)

$$a_{\gamma,k} = \mathrm{Res}_{s=0}\left(\sum_{p=d}^{-\infty}\int_M\int_{S^{m-1}} a^{(k)}_{\gamma,N,p}(L,x,x,\omega)\nu^+_{m-1-s+j}(\varphi_\gamma(L,x,x,\omega))dV d\omega\right)$$

where

$$\nu^+_m(x) = \int_0^\infty \chi(r)r^m e^{irx}dr \,. \tag{25}$$

Here, χ is a smooth cutoff, equal to 1 for $r \geq 1$ and to 0 for $r \leq \frac{1}{2}$, as in [[G.Sh], p. 344, (4.5)]. It is evident that the choice of χ has no effect on the poles or residues above. Since

$$\nu^+_m(x) = e^{\frac{i\pi}{2}(m+1)}(x+iO)^{-m-1} \tag{26}$$

([G.Sh] loc. cit.), we get

$$a_{\gamma,k} = \mathrm{Res}_{s=0}\sum_{p=d}^{-\infty} e^{\frac{i\pi}{2}(m+j)}\int_M\int_{S^{m-1}} a^{(k)}_{\gamma,N,p}(\varphi_\gamma + iO)^{s-m-j}dV d\omega \,. \tag{27}$$

The main point then is to determine the poles and residues of the meromorphic family of distributions $(\varphi_\gamma(L,x,x,\omega)+i0)^\lambda$. Together with a sufficiently concrete construction of the amplitudes $a^{(k)}_{\gamma,N,j}(L,x,x,\omega)$, this will lead to explicit formulae for the wave coefficients.

2. Wave invariants on manifolds without conjugate points

The special feature of manifolds without conjugate points is that the Hadamard-Riesz parametrix is valid globally in space and time. Hence the preceding residue calculation gives geometric formulae for all the wave invariants. In this section we will carry out the details.

The resulting formulae express the wave invariants $a_{\gamma k}$ in terms of geometric quantities such as such as the displacement function $f_\gamma(x) := r(x,\gamma x)^2$ and volume distortian $\Theta(x,\gamma x)$ of the isometry of $\tilde{M}$ associated to γ. For instance, the subprincipal wave invariant $a_{\gamma o}$ is given by the following:

$$\begin{aligned} a_{\gamma o} &= C^o_{n,o,1}\int_\gamma \frac{U_1(x,\gamma x)d\sigma}{\sqrt{\det \mathrm{Hess}(f\gamma)_\sigma}} \\ &+ C^o_{n,o,o}\int_\gamma \frac{\mathrm{Hess}(f_\gamma)^{-1}_\sigma(\frac{\partial}{\partial y},\frac{\partial}{\partial y})[(U_0(x,\gamma x)J_\gamma(y,s)]d\sigma}{\sqrt{\det \mathrm{Hess}(f_\gamma)_\sigma}} \end{aligned} \tag{28}$$

for certain universal constants. Here, U_o, U_1 are the first two Hadamard coefficients (see below), and $J_{[\gamma]}$ is the volume density in certain Morse coordinates. This expression is similar to the one give by Donnelly [D] for $a_{\gamma o}$ on surfaces of negative curvature. In effect, the only condition actually needed for Donnelly's calculation is that the (M, g) be a manifold without conjugate points.

This expression for $a_{\gamma 0}$ is somewhat non-constructive in that it is written in a Morse coordinate system which has not been effectively constructed (although it could be). But there is a simple modification of the calculation which is constructive and leads to effectively computable wave invariants. We will explain below how to carry out both calculations.

Let us observe that the formulae produced here are quite different from those produced in [G.1] [Z.1] by the method of normal form. The latter method yields expressions for $a_{\gamma k}$ in terms of Jacobi fields and their derivatives along γ and in terms of eigenvalues of the Poincare map P_γ. One can show directly that both kinds of formulae agree, but clearly they roll things up in quite different ways.

2.1. **Geometric preliminaries.** Let us continue the table from §1 to encompass the special geometry of manifolds without conjugate points (WCPs). We let:

(i)	$\pi : (\tilde{M}, \tilde{g}) \to (M, g)$	denote the universal riemannian cover;
(ii)	Γ	denote its deck transformation group;
(iii)	$\Theta(x, y)$	denote the volume density in normal coordinates centered at x;
(iv)	$r = r(x, y)$	denote the distance function of $\tilde{M}$;
(v)	$[\Gamma] =$	denote the set of conjugacy classes $[\gamma]$ of Γ
(vii)	Γ_γ	denote the centralizer of γ;
(vi)	For $\gamma \in \Gamma$, $f_\gamma(x) = r(x, \gamma x)^2$	denotes the displacement function
(vii)	$\tilde{M}_\gamma$	denote the critical point set of f_γ;

2.2. Wave equation on a manifold without conjugate points. The wave kernels $\cos(t\sqrt{\Delta})(x,y)$ and $\frac{\sin(t\sqrt{\Delta})}{\sqrt{\Delta}}$ are simplest to construct on a Riemannian manifold (M,g) without conjugate points. By definition, there is a unique geodesic (unit speed) between any two points (x,y) of the universal cover $\tilde{M}$ and the geodesic distance function (squared) is a global smooth function $r^2(x,y)$. The wave kernels can then be globally constructed by the classical Hadamard-Riesz parametrix method and are similar to the well-known heat kernel parametrices. In this section we review the construction and establish notation. We refer to [Be], and [HoI-IV](Volume III) for detailed expositions and to [Su] for parallel constructions for the heat kernel.

On $\tilde{M}$, the wave operator $\tilde{E}$ can be globally constructed (modulo $C^\infty(\mathbb{R}\times M\times M)$) by the Hadamard-Riesz parametrix method ([Be]). That is, the wave kernel $\tilde{E}(t,x,y)=\cos(t\sqrt{\Delta})$ is given modulo smooth kernels by:

$$\tilde{E}(t,x,y)\equiv\int_0^\infty e^{i\theta(r^2-t^2)}\sum_{j=0}^\infty W_j(x,y)\theta^{\frac{n-1}{2}-j}\chi(\theta)d\theta \tag{29}$$

where χ is (as above) a smooth cutoff near 0 and where the W_j are given recursively by the formulae in (8). Note that r^2 and $\Theta^{-\frac{1}{2}}$ are smooth for a metric WCP.

The wave kernel $E(t,x,y)$ on M is obtained by projecting this kernel from $\tilde{M}$, i.e. by summing over the deck transformation group:

$$\begin{aligned}E(t,x,y)&=\textstyle\sum_{\gamma\in\Gamma}\tilde{E}(t,x,\gamma\cdot y)\\ &\equiv\textstyle\sum_{\gamma\in\Gamma}\int_0^\infty e^{i\theta(r^2(x,\gamma y)-t^2)}\sum_{j=0}^\infty W_j(x,\gamma y)\theta^{\frac{n-1}{2}-j}\chi(\theta)d\theta.\end{aligned} \tag{30}$$

2.3. Preliminaries on wave invariants on a manifold WCP. Following the general prescription we determine wave invariants by taking residues.

We first calculate on $\tilde{M}$ that

$$D_t^k\tilde{E}(t,x,y)|_{t=L}=\int_0^\infty e^{i\theta(r^2-t^2)}P_k(t,\theta)\sum_j W_j\theta^{\frac{n-1}{2}-j}\chi d\theta \tag{31}$$

where $P_k(t,\theta)$ are certain polynomials of degree k in θ, given recursively by

$$\begin{aligned}P_1(t,\theta)&=-2t\theta\\ P_{k+1}(t,\theta)&=(D_t-2t\theta)P_k(t,\theta)\,,\end{aligned} \tag{32}$$

so that

$$P_k(t,\theta)=\sum_{\ell=0}^k a_\ell(t)\theta^\ell$$

for some easily calculable coefficients $a_\ell(t)$.

Projecting (31) to $M = \tilde{M}\backslash\Gamma$, gauging as above and applying the standard Selberg rearrangement to the trace, we get

$$\text{res}(D_t^k U(t)|_{t=L}) = \sum_{[\gamma]} \text{Res}_{s=0} I_{[\gamma]}^{(k)}(s) \tag{33}$$

where

$$I_{[\gamma]}^{(k)}(s) = \int_{\tilde{M}\backslash\Gamma_\gamma} \int_0^\infty e^{i\theta(r(x,\gamma x)^2 - t^2)} P_k(t,\theta) \sum_0^\infty W_j(x,\gamma x) \cdot \theta^{\frac{n-1}{2}-j-s} \chi(\theta) d\theta \, dV .$$

Evidently,

$$\begin{aligned} a_{\gamma,k} &= \text{Res}_{s=0} I_{[\gamma]}^{(k)}(s) \\ &= \sum_{j=0}^\infty \sum_{\ell=0}^k a_\ell(L) e^{\frac{i\pi}{2}(\frac{n+1}{2}-j+\ell)} \text{Res}\, I_{[\gamma]}^{(k,\ell,j)}(s) \end{aligned} \tag{34}$$

where

$$I_{[\gamma]}^{(k,\ell,j)}(s) = \int_{\tilde{M}\backslash\Gamma_\gamma} (r(x,\gamma x)^2 - L_\gamma^2 + iO)^{j+s-\frac{n+1}{2}-\ell} W_j(x,\gamma x) dV . \tag{35}$$

Note that, in our notation above,

$$\begin{array}{c} \varphi_\gamma(L_\gamma, x, x, \theta) = \theta(r(x,\gamma x)^2 - L^2) \quad (L = L_\gamma) \\ m = 1\,, \quad d = \frac{n-1}{2}\,, \quad p = \frac{n+1}{2} + \ell - j \\ a_{\gamma,p}^{(k)}(x) = \sum_{j,\ell: \frac{n+1}{2}+\ell-j=p} a_\ell(L) e^{\frac{i\pi}{2}(\frac{n+1}{2}-j+\ell)} W_j(x,\gamma x) . \end{array} \tag{36}$$

To proceed, we have to analyze the poles and residues of the family $(r(x,\gamma x)^2 - L_\gamma^2 + iO)^\lambda$.

We will carry out the analysis by two methods. The first ("ineffective" method) is to change coordinates $x \to y$ to put f_γ into a normal form. Uner the non-degeneracy assumptions, f_γ is a Bott-Morse function, so the normal form is a quadratic form $Q(y)$. This reduces the analysis to the classical case of the family $(Q(y) + iO)^\lambda$, and leads to rather manageable expressions for the residues. However, it has the defect that the coordinates y have not explicitly constructed . The second ("effective" method) involves a Taylor expansion of $r(x,\gamma x)^2$ around its critical points, and is adapted from the analogous effective method of calculating coefficients in the MSP expansion (Hö I, Theorem 7.7.6].

In order to apply the methods, we will need the following.

PROPOSITION **2.1.** *Let (M,g) be a manifold WCP. Fix $[\gamma] \in [\Gamma]$. Then the following are equivalent:*

(i) The metric $\tilde{g}$ on $\tilde{M}/\Gamma_\gamma$ is non-degenerate;

(ii) Fix $(\tilde{G}^L)\imath S^*(\tilde{M}/\Gamma_\gamma)$ is a clean fixed point set for all $L \in LSP(\tilde{M}/\Gamma_\gamma)$;

(iii) f_γ is a Bott-Morse function on $\tilde{M}/\Gamma_\gamma$.

(Here, $\tilde{G}^L$ is the geodesic flow of $\tilde{g}$ on $\tilde{M}/\Gamma_\gamma$.)

2.4. **Calculation of wave invariants on a manifold WCP.** Let us now carry out the calculations. We should note that they specialize to the formulae of Donnelly[D] in the case of negatively curved surfaces and also use some calculations by Sunada [S].

2.4.1. *The Ineffective Method.* Under our assumption that (M, g) is bumpy and WCP, the non-degeneracy conditions are satisfied, and $\tilde{M}_\gamma/\Gamma_\gamma$ is the single closed geodesic, say c_γ, of $\tilde{M}/\Gamma_\gamma$.

By the Morse Lemma for Bott-Morse functions, there exists a tubular neighborhood $T_\epsilon(c_\gamma)$ and coordinates $x = (\sigma, y_1, \dots, y_{n-1})$ such that

(i) $y_1 = \cdots = y_{n-1} = 0$ defines c_γ, and σ is arclength along c_γ (centered at some point p);

(ii) $r(x, \gamma x)^2 - L_\gamma^2 = \text{Hess}(f_\gamma)_\sigma(y, y)$ where $\text{Hess}(f_\gamma)_\sigma$ is the Hessian at $(\sigma, 0, \dots, 0)$;

(iii) The Jacobian $J_\gamma(y, \sigma)$, defined by $dV = J_\gamma(y, \sigma) d\sigma dy$ satisifes $J_\gamma(0, \sigma) \equiv 1$.

Let χ_ε be a smooth cutoff on $\tilde{M}/\Gamma_\gamma$, equal to one on a neighborhood of c_γ and zero outside of $T_\epsilon(c_\gamma)$. We can break up the integral $I^{(k,\ell,j)}_{[\gamma]}(s)$ into two terms corresponding to $1 = \chi_\varepsilon + (1 - \chi_\varepsilon)$. Let us write the two terms as $I^{(k,\ell,j)}_{[\gamma]}(s) + II^{(k,\ell,j)}_{[\gamma]}(s)$, with some abuse of notation. It is obvious that $II^{(k,\ell,j)}_{[\gamma]}(s)$ is entire, so that (the new) $I^{(k,\ell,j)}_{[\gamma]}(s)$ has the same poles and residues as (the old) $I^{(k,\ell,j)}_{[\gamma]}(s)$. Hence, we have:

$$(37) \quad I^{(k,\ell,j)}_{[\gamma]}(s)$$
$$= \int_0^{L_\gamma} \int_{|y|<\varepsilon} (\text{Hess}(f_\gamma)_\sigma(y, y) + iO)^{s+j-\frac{n+1}{2}-\ell} \cdot \chi_\varepsilon(y) \cdot W_j(y, \sigma) J_\gamma(y, \sigma) \cdot d\sigma dy$$

where $W_j(y, \sigma) = W_j(x, \gamma x)$ with $x = (\sigma, y)$.

Now, by [[G.Sh], p. 275-276], the family of distributions

$$(\text{Hess}(f_\gamma)_\sigma(y, y) + iO)^\lambda$$

has a meromorphic continuation to $\mathbb{C}$, with simple poles at the points $\lambda = -\frac{1}{2}(n-1) - q$, $q = 0, 1, 2, \dots$, and with residues

$$\text{res}_{\lambda = -\frac{1}{2}(n-1)-q} (\text{Hess}(f_\gamma)_\sigma(y, y) + iO)^\lambda = \frac{\frac{\pi^{\frac{1}{2}(n-1)}}{4^q q! \Gamma(q+1)} \{\text{Hess}(f_\gamma)^{-1}(\frac{\partial}{\partial y}, \frac{\partial}{\partial y})\}^q \delta_0(y)}{\sqrt{\det\ \text{Hess}(f_\gamma)_\sigma}}.$$

Here,

$$\text{Hess}_\sigma^{-1}(f_\gamma)(\frac{\partial}{\partial y}, \frac{\partial}{\partial y}) = \sum h^{ij}(\sigma, 0) \frac{\partial^2}{\partial y_i \partial y_j},$$

is the inverse Hessian operator, with $\text{Hess}_\sigma(f_\gamma) = [h_{ij}]$. Then, $I^{(k,\ell,j)}_{[\gamma]}(s)$ has a simple pole at $s = 0$ if and only if

$$j - \frac{n+1}{2} - \ell = -\frac{1}{2}(n-1) - q$$

for $q = 0, 1, 2, \ldots$ i.e.

$$j = \ell + 1 - q \quad \text{for some} \quad \begin{array}{rcl} \ell & = & 0, 1, \ldots, k \\ q & = & 0, 1, 2, \ldots . \end{array}$$

Since $j \geq 0$, given (k, ℓ) there are only finitely many values of j for which such a pole occurs. We have:

$$res_{s=0} I^k_{[\gamma]} = \sum_{\ell=0}^{k} \sum_{j=0}^{\ell+1} a_\ell(L) e^{\frac{i\pi}{2}(\frac{n+1}{2} - j + \ell)} \times \frac{\frac{\pi^{\frac{1}{2}(n-1)}}{4^q q! \Gamma(q+1)} \{\text{Hess}(f_\gamma)^{-1}(\frac{\partial}{\partial y}, \frac{\partial}{\partial y})\}^q \delta_0(y)}{\sqrt{\det\ \text{Hess}(f_\gamma)_\sigma}}\Big|_{q=\ell+1-j}.$$

The coefficient in the (k, ℓ, j)th term in dimension n will be written $C^{(k)}_{(n,\ell,j)}$.

PROPOSITION **2.2.** *With the above notations and assumptions, we have:*

$$a_{\gamma k} = \sum_{\ell, j : 0 \geq j \leq \ell+1, 0 \leq \ell \leq k} C^{(k)}_{n,\ell,j} \times \int_\gamma \frac{1}{\sqrt{\det\ Hess\ f\gamma}} \{Hess(f_\gamma)^{-1}_\sigma(\frac{\partial}{\partial y}, \frac{\partial}{\partial y})\}^{\ell+1-j} [W_j(y,s) J_\gamma(y,s)]\, d\sigma$$

for certain universal constants $C^{(k)}_{n,\ell,j}$.

For example, consider the k=0 term. Then $\ell = 0, j = 0, 1$. We have:

$$a_{\gamma o} = C^o_{n,o,1} \int_\gamma \frac{U_1(x, \gamma x) d\sigma}{\sqrt{\det \text{Hess}\ (f\gamma)_\sigma}} + C^o_{n,o,o} \int_\gamma \frac{\text{Hess}(f_\gamma)^{-1}_\sigma(\frac{\partial}{\partial y}, \frac{\partial}{\partial y})[(U_0(x, \gamma x) J_\gamma(y, s)] d\sigma}{\sqrt{\det \text{Hess}\ (f_\gamma)_\sigma}} \tag{38}$$

The Hessian determinants above are easily seen to be constant in σ. We follow the discussion in [Su]. The linear Poincaré map can be written

$$P_\gamma = \begin{pmatrix} A & B \\ C & D \end{pmatrix} = \begin{pmatrix} \langle \nu_i, U_j(1) \rangle, & \langle \nu_i, V_j(1) \rangle \\ \langle \nu_i, \nabla U_j(1) \rangle, & \langle \nu_i, \nabla V_j(1) \rangle \end{pmatrix}$$

where U_i, V_i are Jacobi fields along γ with initial condition $\left\{ \begin{array}{ll} U_i(0)=\nu_i & V_i(0)=0 \\ \nabla V_i(0)=0 & \nabla V_i(0)=v_i \end{array} \right.$ where $\nu, \ldots, v_s$ is an orthonormal basis for the normal bunded class γ.

Then

$$|\det B|^{-\frac{1}{2}} = \Theta^{-\frac{1}{2}}(x, \gamma x) \quad (x \in \gamma)$$

and

$$\Theta^{-\frac{1}{2}}(x, \gamma x) \cdot |\det \text{Hess } f_\gamma|^{-\frac{1}{2}} = |\det(I - P_\gamma)|^{-\frac{1}{2}} .$$

Since $|\det(I - P_\gamma)|^{-\frac{1}{2}}$ is independent of $x \in c_\gamma$, the Hessian determinants

above are constant.

Hence, this simplifies to:

$$(39) \quad a_{\gamma o} = C^o_{n,o,1} \frac{1}{|\det(I - P_\gamma)|^{\frac{1}{2}}} \times \int_0^{L_\gamma} \int_0^{L_\gamma} \Theta^{\frac{1}{2}}((\sigma_1, 0), (\sigma_2, 0))(\Delta_2 \Theta^{-\frac{1}{2}})((\sigma_1, 0), (\sigma_2, 0)) \cdot d\sigma_1 d\sigma_2 + \frac{C^o_{n,o,o}}{|\det(I - P_\gamma)|^{\frac{1}{2}}} \int_\gamma \Theta^{\frac{1}{2}}((0,0), (\sigma, 0)) \text{Hess}(f_\gamma)^{-1}_\sigma (\Theta^{-\frac{1}{2}} J_\gamma)((0,0), (\sigma, 0) d\sigma .$$

These are very preliminary formulae for the wave invariants. As they stand, these expressions involve various objects $\{y, J_\gamma, \text{Hess}(f_\gamma)_\sigma(\frac{\partial}{\partial y}, \frac{\partial}{\partial y})\}$ which are defined in terms of Morse coordinates that have not been constructed explicitly. We now give a modification of the method which is effectively computable.

2.4.2. *Effective Method.* Instead of using the Morse coordinates above, we will carry out the calculation in Fermi normal coordinates along γ. Thus, we consider the exponential map $\exp : N_{[\gamma]} \to \tilde{M}/\Gamma_\gamma$ on the normal bundle along γ; it is a diffeomorphism from some ball bundle $|\nu| < \varepsilon$ to a tubular neighborhood $T_\epsilon(\gamma)$.

We fix a unit speed parametrization $\gamma(\sigma)$ of γ, and let

$$\{e_o(\sigma); e_1(\sigma) \dots e_{n-1}(\sigma)\}$$

be a parallel frame along γ with $e_o = \gamma'$ and with $\{e_1(\sigma) \dots e_{n-1}(\sigma)\}$ a normal frame. The Fermi coordinates $(\sigma, \nu_1, \dots, \nu_{n-1})$ are defined by

$$p = \exp_{\gamma(\sigma)}(\nu_1 e_1(\sigma) + \dots + e_{n-1}(\sigma)).$$

For short we will write $\exp_\sigma \nu = p$.

In these coordinates, the f_γ takes the form $f_\gamma(\exp_\sigma \nu) = r^2(\exp_\sigma \nu, \gamma \exp_\sigma \nu)$. Its Taylor expansion around $\nu = 0$ is given by,

$$(40) \qquad f_\gamma(\exp_\sigma \nu) = L^2_\gamma + \text{Hess}(f_\gamma)_\sigma(\nu, \gamma) + g_\gamma(\sigma, \nu)$$

where $g_\gamma(\sigma,\nu) = O(|\nu|^3)$. By the binomial theorem, we have:

$$(41)\quad \begin{aligned}(f_\gamma - L_\gamma^2 + iO)^\lambda &= (\ \mathrm{Hess}(f_\gamma)_\sigma(\nu,\nu) + iO + g(\sigma,\nu))^\lambda \\ &= \sum_{p=0}^{\infty} \binom{\lambda}{n} (\ \mathrm{Hess}(f_\gamma)_\sigma(\nu,\nu) + iO)^{\lambda-p} g(\sigma,\nu)^p\ .\end{aligned}$$

Substituting (3.21) into (3.12a) and ignoring the II-term, we get (with $\lambda = s + j - \frac{n+1}{2} - \ell$):

$$(42)\quad \begin{aligned} I_{[\gamma]}^{k,\ell,j}(s) = \sum_{p=0}^{\infty} \binom{\lambda}{p} \int_0^{L_\gamma} \int_{|\nu|<\varepsilon} (\ \mathrm{Hess}(f_\gamma)_\sigma(\nu,\nu) + iO)^{\lambda-p} \cdot g_\gamma^p(\sigma,\nu) \\ \cdot \chi_\varepsilon W_j(\exp_\sigma \nu, \gamma \exp_c \nu) \cdot J_{[\gamma]}(\sigma,\nu) d\sigma d\nu\ ,\end{aligned}$$

where d vol $= J_{[\gamma]}(\sigma,\nu)d\sigma d\nu$. As in [Hö I, §7], the remainder term g_γ has been absorbed into the amplitude.

By the above, $(\mathrm{Hess} f_\gamma)_\sigma(\nu,\nu) + iO)^{s+j-\frac{n+1}{2}-p-\ell}$ has simple poles only among the points

$$s + j - \frac{n+1}{2} - \ell - p = -\frac{1}{2}(n-1) - q \qquad (q = 0,1,2,\dots).$$

If $s = 0$, this allows for poles at the infinite number of points

$$j = \ell + p + 1 - q \qquad (p,q \in \mathbb{N}, \quad \ell = 0,1,\dots,k).$$

However, the residue vanishes unless $2j + p \le 2(\ell+1)$. Indeed, writing the p-th term in (3.22) as $I_{[\gamma]}^{k,\ell,j,p}$, we have,

$$(43)\quad \begin{aligned}\mathrm{Res}_{s=0} I_{[\gamma]}^{k,\ell,j,p}(s) = C_{k,\ell,j,p} \cdot \int_0^{L_\gamma} \frac{1}{\sqrt{\det\ \mathrm{Hess}(f_\gamma)_\sigma}} \\ \times (\ \mathrm{Hess}(f_\gamma)_\sigma^{-1}(\frac{\partial}{\partial\nu},\frac{\partial}{\partial\nu}))^{\ell+p+1-j}(g_\gamma^p W_j J_{[\gamma]})ds\end{aligned}$$

for a certain constant $C_{k,\ell,j,p}$. Since g_γ^p vanishes to order $3p$ and the Hessian operator has order $2(\ell+p+1-j)$, it is evident that the residue is zero unless $2(\ell+p+1-j) \ge 3p$. Hence, only a finite number of parameters (j,p) lead to non-zero contributions to $\mathrm{Res}_{s=0} I_{[\gamma]}^{(k)}(s)$.

Proposition 2.3. *With the above notation and assumptions, there are universal constants $C_{k,\ell,j,p,n}$ such that the kth wave invariant is given by:*

$$\begin{aligned} a_{\gamma k} = \sum_{\ell=0}^{k} \sum_{(j,p):p+2j\le 2\ell+2} a_\ell(L) C_{k,\ell,j,p,n} \int_0^{L_\gamma} \frac{1}{\sqrt{\det\ Hess(f_\gamma)_\sigma}} \\ \times (\ Hess(f_\gamma)_\sigma^{-1}(\frac{\partial}{\partial\nu},\frac{\partial}{\partial\nu}))^{\ell+p+1-j}(g_\gamma^p W_j J_{[\gamma]})ds.\end{aligned}$$

Let us use this procedure to recalculate $a_{\gamma 0}$.
We have $k = \ell = 0$, so

$$(44)\quad \begin{aligned} a_{\gamma 0} = \mathrm{Res}_{s=0} I_{[\gamma]}^{(0)}(s) &= \sum_{j=0}^{\infty} e^{\frac{i+1}{2}(\frac{n+1}{2}-j)} \mathrm{Res}_{s=0} I_{[\gamma]}^{(0,0,j)}(s) \\ &= \sum_{j,p=0}^{\infty} C_{j,n,p}\ \mathrm{Res}_{s=0} I_{[\gamma]}^{(0,0,j,p)}(s)\ , \end{aligned}$$

with $C_{j,n,p}$ certain universal constants and with

$$(45)\quad \mathrm{Res}_{s=0} I_{[\gamma]}^{(0,0,j,p)}(s) = \int_0^{L_\gamma} \frac{1}{\sqrt{\det\ \mathrm{Hess}(f_\gamma)_\sigma}} (\ \mathrm{Hess}(f_\gamma)_\sigma^{-1}(\frac{\partial}{\partial \nu}, \frac{\partial}{\partial \nu}))^{p+1-j}(g_\gamma^p W_j J_{[\gamma]}) ds\ .$$

As above we must have $2j + p \leq 2$. With some simplifications noted above, we get that $\mathrm{Res}_{s=0} I_{[\gamma]}^{(0,0,j,p)}(s)$ is a sum of terms of the form $\frac{1}{|I-P_\gamma|^{\frac{1}{2}}} \int_\gamma \Theta_\gamma^{\frac{1}{2}}(\sigma, 0) \cdot P^{(j,p)}(\sigma) d\sigma$ for certain functions $P^{(j,p)}(\sigma)$. Here and below,

$$\Theta_\gamma(\sigma, \nu) = \Theta(\exp_\sigma \nu, \gamma \exp_\sigma \nu).$$

Let us also abbreviate the Hessian operator $\mathrm{Hess}(f_\gamma)_\sigma(\frac{\partial}{\partial \nu}, \frac{\partial}{\partial \nu})^{-1}$ by $\mathcal{H}_\sigma$ and $W_j(\exp_\sigma \nu, \gamma \exp_\sigma \nu)$ by $U_{j\gamma}(\sigma, \nu)$. Then we have (for certain constants $C^{j,p}$):

$$(46)\quad \begin{aligned} &(i)(j = p = 0) : P^{(0,0)}(\sigma) = \mathcal{H}(U_{0\gamma} J_{[\gamma]})|_{\nu=0} \\ &(ii)(j = 0, p = 1) : P^{(0,1)}(\sigma) = C^{(0,1)} \mathcal{H}^2(g_\gamma U_{0\gamma} J_{[\gamma]})|_{\nu=0} \\ &(iii)(j = 0, p = 2) : P^{(0,2)}(\sigma) = C^{(0,2)}(\mathcal{H}^3(g_\gamma^2)) U_{0\gamma}|_{\nu=0} \\ &(iv)(j = 1, p = 0) : P^{(1,0)}(\sigma) = C^{(1,0)} U_{1\gamma}(\sigma, 0). \end{aligned}$$

Note that in (ii) at least three derivates must fall on g_γ, and all of them must in (iii). In the resulting expressions, we can of course replace g_γ by f_γ. Hence we have by this point an effective calculation of $a_{\gamma o}$ in the form:

$$(47)\quad a_{\gamma o} = I + II + III + IV$$

with

$$I := C_{n,o,0,o}^{o} \frac{1}{|\det(I - P_\gamma)|^{\frac{1}{2}}} \int_\gamma \Theta_\gamma^{\frac{1}{2}}(\sigma) \{ \Theta_\gamma^{-\frac{1}{2}} \mathrm{Hess}(f_\gamma)_\sigma^{-1}(J_\gamma)((0,0), (\sigma, 0) \\ + \mathrm{Hess}(f_\gamma)_\sigma^{-1}(\Theta_\gamma^{-\frac{1}{2}}((0,0), (\sigma, 0) d\sigma$$

$$II := C_{n,o,o,1}^{o} \frac{1}{|\det(I - P_\gamma)|^{\frac{1}{2}}} \int_0^{L_\gamma} \int_0^{L_\gamma} \Theta_\gamma^{\frac{1}{2}}(s\sigma_1) \{ \mathrm{Hess}(f_\gamma)_\sigma^{-1} \}^2 (g_\gamma \Theta_\gamma^{-\frac{1}{2}} J_{[\gamma]})|_{\nu=0} ds$$

$$III := C^o_{n,o,o,1} \frac{1}{|\det(I - P_\gamma)|^{\frac{1}{2}}} \int_0^{L_\gamma} \int_0^{L_\gamma} \Theta_\gamma^{\frac{1}{2}}(s\sigma_1)\{\mathrm{Hess}(f_\gamma)_\sigma^{-1}\}^3(f_\gamma)\Theta_\gamma^{-\frac{1}{2}}|_{\nu=0} ds$$

$$IV := C^o_{n,o,1} \frac{1}{|\det(I - P_\gamma)|^{\frac{1}{2}}} \int_0^{L_\gamma} \int_0^{L_\gamma} \Theta_\gamma^{\frac{1}{2}}(\sigma_1)(\Delta_2 \Theta^{-\frac{1}{2}})((\sigma_1, 0), (\sigma_2, 0)) \cdot d\sigma_1 d\sigma_2 .$$

2.4.3. *Extensions.* We observe that:

(1) Essentially the same method is valid if g is allowed to be cleanly degenerate in the sense that the geodesic flow has only clean fixed point sets. In this case, $\tilde{M}/\Gamma_\gamma$ can be a submanifold of $\tilde{M}/\Gamma_\gamma$ of dimension greater than one. With the obvious modifications that γ should be replaced by $\tilde{M}_\gamma/\Gamma_\gamma$, $\mathrm{Hess}(f_\gamma)$ by the normal Hessian to $\tilde{M}/\Gamma_\gamma$ and $n-1$ by the co-dimenison of $\tilde{M}_\gamma/\Gamma_\gamma$, all the previous formulae remain valid. We note that $\tilde{M}_\gamma/\Gamma_\gamma = Ax(\gamma) = \mathcal{G}_{[\gamma]}$, so we could express the results in terms of integrals over any of these spaces [cf.[Su]).

(2) The global hypothesis that the metric is free of conjugate points can also be relaxed: the calculation of the $a_{\gamma k}$'s only requires a microlocal parametrix in a neighborhood of γ and hence only requires microlocal assumptions on γ. The simplest such assumption is that (M, g) has no conjugate points near γ, in a sense made precise below. In this case, a (micro-local) Hadamard-Riesz parametrix can be constructed near γ, and the formulae above for the $a_{\gamma k}$'s remain precisely the same.

3. Birkhoff normal forms: $\partial M = \emptyset$

In this section we will briefly describe the notion of a quantum Birkhoff normal form of $\sqrt{\Delta}$ around a non-degenerate closed geodesic γ. We only explain enough so that the reader may contrast the expressions for the wave invariants in terms of normal coefficients with those of the previous section. We also describe the algorithm for putting $\sqrt{\Delta}$ into normal form. In the following section we will implement the algorithm in the case of a Sturm-Liouville operator $D^2 + q$ on $[0, \pi]$ so that the reader can see how it works in a simple case. For further details we refer to [Z.4].

Roughly speaking, to put $\sqrt{\Delta}$ into microlocal normal form around γ is to express it as a function $F(D_s, \hat{I}_1, \ldots, \hat{I}_n)$ ($n = \dim M - 1$) of the derivative $D_s = \frac{\partial}{i\partial s}$ along γ and of local 'action variables' $\hat{I}_j$ along a transversal to γ. These action operators generate a maximal abelian subalgebra $\mathcal{A}$ of pseudodifferential operators on the local model space. In the case of a manifold without boundary, which may be taken to be the normal bundle $N_\gamma \sim S^1\mathbb{R}^n$ of γ. The definition of the $\hat{I}_j$ depends on the spectrum of P_γ.

3.0.4. *The model.* The wave invariants $a_{\gamma k}$ associated to a closed geodesic γ depend only on the germ of the metric in a neighborhood of γ. We may identify this neighborhood with the normal bundle N_γ by means of the exponential map. Thus, the model space is the cylinder $S^1_L \times \mathbb{R}^n$, where $S^1_L = \mathbb{R}/L\mathbb{Z}$. We use the coordinates $(s, y) \in S^1_L \times \mathbb{R}$ and dual symplectic coordinates (σ, η) for $T^*(S^1_L \times \mathbb{R})$. We henceforth assume the length L of γ equals 2π.

The normal form of $\sqrt{\Delta}$ is a microlocal normal form near γ, i.e. it is only well-defined in a conic neighborhood $V = \{(s, \sigma, y, \eta) : |y| \leq \epsilon, |\eta| \leq \epsilon\sigma\}$ of $\mathbb{R}^+\gamma$. Since $\sigma > 0$ in V the quantization of this cone is a subspace of the Hardy space $H^2(S^1) \otimes L^2(\mathbb{R}^n)$ spanned by functions $e^{iks}f(y)$ with $k \geq 0$.

We now introduce analytic objects on the model space. The algebra $\mathcal{E} :=< Y_1, ..., Y_n, D_1, ..., D_n >$ of polynomial differential operators on the 'tranvserve space' $\mathbb{R}^n$ is generated by the elements y_j = "multiplication by y_j" and by $D_{y_j} = \frac{\partial}{i\partial y_j}$. The symplectic algebra $sp(n, \mathbb{C})$ is represented by homogeneous quadratic polynomials in Y_j, D_j. A choice of action operators is equivalent to a choice of a maximal abelian subalgebra $\mathcal{I} :=< \hat{I}_1, ..., \hat{I}_n >$ of $sp(n, \mathbb{C})$. The appropriate choice of $\mathcal{I}$ depends on type of closed geodesic γ, specifically on the spectrum of P_γ. For each pair $e^{i\pm\alpha}$ of elliptic eigenvalues, one introduces the elliptic harmonic oscillator $\hat{I}_j := \alpha_j(D^2_{y_j} + y_j^2)$. In the real hyperbolic directions, i.e. in real 2-planes where P_γ has a pair of inverse real eigenvalues $e^{\pm\mu_j}$, one introduces hyperbolic action operators $\hat{I}_j = \mu_j(y_j D_{y_j})$. In the complex hyperbolic directions, i.e. in complex 2-planes where P_γ has eigenvalues $e^{\pm\mu + i\pm\nu}$ one introduces actions in two y-variables y_j, y_{j+1} of the form $(y_j D_{y_j} + y_{j+1} D_{y_{j+1}}) + \alpha_j(y_{j+1} D_{y_j} - y_j D_{y_{j+1}})$. In short:

Eigenvalue type	Classical Normal form	Quantum normal form
(i) Elliptic type $\{e^{i\pm\alpha}\}$	$I^e = \frac{1}{2}\alpha(\eta^2 + y^2)$	$\hat{I}^e := \frac{1}{2}\alpha(D^2_y + y^2)$
(ii) Real hyperbolic type $\{e^{\pm\lambda}\}$	$I^h = 2\lambda y\eta$	$\hat{I}^h := \lambda(yD_y + D_y y)$
(iii) Complex hyperbolic (or loxodromic type) $\{e^{\pm\mu+i\pm\nu}\}$	$I^{ch,Re} = 2\mu(y_1\eta_1 + y_2\eta_2)$ $I^{ch,Im} = \nu(y_1\eta_2 - y_2\eta_1)$	$\hat{I}^{ch,Re} = \mu(y_1 D_{y_1} + D_{y_1} y_1 + y_2 D_{y_2} + D_{y_2} y_2)$, $\hat{I}^{ch,Im} = \nu(y_1 D_{y_2} - y_2 D_{y_1})$

A maximal abelian subalgebra of the full pseudodifferential algebra on $S^1 \times \mathbb{R}$ is the subalgebra $\mathcal{A} :=< D_s, \hat{I}_1, \ldots, \hat{I}_n >$ An orthonormal basis of $L^2(S^1 \times \mathbb{R}^n)$ of joint model eigenfunctions of $\mathcal{A}$ is is given by

$$\varphi^o_{kq}(s, y) := e^{iks} \otimes \gamma_q(y), \tag{48}$$

where $\gamma_q, q \in \mathbb{N}^n$ is the nth Hermite function. That is, $\gamma_o(y) = \gamma_{iI}(y) := e^{-\frac{1}{2}|y|^2}$ and $\gamma_q := C_q A_1^{*q_1} ... A_n^{*q_n} \gamma_o(q \in \mathbb{N}^n\})$, with $C_q = (2\pi)^{-n/2}(q!)^{-1/2}$, $q! = q_1!...q_n!$ and where $A_j = (D_j + iq_j)$ is the usual annihilation operator.

The normal form theorem is the following:

THEOREM **3.1.** *There exists a microlocally elliptic Fourier Integral operator W from a conic neighborhood V of $\mathbb{R}^+\gamma$ in $T^*N_\gamma - 0$ to a conic neighborhood of $T^*_+S^1$ in $T^*(S^1 \times \mathbb{R}^n)$ such that:*

$$W\sqrt{\Delta_\psi}W^{-1} :\equiv \mathcal{D} \equiv D_s + \frac{1}{L}H_\alpha + \frac{\tilde{p}_1(\hat{I}_1,\ldots,\hat{I}_n)}{D_s} + \frac{\tilde{p}_2(\hat{I}_1,\ldots,\hat{I}_n)}{D_s^2} + \ldots + \frac{\tilde{p}_{k+1}(\hat{I}_1,\ldots,\hat{I}_n)}{D_s^{k+1}} + \ldots$$

where the numerators $\tilde{p}_j(\hat{I}_1,...,\hat{I}_n)$ are polynomials of degree $j+1$ in the variables $\hat{I}_1,...,\hat{I}_n$, where W^{-1} denotes a microlocal inverse to W in V. The k-th remainder term lies in the space $\bigoplus_{j=0}^{k+2} O_{2(k+2-j)}\Psi^{1-j}$.

The coefficients of the polynomials $\tilde{p}_j(\hat{I}_1,...,\hat{I}_n)$ are the quantum Birkhoff normal form invariants. We denote them by $B_{\gamma jk}$. Here, $O_m\Psi^k$ is the space of transverse pseudodifferential operators of order k (in the y-variables) which vanish to order m at $y = 0$. Thus the error term is bigraded by pseudodifferential order and by order of vanishing. The remainder is small if in some combination it has a low pseudodifferential order or a high vanishing order at γ. This is a useful remainder estimate since a given wave invariant a_γ only involves a finite part of the jet of the metric and only a finite part of the complete symbol of $\sqrt{\Delta}$. Indeed, $a_{\gamma k}$ depends only on $\sqrt{\Delta}$ modulo $\bigoplus_{j=0}^{k+2} O_{2(k+2-j)}\Psi^{1-j}$.

For the reader's convenience we reproduce here the formulae for the wave invariants which follow from the method of normal forms. Undefined notation and terminology will be summarized below.

THEOREM **3.2.** *Let γ be a strongly non-degenerate closed geodesic. Then $a_{\gamma k} = \int_\gamma I_{\gamma;k}(s;g)ds$ where:*
(i) $I_{\gamma;k}(s;g)$ is a homogeneous Fermi–Jacobi–Floquet polynomial of weight $-k-1$ in the data $\{y_{ij}, \dot{y}_{ij}, D^m_{s,y}g\}$ with $m = (m_1,\ldots,m_{n+1})$ satisfying $|m| \leq 2k+4$;
(ii) The degree of $I_{\gamma;k}$ in the Jacobi field components is at most $6k+6$;
(iii) At most $2k+1$ indefinite integrations over γ occur in $I_{\gamma;k}$;
(iv) The degree of $I_{\gamma;k}$ in the Floquet invariants β_j is at most $k+2$.

Let us define the term 'Fermi-Floquet-Jacobi polynomial'. First, we write the metric coefficients g_{ij} relative to Fermi normal coordinates (s, y) along γ. The vector fields $\frac{\partial}{\partial s}, \frac{\partial}{\partial y_j}$ and their real linear combinations will be referred to as Fermi normal vector fields along γ and contractions of tensor products of the $\nabla^m R$'s with these vector fields will be referred to as *Fermi curvature polynomials.* The m-th jet of g along γ is denoted by $j^m_\gamma g$, the curvature tensor by R and its covariant derivatives by $\nabla^m R$. Such polynomials will be called *invariant* if they are invariant under the action of $O(n)$ in the normal spaces. Invariant contractions against $\frac{\partial}{\partial s}$ and against the Jacobi eigenfields $Y_j, \overline{Y}_j$, with coefficients given by invariant polynomials in the components y_{jk}, are called *Fermi–Jacobi polynomials.* We will also use this term for functions on γ given by repeated indefinite integrals over γ of such FJ polynomials. Finally, FJ polynomials whose coefficients are given by polynomials in the Floquet invariants $\beta_j = (1 - e^{i\alpha_j})^{-1}$ are *Fermi–Jacobi–Floquet* polynomials.

The 'weights' referred to above describe how the various objects scale under $g \to \epsilon^2 g$. Thus, the variables g_{ij}, $D^m_{s,y} g_{ij}$ (with $m := (m_1, \ldots, m_{n+1})$), $L := L_\gamma, \alpha_j, y_{ij}, \dot{y}_{ij}$ have the following weights: $\mathrm{wgt}(D^m_{s,y} g_{ij}) = -|m|$, $\mathrm{wgt}(L) = 1$, $\mathrm{wgt}(\alpha_j) = 0$, $\mathrm{wgt}(y_{ij}) = \frac{1}{2}, \mathrm{wgt}(\dot{y}_{ij}) = -\frac{1}{2}$. A polynomial in this data is homogeneous of weight s if all its monomials have weight s under this scaling. Finally, τ denotes the scalar curvature, τ_ν its unit normal derivative, $\tau_{\nu\nu}$ the Hessian $\mathrm{Hess}(\tau)(\nu, \nu)$; Y denotes the unique normalized Jacobi eigenfield, $\dot{Y}$ its time-derivative and δ_{j0} the Kronecker symbol (1 if $j = 0$ and otherwise 0).

For instance, in dimension 2 (where there is only one Floquet invariant β) the residual wave invariant $a_{\gamma 0}$ is given by:

$$a_{\gamma 0} = \frac{a_{\gamma,-1}}{L^\#}[B_{\gamma 0;4}(2\beta^2 - \beta - \frac{3}{4}) + B_{\gamma 0;0}] \tag{49}$$

where:

(a) $a_{\gamma,-1}$ is the principal wave invariant;

(b) $L^\#$ is the primitive length of γ; σ is its Morse index; P_γ is its Poincaré map;

(c) $B_{\gamma 0;j}$ has the form:

$$B_{\gamma 0;j} = \frac{1}{L^\#}\int_o^{L^\#} [a\,|\dot{Y}|^4 + b_1\,\tau|\dot{Y}\cdot Y|^2 + b_2\,\tau \mathrm{Re}\,(\bar{Y}\dot{Y})^2$$
$$+ c\,\tau^2|Y|^4 + d\,\tau_{\nu\nu}|Y|^4 + e\,\delta_{j0}\tau]ds$$
$$+\frac{1}{L^\#}\sum_{\substack{m+n=3;\\ 0\le m,n\le 3}} C_{1;mn}\frac{\sin((n-m)\alpha)}{|(1-e^{i(m-n)\alpha})|^2}\left|\int_o^{L^\#}\tau_\nu(s)\bar{Y}^m\cdot Y^n(s)ds\right|^2$$
$$+\frac{1}{L^\#}\sum_{\substack{m+n=3;\\ 0\le m,n\le 3}} C_{2;mn}\,\mathrm{Im}\left\{\int_o^{L^\#}\tau_\nu(s)\bar{Y}^m\cdot Y^n(s)\left[\int_o^s \tau_\nu(t)\bar{Y}^n\cdot Y^m(t)dt\right]ds\right\}$$

for various universal (computable) coefficients.

The algorithm given in [Z.1] for putting Δ into normal form is somewhat lengthy. In [Z.5] we also give a modified algorithm for putting a Dirichlet Laplacian on a bounded plane domain into normal around an elliptic bouncing-ball orbit. Instead of attempting to summarize the method, we will just illustrate it in the simple case of a Sturm-Liouville operator with Dirichlet boundary conditions on an interval. This case is 'all orbit and no transversal' so we will not see the emergence of transverse harmonic oscillators and the normal form is global rather than microlocal. Despite such exceptional features, this case does convey the flavor of the normal form algorithm in a simple setting. Moreover, we have not seen Sturm-Liouville operators treated in quite this way in the literature.

4. Birkhoff normal form of a Sturm-Liouville operator

The purpose of this section is put a Sturm-Liouville operator $L = D^2 + q$ on a finite interval $[0, \pi]$ (with Dirichlet boundary conditions) into a global Birkhoff normal form. Thus we are interested in the eigenvalue problem:

$$\begin{aligned} &L\,\hat{s}(\lambda_k, x) = \lambda_k^2\,\hat{s}(\lambda_k, x) \\ &\hat{s}(\lambda_k, 0) = \hat{s}(\lambda_k, \pi) = 0, \qquad \langle \hat{s}(\lambda_k, \cdot), \hat{s}(\lambda_j, \cdot)\rangle = \delta_{jk}. \end{aligned} \tag{50}$$

Roughly speaking, we wish to conjugate L by a microlocally elliptic Fourier integral operator W near $[0, \pi]$ into the form

$$W^{-1}LW = D^2 + c_0 + \frac{c_2}{|D|^2} + \cdots, \tag{51}$$

where c_j are constants with $c_0 = \frac{1}{\pi}\int_0^\pi q\,dx$, and where $|D| = \sqrt{D^*D}$ is the self-adjoint square root of the 'Laplacian' with Dirichlet boundary conditions. That is, $|D| \sin ks = |k| \sin ks$. It is a theorem, not a definition, that all of the odd terms vanish. The notation $\hat{s}(\lambda_k, x)$ for normalized eigenfunctions is taken from [M].

This Birkhoff normal form is global because there is only one (unit speed) closed 'geodesic' up to time reversal. Indeed, the bicharacteristic flow of L is the constant speed linear motion along $[0, \pi]$ with elastic reflection at the boundary, i.e. $[0, \pi]$ is a one-dimensional billiard table with precisely two unit speed periodic billiard trajectories.

The conjugation to Birkhoff normal form is closely related (in its results, not in its methods) to the classical work of Gelfand-Levitan and Marchenko, see [LS] [M]. Their work is partly based on the construction of 'transformation operators', which are essentially intertwining operators to normal form. The main result of their work on regular Sturm-Liouville operators (with smooth potentials) is the following:

Theorem 4.1. *([M], Theorem 1.5.1; [LS], Chapter 5) Let $q \in C^\infty$. Then the square roots λ_k of the Dirichlet eigenvalues of $L = D^2 + q$ have the complete asymptotic expansions:*

$$\lambda_k = k + \frac{a_1}{2k} + \frac{a_3}{(2k)^3} + \cdots + \frac{a_{2j+1}}{(2k)^{2j+1}} + \cdots$$

with $a_1 = \frac{1}{\pi}\int_0^\pi q(t)dt$. *Moreover the eigenfunctions* $s(\lambda_k, x)$ *with initial values* $s(\lambda, 0) = 0, s'(\lambda_k, 0) = 1$ *have the form*

$$s(\lambda_k, x) = \frac{\operatorname{Im} y(\lambda_k, x)}{\omega(\lambda_k, 0)}, \qquad y(\lambda, x) = e^{i\lambda x + \int_0^x \sigma(\lambda, t)dt},$$

$$\sigma(\lambda, x) \sim \sum_{k=1}^{\infty} \frac{\sigma_k(x)}{(2i\lambda)^k}, \quad \omega(\lambda, x) = W(y(\lambda, x), y(-\lambda, x))$$

where W *is the Wronskian.*

This leads to the equations

$$\sigma_1(x) = q(x), \quad \sigma_2(x) = -q'(x), \quad \sigma_3(x) = q''(x) - q'(x)^2$$

$$\sigma_{k+1} = -\sigma_k'(x) - \sum_{j=1}^{k-1} \sigma_{k-j}(x)\sigma_j(x).$$

One easily sees that if q is real, then all of the σ_j are real.

We are going to reprove (or re-interpret) this result by means of a conjuation to normal form. Rather than use Marchenko's transformation operators we construct an interwining operator by the method of Birkhoff normal forms. The details are closer to those of Marchenko's construction of the quasimodes $y(\lambda, x)$ and $s(\lambda, x)$ (see [M], sections 4-5; [LS], §5.1). We will compare the two methods below. The main difference is that we use series in k rather than in an unknown λ as in [M][LS]. To relate the various expressions one merely has to substitute in the expansion in Theorem (4.1) for λ_k.

4.1. Conjugation to normal form.

In outline our normal form algorithm involves three steps. First, we 'rescale' the operator so that it admits a semiclassical expansion in powers of a small parameter $h = \lambda^{-1}$. Second, we use semiclassical pseudodifferential operators to conjugate the semiclassical expansion into normal form. Finally, we 'quantize' the parameter h to a discrete set of values and glue together semiclassical intertwining operators into one global intertwining operator to normal form.

To motivate these steps, we observe that the desired normal form is a polyhomogeneous symbol in $|D|$. Hence the eigenfunctions of the normal form are just the sines $\{\sin kx\}$. The intertwining operator W to normal form should then have the properties that

- $W \sin kx = s(\lambda_k, x)$ is an (asymptotic) eigenfunction for L;
- $W \sin kx$ vanishes at $x = 0, \pi$.

Thus, $W \sin kx = C_k \hat{s}(\lambda_k, x)$. To construct W we construct an operator W_k^+ such that $W_k^+ e^{ikx}$ is an eigenfunction and such that $\operatorname{Im} W_k^+ e^{ikx}$ vanishes at $x = 0, \pi$. We then put $W \sin kx = \operatorname{Im} W_k^+ e^{ikx}$. To ensure that $\operatorname{Im} W_k^+ e^{ikx}$ vanishes at $x = 0, \pi$ it suffices that $W_k^+(0), W_k^+(\pi)$ commute with complex conjugation C. In our 1D (one-dimensional) setting, W_k^+ will simply be a multiplication operator and the condition is that

$$\mathrm{Im}\, W_k^+(0) = \mathrm{Im}\, W_k^+(\pi) = 0. \tag{52}$$

4.1.1. *Scaling L.* The first step is to semiclassically scale L, which in the 1D setting simply amounts to rewriting the eigenvalue equation

$$LW_k^+ e^{ikx} = \lambda_k^2 W_k^+ e^{ikx}$$

in the form

$$[e^{-ikx} L e^{ikx}] W_k^+ = \lambda_k^2 W_k^+.$$

We then define the rescaled Sturm-Liouville operator by

$$L_k := [e^{-ikx} L e^{ikx}] = k^2 + 2kD + L. \tag{53}$$

In higher dimensions we also have a transverse space $\mathbb{R}^n_y$ and scaling also involves the sending $y \to \sqrt{k} y$.

4.1.2. *Conjugation.* We then wish to conjugate L_k into semiclassical normal form. At this point, we have to decide what Hilbert space our intertwining operators W_k should act on. One's first reaction might be to define them on the interval $[0, \pi]$ since that is where L and the normal form lives. However, we would like to employ the machinery of pseudodifferential and Fourier integral operators (FIOs), and that machinery is much simpler on manifolds without boundary. Moreover, our very definition of W_k involves $W_k^+ e^{ikx}$ and e^{ikx} does not live (as a smooth function) on $[0, \pi]$. This suggests that we 'double' $[0, \pi]$ and work on the boundaryless manifold $S^1 = \mathbb{R}/2\pi\mathbb{Z}$. For this to work we need $q(x)$ to extend to a smooth function on some larger interval $(-\epsilon, \pi + \epsilon)$. We are only interested in what happens on $[0, \pi]$ so we cut off L to an operator φL where φ is a smooth cutoff on S^1 equalling one on the interval $[0, \pi]$ and supported in $(-\epsilon_0, \pi + \epsilon_0)$ for some $\epsilon_0 > 0$. We will not indicate φ in the notation; later we will use explicit cutoff functions. We also note that the operator $A : L^2[0, \pi] \to L^2(S^1)$ of odd extension acts as the identity on the sines $\{\sin kx\}$ and hence embeds the model Dirichlet problem into the odd suspace of $L^2(S^1)$ (under the 'reflection' $r(e^{ix}) = e^{-ix}$.)

Our object now is to construct a semiclassical pseudodifferential operator W_k^+ on S^1 such that $W_k^+ e^{ikx}$ is an (asymptotic) eigenfunction of L on $[0, \pi]$ and such that $\mathrm{Im}\, W_k^+ e^{ikx} = 0$ for $x = 0, L$. We assume here that $k > 0$ so that $e^{ikx} \in H^+(S^1)$, the Hardy space of smooth functions whose Fourier series only contain positive frequencies. The standard method (e.g. [Sj] [Z.1]) is to construct W_k^+ as an asymptotic product

$$W_k^+ = \Pi_{j=1}^{\infty} e^{ik^{-j} Q_j(x)} \tag{54}$$

of semiclassical pseudodifferential operators $e^{ik^{-j} Q_j(x)}$. In the 1D case, the exponents are simply multiplication operators (and hence, so is W_k^+). In general Q_j can have non-trivial real and imaginary parts. The boundary condition is that $\operatorname{Re} Q_j(0) = \operatorname{Re} Q_j(\pi) = 0$ for all j. In higher dimensions, the Q_j are differential operators on the transverse space.

We now rewrite the desired properties of W_k^+ in operator form:

$$(55)\qquad \begin{aligned} &W_k^{+*}\, L_k\, W_k^+|_0 = k^2 + c_0 + \tfrac{c_2}{k^2} + \dots \\ &\operatorname{Re} Q_k(x) = 0, \quad x = 0, \pi. \end{aligned}$$

In the first line, the notation $|_0$ means the restriction of the operator to constant functions. For instance, $D|_0 = 0$. The reason for this is that we have rewritten the eignvalue problem $L \;\; W_k^+ e^{ikx} = \lambda_k^2 W_k^+ e^{ikx}$ in the form $W_k^{+*}\, L_k\, W_k^+ 1 \sim \lambda_k$, i.e. the operator is only acting on the constant function 1. In higher dimensions, $|_0$ would represent the restriction to functions on the transverse space.

In the 1D case, all the Q_k's commute and we have

$$(56)\qquad W_k^+(x) = e^{i \sum_{k=1}^{\infty} k^{-j} Q_j(x)}.$$

Thus, conjugating L_k to normal form is precisely the same as constructing a function $W_k^+(x)$ satisfying

$$(57)\qquad L\, W_k^+(x)\, e^{ikx} = F(k)^2\, W_k^+(x)\, e^{ikx}, \qquad F(k) \sim k + \frac{a_1}{k} + \frac{a_3}{k^3} + \dots$$

and satisfying the boundary conditions above (55). Clearly the construction of our intertwining operator is essentially the same as the construction of Marchenko's function $e^{\int_0^x \sigma(\lambda,t)dt}$. The essential point to prove is that $F(k) \sim \lambda_k$.

Let us construct the Q_k's by the same inductive method that we would use in higher dimensions. The idea is to successively remove the lower order terms in L_k. There are obstructions to doing so because of the boundary conditions but we can replace the lower order terms by constants $c_j k^{-j}$. The latter essentially constitute the normal form.

We thus begin by determining $Q_1(x)$ so that $L_k^{(1)} := Ad(e^{ik^{-1}Q_1(x)}) L_k$ satisfies $L_k^{(1)}|_0 = [k^2 + c_0 + O(k^{-1})]|_0$. Writing $Ad(e^{ik^{-1}Q_1(x)}) = e^{ik^{-1}ad(Q_1)}$ and expanding the exponential gives the first *homological equation*

$$(58)\qquad [2iD, Q_1(x)] = c_0 - q(x).$$

This simply reads $Q'(s) = \frac{1}{2}(c_0 - q(x))$. Assuming $q(x)$ is real, the boundary condition on Q_1 is that $\operatorname{Re} Q_1(0) = \operatorname{Re} Q_1(\pi) = 0$. The solution is then given by:

$$(59)\qquad Q_1(x) = \frac{1}{2}\int_0^x (c_0 - q(t))dt, \qquad c_0 = \frac{1}{\pi}\int_0^\pi q(t)dt.$$

This is similar to Marchenko's formula $\int_0^x \sigma_1(t)dt = \int_0^x q(t)dt$ except for the constant c_0. This discrepency is due to the fact that our exponent is a series in k and Marchenko's is a series in λ_k and it requires some re-arrangement to align them. We thus have

$$(60)\quad L_k^{(1)} = k^2 + 2kD + c_0 + D^2 + k^{-1}[2Q_1'(x)D + \frac{1}{i}Q_1''(x)] + k^{-2}Q_1'(x)^2.$$

We now repeat the process with $L_k^{(1)}$ in place of L_k. (In the 1D case, it would be simpler to combine the Q_j's into one series and do all the steps at once, but in higher dimensions one does not have this option).

We then seek $Q_2(x)$ so that $L_k^{(2)} := Ad(e^{ik^{-2}Q_2(x)})L_k^{(1)}$ satisfies $L_k^{(2)}|_0 = k^2 + c_0 + O(k^{-2})|_0$. We thus need to kill the k^{-1} term. The homological equation thus reads: $[2D, Q_2] + [2Q_1'(x)D + \frac{1}{i}Q_1''(x)]|_0$ or equivalently

$$2Q_2' = -\frac{1}{i}Q_1''(x).$$

This time Q_2 is purely imaginary, so there is no boundary condition on it. Therefore we do not need to introduce a new constant to kill its boundary values. We may solve by:

$$Q_2(x) = -\frac{1}{2i}\int_0^x Q_1''(t)dt = \frac{1}{4i}\int_0^x q'(t)dt.$$

We observe that the integrand of Q_2 is the same (up to a constant) as σ_2. It is somewhat messy to keep track of the correct constants so to be on the safe side we only only assert that our expressions are correct up to universal constants.

At this point we have:

$$L_k^{(2)} = k^2 + 2kD + c_0 + k^{-1}[2Q_1'(x)D] \\ + k^{-2}[Q_1'(x)^2 + 2Q_2'(x)D + \frac{1}{i}Q_2''(x)] + k^{-3}[2Q_1'Q_2'].$$

Let us carry out the algorithm one more step. We thus seek $Q_3(x)$ so that $L_k^{(3)} := Ad(e^{ik^{-2}Q_3(x)})L_k^{(2)}$ satisfies $L_k^{(3)}|_0 = k^2 + c_0 + \frac{c_2}{k^2} + O(k^{-3})|_0$. The homological equation now reads:

$$2Q_3' = -[Q_1'(x)^2 + \frac{1}{i}Q_2''(x)] + c_2.$$

The solution is:

$$Q_3(x) = (Const.)\int_0^x [q''(t) - q(t)^2 + c_2]dt,$$

$$c_2 = \frac{1}{\pi}\int_0^\pi [q''(t) - q(t)^2]dt.$$

So far we have conjugating L to the normal form $D^2 + c_0 + \frac{c_2}{D^2}$ plus lower order terms. Proceeding in this way, we simeltaneously get a normal form for $D^2 + q$ and a sequence of quasimodes $\operatorname{Im} W_k^+ e^{ikx}$ for the Dirichlet problem. One checks by induction that

- all of the odd Q_j's are real and the even ones are imaginary, hence
- only even powers of $|D|$ occur in the normal form.

These results are equivalent to the details on the expansion for λ_k and the reality of σ_k in Theorem (4.1).

4.1.3. *Intertwining operator.* We now 'glue' together all of these intertwining operators W_k to a 'homogeneous' intertwining operator, i.e. one for which the Planck constant k^{-1} is an internal rather than external parameter. The basic idea is to set $W \sin kx = \operatorname{Im} W_k^+ e^{ikx}$ but we must be careful again to specify the Hilbert space W is to act on. As mentioned above, W_k^+ is only naturally defined on $L^2(S^1)$ so we also define W on that space. Then W is not determined by its values just on $\{\sin kx\}$ but is completely defined by the formula:

$$W = \varphi(x) \sum_{k=1}^{\infty} W_k^+ \Pi_k^+ + W_k^- \Pi_k^- \tag{61}$$

where $W_k^- = CW_k^+C$ (C being the operator of complex conjugation) and where $\Pi_k^\pm : L^2(S^1) \to L^2(S^1)$ is the orthogonal projection onto $e^{\pm ikx}$.

It is natural to consider as well the intertwining operator $\hat{W}$ defined directly on $L^2([0,\pi])$ by $\hat{W} \sin kx := s(\lambda_k, x) := \operatorname{Im} W_k^+ e^{ikx}$. In other words, $\hat{W}$ takes model eigenfunctions of $|D|$ to the (asymptotic) eigenfunctions of L. The functions $s(\lambda_k, x)$ form an orthogonal but not necessarily orthonormal sequence. Hence $\hat{W}$ is not necessarily unitary; to make it so we must compose it with the map $s(\lambda_k, x) \to \hat{s}(\lambda_k, x)$. However this map is not directly constructed by the normal form method. The relation between the intertwining operators is thus given by

$$\hat{W} \sin kx = 1_{[0,\pi]} W \sin kx, \tag{62}$$

where $\hat{W}$ acts on $\sin kx$ as a function on $[0,\pi]$ and W acts on it as an odd function on $[0, 2\pi]$.

Some notational remarks before we proceed. Given two pseudodifferential operators A, B on $L^2(S^1)$ we will write $A \sim B$ in a cone $V \subset T^*S^1 - 0$ if $A - B$ is of order $-\infty$ in that cone, (i.e. if the complete symbol of $A - B$ is of order $-\infty$). In 1D we sometimes abuse notation and write $A \sim B$ on a closed intertval $[-\epsilon, \pi + \epsilon]$ (including $\epsilon = 0$) if $A - B$ has order $-\infty$ in the closure of $T^*(-\epsilon, \pi + \epsilon)$. We also write $f(k,x) \sim g(k,x)$ if $f(k,x) - g(k,x) = O(k^{-N})$ uniformly on $[0,\pi]$.

On $(-\epsilon, \pi + \epsilon)$ we then have

$$LWe^{ikx} \sim F(|k|)^2 We^{ikx}$$

(63) $$L\hat{W} \sin kx \sim F(|k|)^2 \hat{W} \sin kx$$

where $F(|k|) := \lambda_k = |k| + \frac{a_1}{|k|} + \frac{a_3}{|k|^3} + \cdots$

Let us define

(64) $$F(|D|) = |D| + \frac{c_1}{|D|} + \cdots.$$

Here, $|D|$ is defined as $\sqrt{D^*D}$ on $L^2(S^1)$ with $D = \frac{\partial}{i\partial x}$ and in the analogous way on $L^2[0,\pi]$ with the square D^*D is given Dirichlet boundary conditions. These two definitions are compatible if $\sin kx$ is identified with an odd function on S^1. We may then rewrite (63) as:

(65) $$LW \sim WF(|D|)^2, \qquad L\hat{W} \sim \hat{W}F(|D|)^2$$

In addition W and $\hat{W}$ preserve the interior Sobolev space $H_0^1([0,\pi])$ so they take the domain of L into the domain of $F(|D|)$ on $[0,\pi]$.

We would like to invert W and $\hat{W}$ so that (65) can be regarded as conjugation equations. In fact, we only need 'microlocal inverses' near $[0,\pi]$. Again, what we mean by 'inverse' depends on whether we are working microlocally on $[0,\pi]$ as a subset of $[0,2\pi]$ or entirely on $[0,\pi]$. Note that even the definition of the identity operator I depends on which space we are working, e.g. the projection $\Pi_o(x,y) = \sum_{k=1}^{\infty} \sin kx \sin ky$ equals $\frac{1}{2}(I - r)$ on $[0,2\pi]$ (where r is the reflection $e^{ix} \to e^{-ix}$) and $\frac{1}{2}I$ on $[0,\pi]$. To microlocally invert W near $[0,\pi]$ as an operator on $[0,2\pi]$ we first prove:

PROPOSITION **4.2.** *W is an elliptic zeroth order pseudodifferential operator in an open interval $(-\epsilon, \pi+\epsilon)$ around $[0,\pi]$.*

Proof:

To analyse the sum over k, we enlarge S^1 to the 'space-time' $S^1 \times S^1$ and define the operator

(66) $$e^{it|D_x|} : L^2(S^1) \to L^2(S^1 \times S^1), \quad e^{it|D_x|}e^{ikx} = e^{ikt}e^{ikx}.$$

The range of $e^{it|D_x|}$ is contained in the kernel $\mathcal{H}$ of the 'wave operator' $|D_x|^2 - |D_t|^2$ on $S^1 \times S^1$. Let $e^{it|D_x|*} : L^2(S^1 \times S^1) \to L^2(S^1)$ denote the adjoint of $e^{it|D_x|}$. It is easy to see that $e^{it|D_x|*}e^{it|D_x|} = Id$.

We then introduce the further operators $W^{\pm}_{|D_t|}$ and $W_{|D_t|}$ on $L^2(S^1 \times S^1)$. For $k \geq 0$ we put $W^{+}_{|D_t|}(x)\Pi^+ f(x)e^{ikt} = e^{ikt}W_k^+ f(x)$ or equivalently $W^+_{|D_t|}(x) = \Pi_{j=0}^{\infty} e^{i|D_t|^{-j}Q_j(x)}$. (For $k < 0$ put it equal to zero.) We then put

$W^-_{|D_t|}(x) = CW^+_{|D_t|}(x)C$ where C is the operator of complex conjugation. Finally we put $W_{|D_t|}(x) = W^+_{|D_t|}(x)\Pi_+ - W^-_{|D_t|}(x)\Pi_-$. Here, Π_+ (resp. Π_- projects onto positive (resp. negative) frequencies in e^{ikt}.

The following identity is the key to the properties of $W^\pm, W$: Let $j : S^1 \to S^1 \times S^1$ be the inclusion $j(x) = (x, 0)$. Then:

$$W^\pm = j^* W^\pm_{|D_t|} e^{it|D_x|}, \quad W = j^* W_{|D_t|} e^{it|D_x|}. \tag{67}$$

To prove it, just apply both sides to the basis elements e^{ikx}.

These formulae manifestly define Fourier integral operators. In fact, $W^+_{|D_t|} = e^{i\sum_{j=1}^\infty |D_t|^{-j} Q_j(x)}$ is a pseudodifferential operator of order 0. (In higher dimensions it is a Fourier integral operator). The operators $e^{it|D_x|}$ and j^* are clearly Fourier integral operators, with associated canonical relations

$$\begin{aligned} C &= \{((s,\sigma;(s+t,\sigma,t,\sigma))\} \subset T^*(S^1 \times S^1 \times S^1) \\ \Gamma_{j^*} &= \{((s,\sigma);(s,\sigma,0,\sigma)\} \subset T^*(S^1 \times S^1 \times S^1). \end{aligned} \tag{68}$$

The canonical relation underlying $j^* W^+_{|D_t|} e^{it|D_x|}$ is therefore the composite

$$\Lambda = \Gamma_{j^*} \circ C$$

which equals the identity. Hence W is a pseudodifferential operator.

To prove ellipticity we only have to compute the principal symbols of $j^*, W_{|D_t|}$ and $e^{it|D_x|}$. That of $W_{|D_t|}$ is clearly just 1. The other composition similar works out to 1. Hence W has principal symbol equal to 1.

□

Since W is microlocally elliptic we can construct a parametrix for it.

PROPOSITION **4.3.** *There exists a (two-sided) microlocal inverse W_{-1} to W on $(0,\pi)$. That is,*

$$\begin{cases} WW_{-1} \sim W_{-1}W = I + R \ \text{with} \ R \in \Psi^{-\infty}(0,\pi) \\ WF^2 W_{-1} = L + R_1, \quad R_1 \in \Psi^{-\infty} \end{cases}$$

Proof: By the above, WW^* is a zeroth order positive self-adjoint pseudodifferential operator with symbol identically equal to one near $[0,\pi]$. Hence by a standard construction, there exists a positive zeroth order self-adjoint pseudodifferential operator G such that $WW^*G \sim I + R$ with R smoothing on $(-\epsilon_0, \pi + \epsilon_0)$. The operator $W_{-1} = W^*G$ is then a parametrix satisfying:

$$WF^2 W_{-1} \sim LWW_{-1} \mod \Psi^{-K} \tag{69}$$

□

It will be important below to analyse the operator $W\Pi_o 1_{[0,\pi]} W_{-1}$.

PROPOSITION **4.4.** *For any interior pseudodiffererntial operator A (with symbol compactly supported in $T^*(0,\pi)$ we have*

$$AW\Pi_o 1_{[0,\pi]}W_{-1} \sim W\Pi_o 1_{[0,\pi]}W_{-1}A \sim \frac{1}{2}A$$

in the sense that the differences are pseudodifferential operators whose symbols are of order $-\infty$ in $T^*(0,\pi)$.

Proof We have $\Pi_o = \frac{1}{2}(I - r)$ where (as above) r is reflection ($x \to -x$ modulo 2π). So $W\Pi_o 1_{[0,\pi]}W_{-1}A = \frac{1}{2}W1_{[0,\pi]}W_{-1}A - \frac{1}{2}Wr1_{[0,\pi]}W_{-1}A$. The proposition follows if $W1_{[0,\pi]}W_{-1}A \sim A$ and if $Wr1_{[0,\pi]}W_{-1}A \sim 0$ on $[0,\pi]$.

The symbolic construction of W_{-1} is local in the symbol and hence is smooth on the essential support of A, where $1_{[0,\pi]} = 1$. Hence $W1_{[0,\pi]}W_{-1}A \sim A$.

We observe that $Wr1_{[0,\pi]}W_{-1}A$ is a Fourier integral operator on S^1 associated to the canonical transformation $r^*(x,\xi) = (-x,-\xi)$. The graph of this canonical relation does not intersect the essential support of A (which lies in $T^*(0,\pi)$) so the kernel of $Wr1_{[0,\pi]}W_{-1}A$ is C^∞ in the interior of $(0,\pi)\times(0,\pi)$. The case with A on the left is similar.

□

Having defined an inverse we can now assert that there exists a microlocally invertible pseudodifferential intertwining operator W on $(-\epsilon_0, \pi+\epsilon_0)$ which preserves Dirichlet boundary conditions and such that

$$L \sim WF(|D|)^2 W_{-1}. \tag{70}$$

We would then like to conclude that $\lambda_k \sim F(k)$. This was already proved by Marchenko et al [M], and follows from the quasimode consruction since the asymptotic eigenfunctions are close enough approximations to the actual eigenfunctions. In the next section, we are going to prove that $\lambda_k \sim F(k)$ by a wave trace method. Our main interest in this proof is that it generalizes to higher dimensions, whereas the quasimode proof does not. Indeed, in higher dimensions the $s(\lambda_k, x)$ resp. λ_k are replaced by quasimodes, resp. quasi-eigenvalues associated to a closed geodesic. In general, the quasimodes are not close to actual eigenfunctions and it is not even clear that the coefficients in the asymptotic expansions of the approximate eigenvalues $F(k)$ are spectral invariants. However, they are. The path which generalizes to higher dimensions is through the trace of the wave group.

4.2. **Comparison to Marchenko's transformation operator.** Let us pause to compare our construction of W and $\hat{W}$ to that of Marchenko. Since the intertwining operator to normal form is almost uniquely defined – up to phases – as the operator $\sin kx \to \hat{s}(\lambda_k, x)$, the operators $W, \hat{W}$ constructed above must be closely related to Marchenko's transformation operator. Let us briefly review Marchenko's construction and compare the two constructions.

Marchenko's transformation operator is the composition of two operators. The first is constructed in the next result.

THEOREM 4.5. *([M], Theorem 1.2.1 and its Corollary; (1.2.11")) Let* $s(\lambda, x)$ *be the solution of* $L\, s(\lambda, x) = \lambda^2 s(\lambda, x)$ *with initial values* $s(\lambda, 0) = 0$, $s'(\lambda, 0) = 1$. *Then there exists a smooth kernel* $K(x, t, \infty)$ *on* $[0, \pi] \times [0, \pi]$ *such that*

$$s(\lambda, x) = \frac{\sin \lambda x}{\lambda} + \int_0^x K(x, t, \infty) \frac{\sin \lambda t}{\lambda} dt.$$

Moreover, there exists a smooth kernel $L(x, t, \infty)$ *such that*

$$\frac{\sin \lambda x}{\lambda} = s(\lambda, x) + \int_0^x L(x, t, \infty) s(\lambda, t) dt.$$

The basic idea is that $I + K$ intertwines $D^2 + q$ to D^2 and hence satisfies the commutation relation $(D^2 + q)(I + K) = (I + K)D^2$ or equivalently $[D^2, K] = -qK$. This is a hyperbolic equation for $K(x, t)$. Since $\sin \lambda x$ is odd, one can further write the intertwining kernel K in the form $K(x, t) - K(x, t)$ where $K(x, t)$ solves the hyperbolic system ([M], (1.2.9), (1.2.27))

$$((\tfrac{\partial}{\partial t})^2 - (\tfrac{\partial}{\partial x})^2)K = -qK$$

$$K(x, x) = \tfrac{1}{2} \int_0^x q(t) dt, \qquad K(x, -x) = 0.$$

This system can be solved to produce the intertwining operator.

The resulting operator $I + K$ does not take the model Dirichlet eigenfunctions $\sin kx$ to the Dirichlet Sturm-Liouville eigenfunctions $s(\lambda_k, x)$ since it is only devised to take initial values to initial values. Moreover, it conjugates $D^2 + q$ to D^2 which is not the correct normal form. Therefore a further conjugation is needed.

What Marchenko does, essentially, is to define the eigenvalues λ_k by the condition that $(I + K) \sin \lambda_k x$ satisfies Dirichlet boundary conditions at $x = 0, \pi$. Since $\lambda_k = k + \frac{a_1}{k} + \cdots$ we can define a scaling operator S by $S\frac{\sin kx}{k} = \frac{\sin \lambda_k x}{\lambda_k}$. We may then restate Marchenko's theorem as follows:

THEOREM 4.6. *The operator* $(I + K) \circ S : L^2([0, \pi]) \to L^2([0, \pi])$ *defined by* $(I + K) \circ S\frac{\sin kx}{k} = s(\lambda_k, x)$ *preserves Dirichlet boundary conditions and conjuates* $D^2 + q$ *to* $F(|D|)$.

We note here that neither $I + K$ nor S is separately defined on $L^2([0, \pi])$ but rather both are defined on all of $\mathbb{R}$.

It is interesting to note that when Marchenko studies the asymptotic behaviour of the eigenvalues and eigenfunctions, he employs a different method. Namely, he solves the eigenvalue problem $L[y] = \lambda^2 y$ on $[0, \pi]$ (without specifying boundary conditions) by the WKB ansatz in Theorem (4.1). The eigenvalues λ_k^2 are (as mentioned above) determined by the condition that $\operatorname{Im} y(\lambda_k, x)$ satisfies Dirichlet boundary conditions. Because the spectrum is well-separated (consecutive spacings $\lambda_{k+1} - \lambda_k \sim 1$), the asymptotic eigenfunctions are easily seen (by the spectral theorem) to be asymptotic expansions of actual eigenfunctions. The $s(\lambda_k, x)$ are not necessarily of unit norm, their relation to the above $\hat{s}(\lambda_k, x)$'s being $\hat{s}(\lambda_k, x) = \frac{s(\lambda_k, x)}{||s(\lambda_k, x)||}$ where $||\cdot||$ is the L^2-norm. Let us observe that all of the coefficients $\sigma_k(x)$ are real if $q(x)$ is real-valued ([LS], §5.1). It is not hard to see that this implies that the coefficients of the k-series for the exponent of $y(\lambda_k, x)$ are imaginary for odd k and real for even k. This is precisely the even/odd result we saw in the conjugation to normal form.

Because $\lambda_k \sim k + \frac{a_1}{k} + \cdots$, the scaling operator S is easily seen to be a pseudodifferential operator on $\mathbb{R}$. Indeed, one only has to use the addition law for sines to expand $\sin(kx + \frac{a_1}{k}x + \cdots)$ in a k-series, and to rewrite the series in terms of a pseudodifferential operator. We further see that $S = I + S_1$ where S_1 has order -1. It follows that $(I+K)\circ S = I + K + S' + K\circ S'$ is I plus a Volterra operator. As mentioned above, this is essentially due to the fact that the intertwining kernel solves a hyperbolic equation (which would make it supported on $|t| \leq |x|$) and the odd properties of the sines. It follows that the inverse $\hat{W}^{-1}$ is also of the form I plus a Volterra operator.

To understand the relations between $W, W_{-1}, \hat{W}, \hat{W}^{-1}$ it is useful to analyse Marchenko's construction further. We note that just as we defined $W \sin kx = \operatorname{Im} W_k^+ e^{ikx}$ so also Marchenko defines the transformation operator first on the solutions $e(\lambda, x)$ of the eigenvalue problem with initial conditions $e(\lambda, 0) = 1, e'(\lambda, 0) = i\lambda$ ([M], (1.2.2). Then the kernel $K(x,t)$ above may be defined as the kernel satisfying ([M], Theorem 1.2.1):

$$e(\lambda, x) = e^{i\lambda x} + \int_{-x}^{x} K(x,t) e^{i\lambda t} dt. \tag{71}$$

We also need to introduce some operators to take care of the factors of k and $\omega(\lambda_k, 0)$ which normalize the initial values of the eigenfunctions at $x = 0$. Since

$$|D|^{-1} \sin kx = \frac{\sin kx}{k}, \qquad \omega(L, 0) \operatorname{Im} y(\lambda_k, x) = s(\lambda_k, x),$$

$$\omega(|F(|D|), 0) \sin kx = \omega(\lambda_k, 0) \sin kx,$$

we have

$$\begin{aligned} & W \sim (I+K)\circ S\omega(F(|D|),0)|D|^{-1}, \\ & W_{-1} \sim |D|\omega(F(|D|,0)^{-1} S^{-1}(I+L) \\ & W \circ \Pi_o = (I + K_\infty)\circ S\omega(F(|D|),0)|D|^{-1}\circ \Pi_o, \\ & K_\infty \Pi_o f(x) = \textstyle\int_0^x [K(x,t) - K(x,-t)]\Pi_o f(t) dt. \end{aligned} \tag{72}$$

Thus, $W \sin kx \sim \omega(\lambda_k, 0)s(\lambda_k, x)$ on $[0, \pi]$ and $W_{-1}\omega(\lambda_k, 0)s(\lambda_k, x) \sim \sin kx$ on $[0, \pi]$. We further note that

$$(73) \qquad (1_{[0,\pi]}W_{-1})^* \sin ky \sim \omega(\lambda_k, 0)^{-1} \frac{s(\lambda_k, y)}{||s(\lambda_k, \cdot)||^2} \text{ on } (0, \pi)$$

which follows from the inner product formulae

$$\langle (1_{[0,\pi]}W_{-1})^* \sin ky, s(\lambda_j, y)\rangle = \langle W_{-1}^* 1_{[0,\pi]} \sin ky, s(\lambda_j, y)\rangle$$

$$= \langle \sin ky, W_{-1}s(\lambda_j, y)\rangle_{[0,\pi]} \sim \omega(\lambda_k, 0)^{-1}\delta_{jk}$$

where $\langle , \rangle_{[0,\pi]}$ is the inner product on $[0, \pi]$. It follows that

$$W\Pi_o 1_{[0,\pi]}W_{-1}(x, y) = \textstyle\sum_{k=1}^{\infty}(W \sin kx) \otimes (1_{[0,\pi]}W_{-1})^* \sin ky$$

$$(74) \qquad = \textstyle\sum_{k=1}^{\infty} \omega(\lambda_k, 0)s(\lambda_k, x) \otimes (1_{[0,\pi]}W_{-1})^* \sin ky$$

$$\sim \textstyle\sum_{k=1}^{\infty} \omega(\lambda_k, 0)s(\lambda_k, x) \otimes \omega(\lambda_k, 0)^{-1}\frac{s(\lambda_k, y)}{||s(\lambda_k, \cdot)||^2} \sim \delta(x - y) \text{ on } [0, \pi].$$

This corroborates Proposition (4.4).

4.3. **Normal form and microlocal parametrix.** We now prove that the normal form $F(|D|)$ is a spectral invariant of L. The basic idea is to use W and $F(|D|)$ to construct a microlocal parametrix for the Dirichlet wave group of L on $[0, \pi]$. This will allow us to prove that

$$(75) \qquad \sum_{k=1}^{\infty} \cos t\lambda_k \sim \sum_{k=1}^{\infty} \cos tF(k)$$

in the sense that both sides have precisely the same singularities. This is equivalent to $\lambda_k \sim F(k)$ in 1D. We begin by defining the relevant wave groups:

Definition: By $E(t, x, y)$ we denote the kernel of the Dirichlet wave operator $E(t) = \cos tL$ on $[0, \pi]$. It has the eigenfunction expansion:

$$E(t, x, y) = \sum_{k=1}^{\infty} (\cos t\lambda_k)\hat{s}(\lambda_k, x)\hat{s}(\lambda_k, y).$$

It is a classical result that $E(t, x, y)$ is the restriction to $\mathbb{R} \times [0, \pi] \times [0, \pi]$ of a Lagrangean kernel in an open neighborhood of this domain in $\mathbb{R} \times S^1 \times S^1$ (see [GM] for a proof in all dimensions).

The Dirichlet normal form wave group is defined as follows:

Definition: By $F_o(t, x, y)$ we denote the odd (spatial) part

$$(76) \qquad F_o(t, x, y) = \sum_{k=1}^{\infty} \cos t(F(k) \sin kx \sin ky.$$

of the fundamental solution of the wave equation on S^1:

$$\begin{cases} \frac{\partial^2 F_o(t)}{\partial t^2} = F(|D|)^2 F_o(t) & \text{on } (\mathbb{R} \times S^1) \times (S^1) \\ F_o(0,x,y) = \Pi_o(x,y) := \sum_{k=1}^{\infty} \sin kx \sin ky & \frac{\partial F_o}{\partial t}(0) = 0 \\ F_o(t,\cdot,z) = 0 = F_o(t,z,\cdot) & \text{on } \partial[0,\pi]. \end{cases}$$

By odd part we mean odd with respect to r.

We note that Π_o defines half of the identity operator on $[0,\pi]$ and that $F_o(t,x,y)$ is simeltaneously (half) the normal form Dirichlet wave group on $[0,\pi]$ and the odd part of the normal form on $[0,2\pi]$.

We now wish to conjugate the odd normal form wave group on $[0,2\pi]$ to a microlocal parametrix on $[0,\pi]$ for the Dirichlet wave group $E(t)$. We define:

$$(77)\qquad \begin{aligned} &\mathcal{E} = W F_o(t) 1_{[0,\pi]} W_{-1} \\ &\mathcal{E}(t,x,y) = \sum_{k=1}^{\infty} (\cos tF(|k|)) W \sin kx \otimes W_{-1}^* 1_{[0,\pi]} \sin ky. \end{aligned}$$

Here, W_{-1}^* is the adjoint on $L^2([0,\pi])$.

The properties of this kernel are contained in the next lemma:

LEMMA **4.7.** *On* $(0,\pi)$ *we have:*

$$\frac{1}{2}E(t) \sim \mathcal{E}(t) + R_K(t)$$

where $R_K(t)$ *is of order* $-K$.

Proof: $E(t)$ is uniquely characterized as the microlocal solution on $(0,\pi)$ of the Cauchy problem

$$(78)\qquad \begin{cases} ((\frac{\partial}{\partial t})^2 - L)u = 0 \\ u(0) = \Pi_o, \quad \frac{\partial}{\partial t}u(0) = 0 \\ u(t,\cdot)|_{\partial[0,\pi]} = 0. \end{cases}$$

Here we use that Π_o equals the half of the identity operator on $[0,\pi]$. In terms of the eigenfunction expansion we may also write $E(0) = \sum_{k=1}^{\infty} \hat{s}(\lambda_k,x)\hat{s}(\lambda_k,y)$. It suffices to show that $\mathcal{E}$ is a also a microlocal solution to this Cauchy problem modulo errors in the stated class.

We first verify that $\mathcal{E}(t)$ is a microlocal solution of the wave equation. By proposition (4.3) we have:

(79)

$$\begin{cases} \frac{\partial^2}{\partial t^2}\mathcal{E}(t) = WF(|D|)^2 F_o 1_{[0,\pi]} W_{-1} \\ = WF(|D|)^2 W_{-1} W F_o 1_{[0,\pi]} W_{-1} + WF(|D|)^2 (I - W_{-1}W) F_o 1_{[0,\pi]} W_{-1} \\ = L\mathcal{E}(t) + R_1(t), \end{cases}$$

where R_1 has order $-K$ on $(0,\pi)$. Hence $\mathcal{E}(t)$ is a microlocal solution of the forced wave equation with forcing term of the form $R_K\mathcal{E}(t) + R_1(t)$.

The next step is to verify that $\mathcal{E}$ has the same initial condition and zero boundary values as $E(t)$ near $[0,\pi]$. Regarding the boundary condition, we observe that $W\Pi_o(x,y) = 0$ if $x \in \partial[0,\pi]$ and so $\mathcal{E}_\epsilon(x,y) = 0$ if $x \in \partial[0,\pi]$. As for the initial condition, it asserts that $\mathcal{E}(0) = W\Pi_o 1_{[0,\pi]} W_{-1} \sim \frac{1}{2} Id$ on $(0,\pi)$. This was proved in proposition (4.4).

Thus $\mathcal{E}(t)$ is a microlocal solution of the mixed Cauchy problem modulo errors of order $-K$. Since the Cauchy problem is well-posed, we have by Duhamel's principle that

$$(80) \qquad (\mathcal{E}(t) - E(t)) \sim \int_0^t G_o(t-u) R(u) \mathcal{E}(u) du$$

where $R(u)$ is an error in the stated class and where where G_o is the kernel of the mixed problem

$$(81) \qquad \begin{cases} (\partial_t^2 - (L) G_o = 0 & \text{on } [0,\pi] \times [0,\pi] \\ G_o(t,x,y)|_{t=0} = 0 & \frac{\partial}{\partial t} G_o(t,x,y)|_{t=0} = Id \\ G_o(t,x,y) = 0 & \text{for } x = 0, L. \end{cases}$$

Since G_o is well-known to be a Fourier integral operator, the proof is complete. □

4.4. **Wave invariants and normal form.** We now prove the the coefficients a_j of the normal form normal form $F(|k|)$ are spectral invariants of L. We do this by showing that the wave invariants of the normal form are the same as for L. That immediately implies the spectral invariance of the normal form, since the normal form coefficients are easily seen to be in bijection with the wave invariants of the normal form. We use here that $\cos t(F(|D|)$ is simeltaneously the normal form Dirichlet wave group on $[0,\pi]$ and the odd normal form wave group on S^1. We also use the notation γ^m for the mth iterate of the one bouncing ball orbit on $[0,\pi]$. Its length is then $2\pi m$. The normal form has the same broken bicharacterstic flow, so we use the same notation for its closed geodesics.

THEOREM **4.8.** $a_{\gamma^m k}(\sqrt{D^2+q}) = a_{\gamma^m k}(F(|D|)$.

Proof

Using the microlocal parametrix (cf. Lemma 4.7), we have (for $t = 2m\pi$),

$$
\begin{aligned}
a_{\gamma^m k}(\sqrt{L}) &= resD_t^k E(t)|_{t=2m\pi} 1_{[0,\pi]} \\
&= 2resD_t^k \mathcal{E}(t)|_{t=2m\pi} 1_{[0,\pi]} + 2resD_t^k R(t)|_{t=2m\pi} 1_{[0,\pi]} \\
&= 2resD_t^k \mathcal{E}(t)|_{t=2m\pi} 1_{[0,\pi]} \\
&= 2res 1_{[0,\pi]} D_t^k W F_o(t)|_{t=2m\pi} 1_{[0,\pi]} W_{-1}
\end{aligned} \tag{82}
$$

since $R(t)$ of order $-K$ and the residue is unaffected by such a term for large enough K.To evalute the residue let us approximate $1_{[0,\pi]}$ by a smooth cutoff $\tau_\epsilon(x)$, equal to one on $[0,\pi]$ and supported on $(-\epsilon, \pi+\epsilon)$. We have:

$$
\begin{aligned}
& res\ 1_{[0,\pi]} D_t^k W F_o(t)|_{t=2m\pi} 1_{[0,\pi]} W_{-1} \\
&= \lim_{\epsilon\to 0} res\ \tau_\epsilon(x) W D_t^k F_o(t)|_{t=2m\pi} 1_{[0,\pi]} W_{-1} \\
&= \lim_{\epsilon\to 0} res\ 1_{[0,\pi]} W_{-1} \tau_\epsilon(x) W D_t^k F_o(t)|_{t=2m\pi}
\end{aligned} \tag{83}
$$

where in the last line we used the tracial property of res. We further note that $W_{-1}\tau_\epsilon D_t^k W F_o(t)|_{t=2m\pi}$ is a (standard) Fourier integral operator on the boundaryless manifold S^1. The residue of $1_{[0,\pi]}$ times this operator can be calculated as in the boundaryless case by integrating the residue density over the interval. This follows as in the proof of the Poisson relation for manifolds with boundary [GM] and ultimately comes down to the method of stationary phase on a manifold with boundary (loc. cit.) To simplify the calculation, we observe that $F_o(t) = \cos(tF(|D|)\Pi_o$ where $|D|$ denotes the operator $|D|e^{ikx} = |k|e^{ikx}$ on S^1. It follows that

$$
\begin{aligned}
D_t^k F_o|_{t=2m\pi} = \ & F(|D|)^k(-1)^{k/2} \cos 2m\pi F(|D|)\Pi_o \qquad k \text{ even} \\
& = F(|D|)^k(-1)^{\frac{k+1}{2}} \sin 2m\pi F(|D|)\Pi_o \quad k \text{ odd.}
\end{aligned}
$$

Since both cases are similar we will assume k is even. Thus we have

$$
a_{\gamma^m k}(L) = \lim_{\epsilon\to 0} res\ 1_{[0,\pi]} W_{-1}\tau_\epsilon W F(|D|)^k(-1)^{k/2} \cos 2m\pi F(|D|)\Pi_o 1_{[0,\pi]}, \tag{84}
$$

We also have that $\cos 2m\pi F(|D|) = \cos 2m\pi(\frac{c_1}{|D|} + \dots)$ and that $\sin 2m\pi|D| = \sin 2m\pi(\frac{c_1}{|D|} + \dots)$ are pseudodifferential operators so the operator whose residue we are taking is Π_o composed with a pseudodifferential operator. The residue of a Fourier integral operator is obtained by integrating a density over the fixed point set of its underlying canonical relation. The canonical relation underlying $\Pi_o = \frac{1}{2}(I - r)$ has two components, the identity graph and the graph of the map $r(x,\xi) = (-x, -\xi)$. The only fixed points of r occur at $x = 0, \pi \pmod{2\pi}$ and there are no fixed vectors, so the r term contributes zero to the residue and we may replace Π_o by $\frac{1}{2}I$. Then we are reduced to calculating the residue of a pseudodifferential operator, which in the 1D case equals the integral over γ_o of the term of order -1 in its complete symbol exapansion.

The main point now is that due to the leftmost factor of $1_{[0,\pi]}$, the residue density may be calculated as if $\tau_\epsilon \equiv 1$. Since this is an important point, we carry out the calculation in some detail using the Kohn-Nirenberg calculus. We recall that the formula for the symbol of the composition $a \circ b$ of two pseudodiffererential symbols in that calculus is given by $a \circ b \sim \sum_\alpha \frac{i^{|\alpha|}}{\alpha!} D_\xi^\alpha a(x,\xi) D_x^\alpha b(x,\xi)$ (see e.g. [T], Prop. 3.3). For notational simplicity we use the same notation for operators and their symbols.

We further observe that the complete symbol of

$$F(|D|)^k(-1)^{k/2}\cos 2m\pi F(|D|)$$

equals

$$F(|\xi|)^k(-1)^{k/2}\cos 2m\pi F(|\xi|)$$

and hence is independent of x. Also using the tracial property of res we have:

(85)

$$a_{\gamma^m k}(\sqrt{L})$$

$$= \sum_\alpha \frac{i^{|\alpha|}}{\alpha!} \lim_{\epsilon\to 0} res\, 1_{[0,\pi]}\{(D_\xi^\alpha W_{-1})(D_x^\alpha \tau_\epsilon W)(F(|\xi|)^k(-1)^{k/2}\cos 2m\pi F(|\xi|))\}$$

$$= \sum_\alpha \frac{i^{|\alpha|}}{\alpha!} \lim_{\epsilon\to 0} Res_{s=0} \int_0^\pi \int_0^\infty \{(D_\xi^\alpha W_{-1}) D_x^\alpha(\tau_\epsilon W)(F(|\xi|)^k \times(-1)^{k/2}\cos 2m\pi F(|\xi|)\}|\xi|^{-s} dx d\xi$$

$$= \sum_\alpha \frac{i^{|\alpha|}}{\alpha!} Res_{s=0} \int_0^\pi \int_0^\infty \{(D_\xi^\alpha W_{-1}) D_x^\alpha(W)(F(|\xi|)^k \times(-1)^{k/2}\cos 2m\pi F(|\xi|)\}|\xi|^{-s} dx d\xi$$

$$= Res_{s=0} \int_0^\pi \int_0^\infty \{(F(|\xi|)^k(-1)^{k/2}\cos 2m\pi F(|\xi|)\}|\xi|^{-s} dx d\xi.$$

Above we used the fact that $WW_{-1} \sim 1$ on $(-\epsilon_0, \pi + \epsilon_0)$ to replace

$$\sum_\alpha \frac{i^{|\alpha|}}{\alpha!} D_\xi^\alpha W_{-1}(x,\xi) D_x^\alpha W(x,\xi)$$

by 1. We also used the fact that $D_x^\alpha \tau_\epsilon = 0$ on $[0,\pi]$. The final integral is precisely the residue of the normal form claimed in the statement of the theorem.

□

In higher dimensions, the fact that the normal form is a spectral invariant leads to a number of concrete inverse results. We discuss these and related open problems in the next section.

5. Principal results and open problems

The purpose of this section is simply to state the inverse spectral results proved to date by the method of normal forms, to list some obvious problems they suggest, and to correct some errors from earlier papers.

5.1. **Principal results and open problems.** We begin with a general inverse result valid on all bumpy compact Riemannian manifolds. In the elliptic case the theorem was conceived and proved by V. Guillemin. The author developed a somewhat different approach and generalized it to any non-degenerate closed geodesic.

THEOREM **5.1.** *([G.1-2]; see also [Z.1-2]) Suppose that (M, g) is a compact Riemannian manifold without boundary, and suppose that γ is a non-degenerate closed geodesic whose length L_γ is of multiplicity one in the length spectrum of (M, g). Then the quantum Birkhoff normal form of $\sqrt{\Delta}$ at γ is a spectral invariant.*

Problem 1: To what degree is the metric g determined by the Birkhoff normal form invariants $B_{\gamma k j}$ of $\sqrt{\Delta}$ at one closed geodesic γ? Do there exist 'iso-normal form' deformations, i.e. deformations g_t of g which preserves the normal form at γ?

It would not be surprising if the answer were yes in many cases. Then there is no hope, even for analytic metrics, of determing the whole of (M, g) from the normal form at one γ and one would have to put together information from different γ. It is not clear how to do this.

Problem 2: Use the spectral invariance of the normal form to investigate new cases of spectral rigidity or compactness of isospectral sets.

A simpler avenue of attack is to find special classes of metrics where the normal form at just one closed geodesic gives a lot of information. Such metrics should only involve one unknown function of one variable.

The simplest situation is that of real analytic surfaces of revolution of 'simple type' i.e. surfaces in $\mathbb{R}^3$ with just one closed rotationally invariant geodesic, and additionally whose Poincare map of twist type. In this case there is a global Birkhoff normal form and we proved:

THEOREM **5.2.** *([Z.3]) Let (S^2, g) be an analytic simple surface of revolution with simple length spectrum. Then the normal form is a spectral invariant.*

This allowed us to prove the inverse result:

THEOREM **5.3.** ([Z.3]) *Suppose that g_1, g_2 are two real analytic metrics on S^2 such that (S^2, g_i) are simple surfaces of revolution with simple length spectra. Then $Sp(\Delta_{g_1}) = Sp(\Delta_{g_2})$ implies $g_1 = g_2$.*

Problem 3: Are analytic metrics of revolution of simple type on S^2 spectrally determined? (At the present time, the only metric on S^2 known to be spectrally determined is the round metric g_{can}). If g is such a metric and if h is isospectral to g must h be a metric of revolution of simple type? (G. Forni and the author have proved that its geodesic flow must be C^0-integrable).

Another natural class of metrics is analytic bounded plane domains. Of course it is quite a famous problem to prove (or disprove) that they are spectrally determined. In [Z.5] we studied metrics in the class $\mathcal{D}$ of domains Ω satisfying

$$\begin{cases} \bullet \ \Omega \text{ is real analytic} \\ \bullet \ \Omega \text{ is } \mathbb{Z}_2 \times \mathbb{Z}_2 - \text{ symmetric} \\ \bullet \text{ the axis of } \Omega \text{ is a non} - \text{degenerate elliptic bouncing ball orbit } \overline{AB} \\ \bullet \ Lsp(\Omega) \ \text{ is simple.} \end{cases}$$

We proved:

THEOREM **5.4.** *The Dirichlet spectrum Spec:* $\mathcal{D} \mapsto \mathbb{R}_+^{\mathbb{N}}$ *is 1-1.*

Thus, bi-axisymmetric analytic domains with non-degenerate elliptic bouncing ball orbits and with simple length spectra are spectrally determined.

Problem 4: Can one remove the symmetry assumption? In [Z.5] the domain was determined just from the principal symbol data of the quantum Birkhoff normal form, i.e. from the classical normal form. Probably one can get more out of the quantum normal form. If one expresses the two components of the boundary at the ends of the bouncing ball orbits as graphs of functions f_1, f_2 over the x-axis, one sees why the symmetry assumption is useful: namely, with the up/down symmetry one has $f_1 = f_2$ and with the right/left symmetry one has only to determine the even Taylor coefficients of f_1. It is quite likely that the second symmetry assumption is unnecessary. This is because there should be 'enough' quantum Birkhoff normal form coefficients to determine the whole Taylor expansion of f_1. However it seems rather difficult to drop the up/down symmetry assumption.

A general problem:

Problem 5: In view of the fact that one can determine the Birkhoff normal form at each closed geodesic γ, can one determine the 'dynamical type' of g from the spectrum of Δ? E.g. can one determine if the geodesic flow of g is hyperbolic?

It is curious to compare the expressions for the wave invariants produced by the method of normal forms versus those produced by the Hadamard-Riesz parametrix method. The only part of the normal form invariants which change under iteration are the Floquet invariants, which go to their multiples. This observation motivates:

Problem 6: In the formula for the wave invariants of manifolds WCP's, how do the wave invariants of γ change under iteration $\gamma \to \gamma^k$.

REFERENCES

[B.B] V.M.Babic, V.S. Buldyrev: *Short-Wavelength Diffraction Theory*, Springer Series on Wave Phenomena 4, Springer-Verlag, New York (1991).

[Be] P. Berard, On the wave equation without conjugate points, Math. Zeit. **155** (1977), 249–276.

[CV] Y. Colin de Verdiere, Parametrix de l'equation des ondes et integrales sur l'espace des chemins, Seminaire Gaoulaonic–Schwartz, expose n^{O} 20, 1974–75.

[CV.1] Y.Colin de Verdiere, Sur les longuers des trajectoires periodiques d'un billard, In: P.Dazord and N. Desolneux-Moulis (eds.) *Geometrie Symplectique et de Contact: Autour du Theoreme de Poincare-Birkhoff. Travaux en Cours, Sem. Sud-Rhodanien de Geometrie III* Pairs: Herman (1984), 122-139.

[D] H. Donnelly, On the wave equation asymptotics of a compact negatively curved surface, Inv. Math. **45** (1978) 115-137.

[D.G] J.J.Duistermaat and V.Guillemin, The spectrum of positive elliptic operators and periodic bicharacteristics, Inv.Math. 24 (1975), 39-80.

[F.G] J.P. Francoise and V. Guillemin: On the period spectrum of a symplectic mapping, J. Fun. Anal. **100**, (1991) 317–358.

[G.Sh] I.M.Gel'fand and G.E.Shilov, *Generalized Functions, Volume 1*, Academic Press, New York (1964).

[G.1] V.Guillemin, Wave trace invariants, Duke Math. J. 83 (1996), 287-352.

[G.2] V. Guillemin, Wave-trace invariants and a theorem of Zelditch, Duke Int.Math.Res.Not. 12 (1993), 303-308.

[G.3] V. Guillemin, A new proof of Weyl's law on the asymptotic distribution of eigenvalues, Adv. in Math. **55** (1985), 131–160.

[GM] V.Guillemin and R.B.Melrose, The Poisson summation formula for manifolds with boundary, Adv.in Math. 32 (1979), 204 - 232.

[G.St] V. Guillemin, S. Sternberg, *Geometric Asymptotics*, AMS Math. Surveys **14** (1977).

[HoI-IV] L. Hörmander, *Theory of Linear Partial Differential Operators I-IV*, Springer-Verlag, New York (1985).

[K] C. Kassel, Le residu non commutatif, Seminaire Bourbaki n^{O} 708, Asterisque 177-178 (1989), 199-229.

[Kl] W. Klingenberg, *Lectures on Closed Geodesics*, Grundlehren der. math. W. **230**, Springer-Verlag (1978).

[L] V.F.Lazutkin, Construction of an asymptotic series of eigenfunctions of the "bouncing ball" type, Proc.Steklov Inst.Math. 95 (1968), 125- 140.

[LS] B.M.Levitan and I.S. Sargsjan, *Sturm-Liouville and Dirac Operators*, Math.and its Appl., Kluwer, London (1991).

[M] V.A.Marchenko, *Sturm-Liouville Operators and Applications*, Operator Theory: Adv.and Appl. 22, Birkhauser, Basel (1986).

[P] G.Popov, Length spectrum invariants of Riemannian manifolds, Math.Zeit. 213 (1993), 311-351.

[Sj] J.Sjostrand, Semi-excited states in nondegenerate potential wells, Asym.An. 6(1992) 29-43.

[Su] T. Sunada, Trace formula and heat equation asymptotics for a non-positively curved manifold, Am. J. Math., vol. 104 (1982), 795–812.

[T] M.E. Taylor, *Partial Differential Equations II*, Applied Math.Sci. 116, Springer-Verlag (1996).

[Z.1] S.Zelditch, Wave invariants at elliptic closed geodesics, Geom.Anal.Fun.Anal. 7 (1997), 145-213.

[Z.2] ———, Wave invariants for non-degenerate closed geodesics, GAFA 8 (1998), 179-217.

[Z.3] ———, The inverse spectral problem for surfaces of revolution, J. Diff. Geom. 49 (1998), 207-264.

[Z.4] ———, Normal form of the wave group and inverse spectral theory, *Journees Equations aux Derivees Partielles*, Saint-Jean-de-Monts (1998).

[Z.5] ———, Spectral determination of analytic axi-symmetric plane domains (preprint 1998).

[Z.6] ———, Kuznecov sum formulae and Szego limit formulae on manifolds, Comm. PDE **17** (1&2) (1992), 221–260.

Department of Mathematics, Johns Hopkins University, Baltimore, MD 21218, USA

E-mail address: zel@math.jhu.edu

Printed in the United States
By Bookmasters